Aviation Security Law

Ruwantissa Abeyratne

Aviation Security Law

Springer

Dr. Ruwantissa Abeyratne
International Civil Aviation Organization
999 University Street
Montreal H3C 5H7, Quebec
Canada
tabeyratne@icao.int

ISBN 978-3-642-11701-5   e-ISBN 978-3-642-11703-9
DOI 10.1007/978-3-642-11703-9
Springer Heidelberg Dordrecht London New York

Library of Congress Control Number: 2010928430

© Springer-Verlag Berlin Heidelberg 2010

This work is subject to copyright. All rights are reserved, whether the whole or part of the material is concerned, specifically the rights of translation, reprinting, reuse of illustrations, recitation, broadcasting, reproduction on microfilm or in any other way, and storage in data banks. Duplication of this publication or parts thereof is permitted only under the provisions of the German Copyright Law of September 9, 1965, in its current version, and permission for use must always be obtained from Springer. Violations are liable to prosecution under the German Copyright Law.

The use of general descriptive names, registered names, trademarks, etc. in this publication does not imply, even in the absence of a specific statement, that such names are exempt from the relevant protective laws and regulations and therefore free for general use.

Cover design: WMXDesign GmbH, Heidelberg, Germany

Printed on acid-free paper

Springer is part of Springer Science+Business Media (www.springer.com)

Preface

Aviation is an important global business and a significant driver of the global economy. It is vital, therefore, that stringent measures are taken to counter acts of unlawful interference with civil aviation. The *Convention on International Civil Aviation* signed at Chicago on 7 December 1944, states in its *Preamble* that whereas the development of civil aviation may help preserve friendship and understanding among the people of the world, yet, its abuse could become a threat to general security.

The genealogy of the term "Terrorism" lies in Latin terminology meaning "to cause to tremble" (*terrere*). Since the catastrophic events of 11 September 2001, we have seen stringent legal measures taken by the United States to attack terrorism, not just curb it. The famous phrase "war on terror" denotes pre-emptive and preventive strikes carried out through applicable provisions of legitimately adopted provisions of legislation. The earliest example is the *Air Transportation Safety and System Stabilization Act* (ATSAA) enacted by President Bush less than two months after the 9/11 attacks. Then, two months after the attacks, in November 2001, Congress passed the *Aviation and Transportation Security Act* (ATSA) with a view to improving security and closing the security loopholes which existed on that fateful day in September 2001. The legislation paved the way for a huge federal body called the Transportation Security Administration (TSA) which was established within the Department of Transportation. The Homeland Security Act of 2002 which followed effected a significant reorganization of the Federal Government.

All this goes to show that the law plays a significant role in ensuring aviation security. This book addresses new and emerging threats to civil aviation; evaluates security tools now in use such as the Public Key Directory, Advance Passenger Information, Passenger Name Record and Machine Readable travel documents in the context of their legal and regulatory background; and discusses applicable security treaties while providing an insight into the process of the security audits conducted by the International Civil Aviation Organization (ICAO).

The book also examines issues of legal responsibility of States and individuals for terrorist acts of third parties against civil aviation and discusses from a legal perspective the latest liability Conventions adopted at ICAO. The Conclusion of the book provides an insight into the application of legal principles through risk management. Since the writing of this book, the author published three feature articles entitled, The NW Flight 253 and the Global Framework of Aviation Security (*Air and Space Law*, Volume 35 Issue 2 April 2010 167–182); The Use of Full Body Scanners and Their Legal Implications; and The Use of Forged Passports for Acts of Criminality (both of which could be accessed through the web page of the Journal of Transportation Security (Springer). These three articles form a useful adjunct to this book.

Montreal, CA   Ruwantissa Abeyratne

Contents

1 A Security Culture 1
  A. A Risk-Based Approach 1
  B. The ICAO Response 2
    I. The ICAO High-Level Ministerial Conference 2
    II. Post Conference Work 7
  C. Emerging Threats 9
    I. Probability 9
    II. Reacting to Probability 10
    III. Deterrence 13
    IV. Problems of Deterrence 14
    V. Threat Assessment in ICAO 16
    VI. The AVSEC Panel 19
    VII. Bioterrorism 21
    VIII. Cyber-Terrorism 24
    IX. MANPADS 25
    X. The Diverse Nature of Missile Attacks 29
    XI. Installation of an Anti-missile System 32
    XII. The Perimeter Guard 32
    XIII. International Accord 33
    XIV. Other Current Threats 36
  References 36

2 Principles of Responsibility 39
  A. State Responsibility 39
    I. Principles of State Responsibility 42
    II. The Theory of Complicity 42
    III. Mechanisms for Extradition of Offenders: The Lockerbie Case 43
    IV. The Condonation Theory 48
    V. The Role of Knowledge 51

		VI. Profiling of Passengers	54
		VII. Airport Profiling	55
		VIII. Profiling and the Right of Privacy	58
	B.	Other Aspects of Responsibility	61
		I. Prelude to the Rome Convention of 1952	61
		II. The Rome Convention of 1952	66
	C.	The Rome Convention of 1952	70
		I. Background	70
		II. Insurance	71
		III. Provisions of the Convention	77
		IV. The Montreal Protocol of 1978	80
		V. Modernizing the Rome Convention	81
	References		91
3	**Initiatives of the Early Twenty-first Century**		**93**
	A.	The Two Liability Conventions	93
		I. The General Risks Convention	93
		II. The Unlawful Interference Compensation Convention	96
	B.	Innovative Security Tools	98
		I. Biometric Identification	109
		II. Public Key Directory	109
	C.	Advance Passenger Information	121
	D.	The Passenger Name Record	122
		I. Definition and Application of PNR	123
		II. The Importance of PNR Data to States	125
		III. Advantages of Unified Guidelines	126
		IV. Advance Passenger Information Guidelines	138
		V. Contracting States' Positions	140
	E.	Machine Readable Travel Documents	151
		I. Some Problem Areas	156
	F.	Unmanned Aerial Vehicles	157
		I. Legal and Regulatory Issues	162
		II. Operations Over the High Seas	163
		III. Air Traffic Services	166
		IV. UAVs as State Aircraft	168
	References		174
4	**Narco-terrorism**		**177**
	A.	Introduction	177
	B.	United Nations Initiatives	179
		I. The United Nations Convention Against Illicit Traffic in Narcotic Drugs and Psychotropic Substances	186
		II. Some Recent Efforts of the United Nations	187

Contents ix

	C.	ICAO Initiatives	189
		I. Basic Principles of Aeronautics on International Narcotic Control	189
	D.	Other Regulatory Provisions	198
		I. Article 4 of the Convention on International Civil Aviation	198
		II. Article 3 *bis*	199
		III. Other Legal Aspects	201
		IV. ICAO Assembly Resolution A 27-12	202
	References		204
5	**The Unlawful Interference Conventions**		205
	A.	United Nations General Assembly Resolutions on Unlawful Interference with Civil Aviation	205
	B.	International Conventions	210
		I. Convention for the Prevention and Punishment of Terrorism (1937)	210
		II. Convention on International Civil Aviation (Chicago Convention of 1944)	210
		III. United Nations Charter	212
		IV. The Geneva Convention on the High Seas (1958)	213
	C.	Concerted Action Under the Auspice of the International Civil Aviation Organization: The Tokyo Convention (1963)	217
		I. The Powers Given to Aircraft Commander and Others in Order to Combat Hijackings	221
		II. Jurisdiction to Punish the Terrorists	223
		III. Powers and Duties of States	224
		IV. Extradition	225
		V. Responsibilities of States	226
		VI. An Answer?	228
	D.	The Hague Convention on Hijacking 1970	230
		I. The Scope of the Convention	231
		II. Powers and Duties Imposed Upon States in Order to Combat Hijacking	232
		III. Other Provisions	236
	E.	The Montreal Convention (1971)	237
		I. Definition of In Service	237
		II. Definition of the Offence	238
		III. Penalties and the Scope of the Convention	241
		IV. Jurisdictional Powers Given to States Under the Montreal Convention (1971)	242
	F.	The Bonn Declaration	246
		I. The Legal Status of the Bonn Declaration	247
		II. Incompatibility of the Declaration with the Vienna Convention on the Law of Treaties	249

		III.	The Incompatibility of the Declaration with the Convention on International Civil Aviation (Chicago Convention 1944) and the International Air Services Transit Agreement	250
		IV.	Problem of Prosecution or Extradition	252
	G.	\multicolumn{2}{l}{A New Convention on the Marking of Plastic Explosives for the Purpose of Detection}	253
		I.	Scope of the Convention	255
		II.	Obligations of States	257
		III.	Technical Annex	258
		IV.	International Explosives Technical Commission	260
		V.	Final Clauses and Final Act	261
	\multicolumn{3}{l}{References}	262

6 Aviation Security Audits ... 265
   A. Security Oversight ... 268
   B. The Role of the ICAO Council ... 270
   References ... 276

7 Conclusion ... 277

Index ... 281

Table of Cases

1969 Case of *Walls v Mussens Ltd* P. 85
1979 Case of *Mannington Mills v. Congoleum Corporation* 595 F.2d 1287 P. 127–128
1988 Cases *International Tin Council v. Amalgamet Inc.* P. 119
1991 Case of *EEOC v. Arabian American Oil Company and ARAMCO Services* 113 L E 2d 274 P. 127
Air India v.Wiggins [1980] 1 WLR 815 at 819 P. 127
Anns v. Merton London Borough Council [1978] A.C. 728 (H.L.) P. 83
Arab Banking Corporation v. International Tin council and Algemene Bank Nederland and Others (Interveners) and Holo Trading Company Ltd. Interveners) (1988) 77 ILR 1–8 P. 120
Barboni v. Cie Air-France (1982) 36 RFDA 358 P. 82
Buchbinder v. American Airlines, P. 87
Case of *Palsgraf v. Long Island Railway Co.* P. 84
Chartered Bank v. International Tin Council and others P. 120
Cork v. Kirby Maclean Ltd., [1952] 2 All.E.R. 402 (C.A.). P. 84
El Al Israel Airlines, Ltd. v. Tseng 525 U.S. 155 (1999) P. 87
Gibbs v. American Airlines, Inc., 1999 P. 87
Haddad v. Cie Air France (1982) 36 RFDA 355 P. 82
Holmes v. Bangladesh Biman Corporation, [1989] 1 AC 1112 at 1126 P. 127
Laura M.B. Janes (USA) v. United Mexican States (1925) 4 R Intl Arb Awards 82 P. 43
Lutcher SA Cellulose e Papel v. Inter-American Development Bank, 382 F.2d. 454 (DC Cir.1967) P. 120
M'Alister (or Donoughue) v. Stevenson [1932] A.C. 562 (H.L.) P. 83
Naziranbai v. the state, 1957 Madhya Bharat Law Reporter, P. 109
Neilson v. Kamloops (City of), [1984] P. 84
Nicaragua v. the United States, ICJ Reports 1986, P. 52
Palsgraf v. Long Island Railway Co. 162 N.E. 99 (N.Y. 1928) P. 84
Schenk v. US, 249 US 47 (1919) P. 9

Smith v. Socialist Peoples Libyan Arab Jamahiriya, 866 F. Supp 306 (1995) P. 107
Standard Chartered Bank v. International Tin Council and others [1986] 2 All ER 257 P.120
Timberlane Lumber Company v. Bank of America, 549 F. 2d 597 (1976) P. 128
UK v. Albania, [1949] ICJ Rep. 4 (9 April) at 22 P. 108

Chapter 1
A Security Culture

A. A Risk-Based Approach

Since the events of 11 September 2001, there have been several attempts against the security of aircraft in flight. These threats have ranged from shoe bombs to dirty bombs to explosives that can be assembled in flight with liquids, aerosols and gels. In every instance the global community has reacted with pre emptive and preventive measures which prohibit any material on board which might seemingly endanger the safety of flight. Some jurisdictions have even gone to extremes in prohibiting human breast milk and prescriptive medications on board.

New and emerging threats to civil aviation are a constant cause for concern to the aviation community. Grave threats such as those posed by the carriage of dangerous pathogens on board, the use of cyber technology calculated to interfere with air navigation systems, and the misuse of man portable air defence systems are real and have to be addressed with vigour and regularity. The International Civil Aviation Organization has been addressing these threats for some time and continues to do so on a global basis.

Since the events of 11 September 2001 took place, the most critical challenge facing international civil aviation remains to be the compelling need to ensure that the air transport industry remains continuous and its consumer is assured of sustained regular, safe and secure air transport services. The Air Transport Association (ATA), in its 2002 State of the United States Airline Industry Statement, advised that, in the United States, the combined impact of the 2001 economic downturn and the precipitous decline in air travel following the 11 September 2001 attacks on the United States resulted in devastating losses for the airline industry which are likely to exceed $7 billion and continue through 2002.[1] Of course, the overall picture, which portended a certain inevitable gloom for the air transport industry, was not the exclusive legacy of United States' carriers. It applied

---

[1] State of the United States Airline Industry, *A Report on Recent Trends for United States Carriers*, Air Transport Association: 2002, Statement by Carol B. Hallett, President and CEO, ATA.

worldwide, as was seen in the abrupt downfall of air traffic globally during 2001. The retaliation by the world community against terrorism, which is an ongoing feature in world affairs, increased the airline passenger's fear and reluctance to use air transport. In most instances in commercial aircraft purchasing, air carriers cancelled or postponed their new aircraft requisition orders. Many carriers, particularly in developing countries, were seen revisiting their cost structures and downsizing their human resource bases. It is incontrovertible that another similar event or series of events will inevitably plunge the aviation industry into similar despair and destitution.

In order to arrive at where we are at the present time with regard to the results of the global measures taken by the International Civil Aviation Organization (ICAO), it is necessary to discuss the various steps taken from a regulatory perspective by ICAO in its role as regulator and mentor of international civil aviation, in countering imminent threats posed to the sustainability of the air transport industry.

B. The ICAO Response

I. The ICAO High-Level Ministerial Conference

At the 33rd Session of the Assembly, held from 25 September to 5 October 2001, ICAO adopted Resolution A33-1 entitled "Declaration on misuse of civil aircraft as weapons of destruction and other terrorist acts involving civil aviation".[2] This Resolution, while singling out for consideration the terrorist acts which occurred in the United States on 11 September 2001, and, *inter alia*, recognizing that the new type of threat posed by terrorist organizations requires new concerted efforts and policies of cooperation on the part of States, urged all Contracting States to intensify their efforts in order to achieve the full implementation and enforcement of the multilateral conventions on aviation security, as well as of the ICAO Standards and Recommended Practices and Procedures (SARPs) relating to aviation security, to monitor such implementation, and to take within their territories appropriate additional security measures commensurate to the level of threat in order to prevent and eradicate terrorist acts involving civil aviation. The Resolution also urged all Contracting States to make contributions in the form of financial or human resources to ICAO's aviation security mechanism to support and strengthen the combat against terrorism and unlawful interference in civil aviation; called on Contracting States to agree on special funding for urgent action by ICAO in the field of aviation security; and directed the Council to develop proposals and take

---

[2]*Assembly Resolutions in Force* (as of 5 October 2001), ICAO Doc 9790, at p. VII-1. Also of general interest is UN General Assembly Resolution 56/88, *Measures to Eliminate International Terrorism*, adopted at the 56th Session of the United Nations which calls upon States to take every possible measure in eliminating international terrorism. See A/RES/56/88, 24 January 2002.

B. The ICAO Response

appropriate decisions for a more stable funding of ICAO action in the field of aviation security, including appropriate remedial action.

Resolution A33-1 also directed the Council to convene, at the earliest date, an international high-level, ministerial conference on aviation security in Montreal with the objectives of preventing, combating and eradicating acts of terrorism involving civil aviation; of strengthening ICAO's role in the adoption of SARPs in the field of security and the audit of their implementation; and of ensuring the necessary financial means to strengthen ICAO's AVSEC Mechanism, while providing special funding for urgent action by ICAO in the field of aviation security.

On 19 and 20 February 2002, in keeping with the requirement of Assembly Resolution A33-a high level ministerial conference on aviation security was held in the Headquarters of the International Civil Aviation Organization, Montreal. In the words of Dr. Assad Kotaite, President of the ICAO Council who opened the Conference (and later served as the Chairman of the Conference), the Conference was being held "...at a critical juncture for civil aviation and for society at large... and would review and develop global strategy for strengthening aviation security with the aim of protecting lives both in the air and on the ground, restoring public confidence in air travel and promoting the health of air transport in order that it can renew its vital contribution to the world economy..."[3] Dr. Kotaite stated that this was a historic moment in the evolution of civil aviation.

At this Conference, attended by Member States of the International Civil Aviation Organization, Some 714 participants from 154 Contracting States and observers from 24 international civil aviation organizations endorsed a global strategy for strengthening aviation security worldwide and issued a public declaration at the conclusion of their two-day meeting.

The High Level Ministerial Conference came to several conclusions and adopted numerous recommendations containing guidance for follow up action. The Conference concluded that the events of 11 September 2001 have had a major negative impact on world economies and an impact on air transport which is unparalleled in history and restoration of consumer confidence in air transport and assurance of the long-term health of the air transport industry are both vital, and many States have already initiated a range of measures to this effect. It was also the view of the Conference that the effective application of enhanced uniform security measures, commensurate with the threat, will help to restore confidence in air transport, but these measures will need to be passenger and cargo user-friendly and not overly costly for the industry and its consumers if traffic growth is to be regenerated. Accordingly, the Conference recommended that consistent with Assembly Resolution A33-1, States should intensify their efforts to achieve the full implementation and enforcement of the multilateral conventions on aviation security as well as of the ICAO Standards and Recommended Practices (SARPs) relating to aviation security and take within their territories appropriate additional security measures

---

[3]ICAO News Release *PIO 02/2002*.

which are commensurate with the level of threat and are cost effective. Since restoration of confidence in air transport is a collective responsibility, the Conference called upon States to enhance international cooperation in aviation security and assist developing countries to the extent possible.

With regard to the compelling need to strengthen aviation security worldwide, the Conference concluded that a strong and viable aviation security (AVSEC) programme was indispensable and that a global uniform approach to the implementation of the international aviation security standards is essential, while leaving room for operational flexibility. It was also considered useful to establish regional and sub-regional approaches which could make a significant contribution to ICAO's aviation security activities. The Conference concluded that aviation security was a responsibility of Contracting States, and States which outsource aviation security programmes should therefore ensure that adequate governmental control and supervision are in place. The Conference also observed that, since gaps and inadequacies appear to exist in international aviation security instruments with regard to new and emerging threats to civil aviation, further study was needed in this regard. There was a need for a comprehensive ICAO *Aviation Security Plan of Action* for strengthening aviation security, through a reinforced AVSEC mechanism, an ICAO aviation security audit programme, technical cooperation projects, promotion of aviation security quality control functions and appropriate performance indicators.

Based on the above conclusions the Conference recommended that States take immediate action to lock flight deck doors for aircraft operated internationally, while maintaining measures on the ground to provide the highest level of aviation security. States were also requested to actively share threat information in accordance with Standards in Annex 17 and employ suitable threat assessment and risk management methodologies appropriate to their circumstances, based on a template to be developed by ICAO and ensure that aviation security measures are implemented in an objective and non-discriminatory manner.

As for ICAO's role in this process, the Conference recommended that the Organization develop, as a matter of high priority, amendments to the appropriate Annexes to require protection of the flight deck door from forcible intrusion; continue its efforts to identify and analyze the new and emerging threats to civil aviation with the purpose of assisting in the development of security measures and to actively collaborate with other associated agencies; carry out a detailed study of the adequacy of the existing aviation security conventions and other aviation security-related documentation with a view to proposing and developing measures to close the existing gaps and remove the inadequacies, including amendment where required, so as to deal effectively with the existing, as well as the new and emerging, threats to international civil aviation; develop and take action to deal with the problem of aviation war risk insurance; and develop and implement a comprehensive *Aviation Security Plan of Action* and any additional actions approved by the Council, including a clear identification of priorities.

One of the key conclusions of the Conference was that, in order to further enhance safety and security and to ensure the systematic implementation of the

B. The ICAO Response

critical elements of a State's aviation security system, there was an urgent need for a comprehensive ICAO programme of aviation security audits and that such a programme should audit national level and airport level compliance with Annex 17 and with aviation security related provisions of other Annexes on a regular, mandatory, systematic and harmonized basis. It was the view of the Conference that the ability to determine whether an airport or State is in compliance will require that auditors have a solid aviation security background and be sufficiently trained and certified by ICAO to ensure that auditing is conducted in a consistent and objective manner. The Conference was strongly convinced that such an audit programme should be undertaken under the auspices of ICAO's AVSEC Mechanism which could be guided by proven and successful concepts used in viable programmes already developed by the European Civil Aviation Conference (ECAC), the United States and other States in the development of the framework for a security audit programme.

It was considered that the regional approach would have many benefits and was to be considered as supplementary to local initiatives, in particular by promoting regional partnership and the activities of the ICAO Regional AVSEC Training Centres. The AVSEC Panel, which is an instrumentality of the ICAO Council should assist in the development of technical requirements and guidance materials needed to administer the audits and assist in the development of an effective quality assurance programme to maintain standards of audit performance; and since an audit programme could provide only security levels of audited airports at the time of the audit, a permanent mechanism based on quality control and the regular conduct of exercises and inspections could guarantee the continuity and improvement of security levels determined by the audits.

Arguably, the most significant and seminal recommendation of the Conference was that ICAO establish a comprehensive programme of a universal, regular, mandatory, systematic and harmonized aviation security audits, with implementation beginning in 2003 based on the final work plan established by the Council. It was also decided that, in order to be effective, the programme should be based on an audit process that uses ICAO trained and certified audit teams which are headed by an ICAO staff member and which consistently apply fair and objective methods to determine compliance with Annex 17 by observing measures at airports and assessing the State's capabilities to sustain those measures.

The Conference was of the view that of singular importance to the audit process was the need for the audit programme to be established under the auspices of ICAO's AVSEC Mechanism. It recommended that, in developing the audit programme, which should be transparent and autonomous, ICAO should ensure the greatest possible coordination and coherence with audit programmes already established at a regional or sub-regional level, taking into account aviation security situation in these States. For this to be a reality, a compliance mechanism has to be built into the programme, which will delineate between minor and serious areas of improvement, ensure that immediate corrective action is taken for serious deficiencies and provide to developing States the necessary assistance to measurably improve security.

With regard to funding an aviation security audit programme to be run by ICAO, an adequate and stable source of funding was to be sought for the AVSEC Mechanism through increased voluntary contributions until such time that an allocation of funds can be sought through the Regular Programme Budget, which was envisioned to be as soon as possible. It was recommended that all States be notified of a completed audit, that ICAO Headquarters be the repository for full audit reports and that the sharing of audit reports between States take place on a bilateral or multilateral basis. States were required, under such a programme, to commit to provide ICAO with national AVSEC findings based on a harmonized procedure to be developed by ICAO as early as possible. Of course, those States – in particular developing countries – should be provided with technical and financial assistance under technical cooperation, so that they may take remedial actions to rectify the deficiencies identified during the audit. States should also utilize the ICAO audits to the maximum extent possible and could always approach ICAO with regard to the audit findings for other States.

The Conference also concluded that, in order to execute the ICAO *Aviation Security Plan of Action*, an indicative additional funding requirement was for a minimum of US $15.4 million through voluntary contributions for the current triennium 2002–2003–2004, these figures to be used as a basis for further study by the Council. However, for the longer term a more stable means of funding the ICAO *Aviation Security Plan of Action* would be either through an increase of the assessment to the ICAO General Fund for the following triennia, or by a long-term commitment, on a voluntary basis, of systematic contributions according to an approved suggested level of contribution, to be determined by the Council, by all States. With regard to recouping policies of States, the Conference observed and confirmed that ICAO's policy and guidance material on cost recovery of security services at airports in *ICAO's Policies on Charges for Airports and Air Navigation Services* (Doc 9082/6) and the *Airport Economics Manual* (Doc 9562) remained valid, although there was a need for development of additional policy and guidance material on cost recovery of security measures with regard to air navigation services complementary to that which already exists with respect to airport security charges. There was also a need for further improvement of human resources, utilizing the existing training centres and the standardization of instruction materials, where appropriate based on ICAO's TRAINAIR methodology.

On the above basis, States were called upon by the Conference to commit to provide adequate resources, financial, human and/or otherwise in kind, for the time being on a voluntary basis through the AVSEC Mechanism, for the ICAO *Aviation Security Plan of Action* for the triennium 2002–2003–2004 as a matter of priority, and be aware of the continuing needs for subsequent triennia. They were also called upon to agree to remove the existing ties they individually imposed on the expenditures of AVSEC Mechanism contributions in order for ICAO to immediately utilize all funds available in the AVSEC Mechanism Trust Funds. The Conference observed that States might wish to use the Technical Co-operation Programme of ICAO as one of the main instruments to obtain assistance in advancing implementation of their obligations under relevant international conventions, Standards and

B. The ICAO Response

Recommended Practices (SARPs) of Annex 17 – *Security* and related provisions of other Annexes, as well as adherence to ICAO guidance material.

As for ICAO's involvement and contribution, the Organization was requested to establish an ICAO *Aviation Security Follow-up Programme* and seek additional resources, similar to the USOAP Follow-up Programme of the Technical Co-operation Bureau, to enable States to obtain technical cooperation in the preparation of necessary documentation and in resource mobilization for aviation security. It was felt that one of the ways in which this could be achieved was by ICAO promoting the use of the ICAO Objectives Implementation Mechanism as a means for States to obtain technical cooperation, as required for the rectification of deficiencies identified during aviation security evaluations and audits and urgently pursuing the development and implementation of an International Financial Facility for Aviation Safety (IFFAS),[4] to encompass not only safety but also security. Another significant function of ICAO was to elaborate on its policy and guidance material on cost recovery of security services, notably to include development of policy and guidance material on cost recovery, through charges, of security measures with regard to air navigation services and explore the issue of using security charges as a means of recovering the cost of ICAO assistance when it is provided to States for security development projects.

II. Post Conference Work

In furtherance to the recommendations of the Conference, the ICAO Secretariat initiated an aviation security plan of action which was aimed at reviewing legal instruments, in particular the enhancement of Annex 17 – *Security* – *Safeguarding International Civil Aviation against Acts of Unlawful Interference* to the Convention on International Civil Aviation (the work undertaken by the AVSEC Panel and the latest Amendment 10 to Annex 17) and introduction or strengthening of security-related provisions in other Annexes to the Convention (Annex 1 – *Personnel Licensing*, Annex 6 – *Operation of Aircraft*, Annex 8 – *Airworthiness of Aircraft*, Annex 9 – *Facilitation*, Annex 11 – *Air Traffic Services*, Annex 14 – *Aerodromes* and Annex 18 – *The Safe Transport of Dangerous Goods by Air*). The plan of action also envisioned reinforcing AVSEC Mechanism activities, notably in the preparation of security audits and in undertaking immediate/urgent assistance to States, and expediting work on improving technical specifications relating to and further implementing the use of Machine Readable Travel Documents (MRTDs), biometric identification and travel document security and improving border security systems. The reviewing of certain Procedures for Air Navigation Services (PANS) and revision of relevant ICAO manuals and other guidance

---

[4]For detailed information on the proposed International Financial Facility for Aviation Safety, see, Abeyratne (2000, pp. 383–407).

material including further development of Aviation Security Training Packages (ASTPs), training programmes, workshops, seminars, as well as assistance to States through ICAO's technical co-operation programme are also on the programme of implementation.

At that time, ICAO considered the development and execution of a comprehensive and integrated ICAO AVSEC Plan of Action as its highest priority. It is no less important to ICAO at the present time. The success of this Plan of Action was to be measured over a long period as the improvements expected in Contracting States would require an intensive and continuous worldwide commitment. It was expected that the full and active participation of all Contracting States, as well as all technical and deliberative bodies of ICAO, was essential for the achievement of concrete results within an acceptable period of time.

The aviation security plan of action of ICAO was to focus on the development of new training and guidance material on National Quality Control (NQC), System Testing, Auditors, audit guidelines and forms, including urgent distribution to all States, including training and certification of international auditors through the existing ICAO Aviation Security Training Centres (ASTCs) network, which was to be reinforced and expanded where required. It was also expected to include undertaking universal, mandatory and regular AVSEC audits to assess the level of implementation and enforcement by States of SARPs contained in Annex 17, together with the assessment of security measures undertaken and, on a sample basis, at airport level for each State. ICAO would maintain an ICAO AVSEC findings database would be maintained. The creation of Aviation Security Regional Units (ASRUs) functionally linked to the AVSEC Mechanism, to be urgently implemented in Africa, the Middle East, Eastern Europe, the Americas and Asia and Pacific, in order to coordinate the execution of AVSEC Mechanism activities and provide direct assistance to States was also be a feature of the plan.

The seminal consideration regarding ICAO's role in sustaining the aviation industry lies in the mandate of the Organization, as contained in Article 44 of the Convention on International Civil Aviation.[5] In this context, ICAO's role throughout the past 63 years has been one of adapting to the trends as civil aviation went through three distinct phases of metamorphosis. The first phase was the modernist era as it prevailed when the Convention on International Civil Aviation was signed at Chicago on 7 December 1944, which was centred on State sovereignty[6] and the widely accepted post-war view that the development of international civil aviation can greatly help to create and preserve friendship and understanding among the nations and peoples of the world, yet its abuse can become a threat to general security.[7] This essentially modernist philosophy focussed on the importance of the

---

[5]Convention on International Civil Aviation (also called the Chicago Convention), signed at Chicago on 7 December 1944. See ICAO doc 7300/9 Ninth Edition, 2006.

[6]Article 1 of the Chicago Convention provides that the Contracting States recognize that every State has complete and exclusive sovereignty over airspace above its territory.

[7]Preamble to the Chicago Convention.

State as the ultimate sovereign authority which can overrule considerations of international community welfare if they clashed with the domestic interests of the State. It gave way, in the 1960s and 1970s to a post-modernist era of recognition of the individual as a global citizen whose interests at public international law were considered paramount over considerations of individual State interests.

The 11 September 2001 events led to a new era that now calls for a neo-post modernist approach. This approach, as has been demonstrably seen after the occurrence of the events of 11 September 2001, admits of social elements and corporate interests being involved with States in an overall effort at securing world peace and security. The role of ICAO in this process is critical, since the Organization is charged with regulating for safe and economic air transportation within the broad parameters of the air transport industry. The industry remains an integral element of commercial and social interactivity and a tool that could be used by the world community to forge closer interactivity between the people of the world.

In the above sense, ICAO's initiatives in the fields of aviation security in the immediate aftermath of the 9/11 events have not been mere reactive responses but a visionary striving to ensure the future sustainability of the industry. Of course, this responsibility should not devolve upon ICAO alone. ICAO's regulatory responsibility can only be fulfilled through active regulatory participation by States.

C. Emerging Threats

I. Probability

Blaise Pascal, in his book *Ars Cogitandi* states that fear of harm ought to be proportional not merely to the gravity of the harm but also to the probability of the event.[8] It is also a fact of risk management that, under similar conditions, the occurrence (or non-occurrence) of an event in the future will follow the same pattern as was observed in the past.[9] Based on these premises one is confronted with the terrifying possibility that there could be a nuclear 9/11 sometime in the future.[10]

In the 1919 decision of *Schenk* v. *US*,[11] Justice Oliver Wendell Holmes used the words *clear and present danger* when the US Supreme Court adjudicated the case of Charles Schenk who had distributed leaflets allegedly calculated to incite and cause insubordination and obstruction in recruits of the American Socialist Party.

---

[8]Ferguson (2008, p. 188).
[9]Ferguson (2008, p. 188).
[10]Bobbitt (2008, p. 98–179).
[11]249 US 47 (1919).

The actions of Schenk were considered to constitute an offence under the *Espionage Act* of 1917. Justice Holmes stated:

> The question in every case is whether the words used are used in such circumstances and are of such a nature as to create a clear and present danger that they will bring about the substantive evils that Congress has a right to prevent. It is a question of proximity and degree. When a nation is at war many things that might be said in times of peace are such a hindrance to its effort that their utterance will not be endured so long as men fight and that no court could regard them as protected by any constitutional right.[12]

One commentator[13] is of the view that the words *clear and present danger* have come to mean that arguably, in times of emergency, usually operative legal norms may be disregarded. Implicit in this statement is the axiom *Necessitat non habet legem* (necessity has no law). The moral foundation of Justice Holmes' statement brings to bear the philosophical discourse of justification in responding to threats. This in turn revolves round the basic consideration as to what the greater harm is: application of the legal principles to the letter; or suspending them to prevent an evil greater than the disregard of the law.

II. Reacting to Probability

In every instance of terrorism the focus revolves round those who are harmed by such acts. Therefore, it is difficult not to discuss the merits and demerits of strategy that would bring about the least damage based on a balance of probability. At one of my classes on international law I asked the students "suppose you are law enforcers and you have clear evidence that a certain person will plan and carryout a bomb attack that would kill an entire village in your jurisdiction, and suppose you know that if you apprehend him, there is a danger of many innocent bystanders being killed, would you go ahead and apprehend him?"

We were in the process of discussing the right of a country to exercise self defence against a possible armed attack. In particular, my class was discussing the fact that international law allows, by virtue of Article 51 of the United Nations Charter, a State to defend itself against an armed attack that occurs against its territory and people, while the attack occurs. This prohibition implicitly precludes pre-emptive or preventive attacks on an aggressor based on evidence gathered beforehand.

The United Nations High Level Panel on Threat, Challenges and Change, issued in December 2004 a report which acknowledged that a threatened State can take necessary action as long as the threatened attack is imminent, no other means would deflect it and the action taken to respond to the threat is proportionate. This statement recognizes that the right of self defence extends beyond an actual attack to an imminently threatened one, provided there is credible evidence of such a threat and the threatened State has no obviously alternative recourse available.

---

[12]249 US 47 (1919), 54.
[13]Keith (2005, pp. 185–196).

C. Emerging Threats

Built into my question was the issue of collateral damage, or as we lawyers call it *the law of unidentified consequences*. A case in point is the 2002 targeted killing of a military wing leader of Hamas who was known to be planning and ordering numerous successful bombings against civilians. He was also known to have been planning attacks that were unprecedented in size and consequences. In the process, he was using young children as human shields to carry out suicide attacks against Israel. The Israelis believed that killing the Hamas military leader would thwart the planned attacks and save hundreds of innocent lives. The difficulty in killing this terrorist was that he was constantly changing his living quarters. Often his wife slept beside him, exposing herself, an "uninvolved" person, to the possibility of being killed in an attack against the terrorist. The decision was therefore taken to order a hit only when the terrorist was alone. In one instance, a strike was called off when it was discovered he was with family members. However, when the strike was eventually carried out, the rocket killed not only the terrorist but also his wife, 14-year-old daughter and several others. Israel later issued a statement saying that if they had known the strike action would have resulted in collateral damage the attack would never have taken place.

The law of unidentified consequences dictates that under no circumstances should innocent bystanders be adversely affected intentionally. At this point, I added a new dimension to my question and asked my class: "suppose in killing the terrorist you have no alternative but to kill his 8-year-old daughter first, whom you know he will use as a human shield while carrying out his attack and this is the only opportunity you will get to save hundreds of lives. How would you weigh the one innocent life against hundreds of others?"

Alan Dershowitz, Professor at Harvard Law School and leading criminal lawyer and constitutional scholar, asks the question differently. In the context of the Holocaust of the Second World War, Dershowitz asks: "what if the Jewish Underground had credibly believed that by blowing up German Kindergartens in Berlin, they could force the closure of death camps – that the killing of a hundred innocent German children could save the lives of one million innocent Jewish children and adults, would this be a morally permissible choice of evils?"[14] It will not be difficult to surmise that most people, in considering this dilemma, will agree that the wilful killing of innocent people crosses a certain moral line that should be crossed, if ever, in the most extreme and compelling circumstances.

The same question was asked by Fyodor Dostoyevski in his monumental work *The Brothers Karamazov* where one brother (Ivan) asks the other (Alyosha) whether the latter would, if it were in his power, build an edifice of human destiny that brings happiness to all mankind, but for that he must inevitably and unavoidably torture just one tiny creature, a child and build the edifice upon the unrequited tears of that child. Alyosha vehemently says he will not agree to such a condition.

In theory and in fiction, Alyosha's position is both noble and admirable. However, it becomes a legislative nightmare when put into practice. In his book

---

[14]Dershowitz (2006, p. 26).

Dershowitz, with his characteristic intellectual dexterity, offers guidance for the development of an appropriate jurisprudence for the international community to follow with regard to pre-emptive action. But first, commonsense would dictate that one consider the uncanny pliability of the English language in identifying two types of strikes against terrorism: Preventive strikes; and pre-emptive strikes. The first characterises a strike against an aggressor who is likely to attack sometime in the future. A Pre-emptive strike on the other hand responds to circumstances that already show action taken toward launching an attack.

An example for a pre-emptive strike is the six day war launched by Israel against Egypt and Syria in 1967. The Israeli attack took place after Egypt and Syria had already closed the Strait of Tiran, expelled United Nations plenipotentiaries, massed their regular armies on the order and threatened a genocidal war. Israeli attacked pre-emptively, destroying Egyptian and Syrian forces on the ground and went onto to carve out a comprehensive and decisive victory in just six days. In this sense, it is arguable along similar lines whether the United States action in Afghanistan can be termed "preventive or pre-emptive" in removing the threat of further action on the United States after the events of 11 September 2001. The preventive part of United States action would have been to remove the ruling government which was allegedly harbouring those who could cause further harm to the United States.

Be that as it may, the words "preventive" and "pre-emptive" are unique to the English language where they are used in separate contexts while most other languages use the words inter-changeably. The result is that an explicit distinction between the two words could often be tenuous.

Certain circumstances over the past decade has made the world more cautious, leading it to guide its philosophy of mutual trust along a path which is now called "the precautionary approach". The world would no longer sit and wait, reacting only when a crisis causes massive human suffering and loss of lives. A new doctrine, propounded by a group of scholars at the behest of the United Nations Secretary General Kofi Annan in 2001 has come into being. Called "the responsibility to protect", this doctrine embraces the principle that all member States of the United Nations have a responsibility to protect the lives, liberty and basic human rights of their citizens, and that if they fail or are unable to carry it out, the international community has a responsibility to step in.

All this is well and good. But it does not give me an answer to my question on *the law of identified consequences* – the same question posed by Dershowitz and Dostoyevski. Can we sacrifice one known innocent person to save the lives of a hundred unknown innocent humans? This is where I go back sheepishly to my undergraduate law class on Jurisprudence. I am bound to find in my third year law notes that Bentham, one of the most influential utilitarians, who argued that the right act or policy was that which would cause "the greatest happiness of the greatest number" might have favored sacrificing an innocent life to save more lives, and that Kant, with his categorical imperative as the central philosophical concept of his moral beliefs – that human beings occupy a special place in creation and that morality can be summed up in one, ultimate commandment of reason, or *imperative*, from which all duties and obligations derive, might have totally rejected the idea of sacrificing even one innocent life.

Thomas Hobbes, on the other hand, is an enigmatic source on this issue. His philosophy, found in his *Leviathan*, is, "do not that to another, which thou wouldst not have done to thyself". Does he mean, do not sacrifice the innocent child as you would not like to be sacrificed in a similar manner? Or would he say, as he has, that a human's primary right is self defense against a violent death and because man is constantly at war and his life is "Solitary, poor, nasty, brutish and short, one human sacrifice could be natural consequence of the exercise of self defense?"

III. Deterrence

Studies have shown that stringent measures, when adopted against a particular type of crime belonging to a generic group (such as hijacking in the spectrum of unlawful interference against civil aviation) would be effective enough to reduce that particular type of crime. However, it might give rise to increase in other forms of crime belonging to that generic group. Called the *displacement theory*, this pattern has applied in particular to civil aviation, as seen in the decrease in offences against aircraft after the events of 11 September 2001.

In order that basic strategies are employed for preventing crime and to combat crime when prevention is impossible, crime prevention strategies adopt two methods of combating crime. The first method is to prevent or stop potential criminal acts. The second method is to apprehend and punish anyone who commits a criminal act. These methods follow the philosophy that the prevention of crime can be achieved by increasing the probability of apprehension and applying severe penal sanction to a crime. For example, installation of metal detectors at airports increases the probability of detecting and apprehending potential hijackers or saboteurs. Theoretically the high risk of being apprehended decreases the potential threat and the stringent penal sanction that may apply consequent to such apprehension compound the ominous quality of the preventive means taken.

Many studies focus on aspects of the deterrence theory with the application of the theory to the varied effects of criminology applications on various modes of crimes. These studies relating to the prevention of crime attracted interesting conclusions which went on to reflect that increasing certainty and the severity of punishment reduced the rate of homicide in the United States. It was found by one study that the effect of severity was greater than that of certainty. There were negative correlations between certainty of imprisonment and total felonies. Another observation was that increasing certainty of punishment decreased the incidence of homicide, robbery, assault, burglary, larceny and auto theft. It was also found that certainty appeared to have an independent effect separate from severity of punishment. In view of the fact that effects of severity decreased as certainty of punishment decreased, it would be reasonable to conclude that it is better for policy to concentrate on increasing certainty in order that such an approach would be more effective.

The application of the above approach within U.S. resulted in an increased number of police on patrol, which resulted in a decrease in the number of robberies in New York City subways. Also, increasing the number of police on patrol decreased the number of outdoor felonies in the Twentieth Precinct of New York. Increasing the certainty and severity of punishment for drunk driving has similarly been is effective in reducing drunk driving.

Generally, these studies support the hypothesis that two factors lead a criminal to perceive a greater risk of punishment. These factors are: first, certainty, or a high probability of being arrested and convicted, second, the severity of harshness of the punishment. Certainty and severity of punishment an each have an individual effect on crime prevention; but there is a greater impact when certainty and severity are combined.

Deterrence as a theoretical concept that can be applied in most instances of criminology with practical results is based on the basic assumption that individuals are rational beings. Rationality promotes benefit maximizing behaviour that appears to human beings even against constraints. This means that individuals as rational beings pursue their maximizing goals by making the best choices they can. The underlying concept of the deterrence theory supports the hypothesis that rationalists consider potential criminals as rational decision-makers faced with constraints and uncertainty in their decision making process. The explanation of governmental actions follows the same pattern of the rational model. Therefore, an analyst could conclude that criminals and governmental officials are engaged in a "game" where criminals try to maximize their illegitimate goals through the "least expensive" (apprehension and punishment) approaches and governmental policy makers try to prevent crimes by increasing the probability of apprehension and creating a punishment measure which will serve as a deterrent.

The model that when other variables are held constant, an increase in the person's probability of conviction would decrease the number of offences he or she commits was first introduced in 1968, where a model was developed based on an individual's participation in criminal activity. This theory holds the view that changing the probability of conviction (certainty) has a greater effect on the number of offences than a change in punishment (severity).

In 1973, a formulation of a more comprehensive model of the decision to engage in unlawful activities, based on available empirical evidence, tested the earlier theory and revealed that the rate of specific felonies was positively related to estimates of relative gains, and negatively related to estimates of costs associated with criminal activity. One wonders, however, what the effect would be of the above findings on a criminal who does not intend living after the perpetration of the crime.

IV. *Problems of Deterrence*

The only deterrence that would be effective against terrorism of any nature is broadly based on the success of convincing the terrorist that the risk he takes

C. Emerging Threats

outweighs the benefits which may accrue to his cause by his act. The futility of attempting to wipe out terrorism by the use of military force or the threat of general sanction on an international level is apparent. The terrorist has to be shown that any attempt at terrorist activity would cause him and his cause more harm than good. Deterrence in this context attains fruition when effective punitive sanctions are prescribed and carried out whilst simultaneously denying the terrorist his demands. In both instances the measures taken should be imperatively effective. It is not sufficient if such measures are merely entered into the statute books of a State or incorporated into international treaty. The international community has to be convinced that such measures are forceful and capable of being carried out.

However, deterrence does not stop at the mere imposition of effective sanction nor does it complete its task by the denial of terrorist demands. Arguably, the most effective method of countering terrorism is psychological warfare. The terrorist himself depends heavily on psychology. His main task is to polarize the people and the establishment. He wants popular support and a sympathetic ear. He wants a lot of people listening and watching, not a lot of people dead. Counter measures taken against a terrorist attack, be it hostage taking, kidnapping or a threat of murder, should essentially include an effective campaign to destroy the terrorist's credibility and sincerity in the eyes of the public. Always, the loyalty of the public should be won over by the target and not by the terrorist. It is only then that the terrorist's risk outweighs the benefits he obtains. To achieve this objective it must be ensured that the terrorist receives publicity detrimental to him, showing the public that if the threatened person, group of persons or State comes to harm, the terrorist alone is responsible. Therefore, the most practical measures that could be adopted to deter the spread of terrorism can be accommodated in two chronological stages:

(1) Measures taken before the commission of an offence such as the effective imposition and carrying out of sanctions and the refusal to readily comply with the demands of the terrorist
(2) Measures taken after the commission of the act such as the skilful use of the media to destroy the credibility of the terrorist cause and to convince the people that the responsibility for the act devolves at all stages solely upon the terrorist

One difficulty in exercising deterrence against terrorism in general and international terrorism in particular is that often, the measures taken are not effective enough to convince the terrorist that in the end, more harm would be caused to him than good. Negotiation with the terrorist in particular has to be done by professionals specially trained for the task. *A fortiori*, the media has to be handled by specialists with experience. Things would be much more difficult for the terrorist if these were done. The greatest problem of deterrence is the pusillanimity of the international community in the face of terrorism and the feeble response offered by States as a composite body. The reasons for this hesitation on the part of the international community to adopt effective measures against international terrorism is by no means inexplicable. When one State supports a revolutionary cause which is aimed against another, it is quite natural that the terrorist is aware of the support he is capable of obtaining from at least one part of the already polarized world. Therein lies the problem.

V. Threat Assessment in ICAO

It is incontrovertible that the underlying philosophical enigma involving the spectre of potential collateral damage should not stop the international community from reacting to terrorism. This is particularly so in civil aviation where weapons of mass destruction accounted for 1,993 lives in the Twin Towers on 11 September 2001 and the threat of a similar repeat offence has not entirely disappeared. Against this backdrop, the three grave and emerging threats to civil aviation are bioterrorism, cyber-terrorism and the misuse of shoulder launched surface to air missiles.[15]

At its 33rd session held in Montreal from 25 September – 5 October 2001, the ICAO[16] Assembly adopted Resolution A33-1[17] which was a direct response to the terrorist acts of 9/11. The Resolution recognized that a new type of threat was posed to civil aviation which required new concerted efforts and policies of cooperation on the part of States. The Resolution also urges all ICAO member States to ensure, in accordance with Article 4 of the Chicago Convention,[18] that civil aviation is not used for any purpose inconsistent with the aims of the Convention, and to hold accountable and punish severely those who misuse civil aircraft as weapons of destruction, including those responsible for planning and organizing such acts or for aiding, supporting or harbouring perpetrators. It also called upon States to cooperate with each other in this endeavour and to ensure that ICAO Standards and Recommended Practices (SARPs) relating to aviation security are adhered to.

---

[15]There is also a new dimension in the sabotage of aviation which results in damage caused by the hostile use of dirty bombs, electromagnetic pulse devices or biochemical materials. Dirty bombs are devices which cause damage through nuclear detonation involving the spread of radioactivity to undetermined areas. This article will not address this threat. However, for a detailed discussion on this threat see Abeyratne (2005, pp. 117–129).

[16]The International Civil Aviation Organization, a specialized agency of the United Nations, was established by Article 44 of the *Convention on International Civil Aviation* (Chicago Convention), signed at Chicago on 7 December 1944 (ICAO Doc 7300/9, Ninth Edition, 2006). The main objectives of ICAO are to develop the principles and techniques of international air navigation and to foster the planning and development of air transport. ICAO has 190 Contracting States. ICAO's Mission and Vision Statement is "to achieve its mission of safe, secure and sustainable development of civil aviation through cooperation amongst its member States". In December 2004, following a decision by the 35th Session of the ICAO Assembly, the Council of ICAO approved six Strategic Objectives for 2005–2010: They are: safety; security; environmental protection; efficiency; continuity; and rule of law. The Strategic Objective applicable to this article is security.

[17]Resolution A33-1, Declaration on misuse of civil aircraft as weapons of destruction and other terrorist acts involving civil aviation, Assembly Resolutions in Force (as of 8 October 2004) ICAO Doc. 9848. at VII-1.

[18]Convention on International Civil Aviation, signed at Chicago on 7 December 1944. See ICAO Doc 7300/9 Ninth Edition 2006.

C. Emerging Threats

Finally the Resolution directed the Council of ICAO and the Secretary General to act urgently to address new and emerging threats to civil aviation, in particular to review the adequacy of existing aviation conventions on security.

In response to the requirement of A33-1, that ICAO act with some urgency to address new and emerging threats to civil aviation, an ICAO Special Sub Committee meeting of the Legal Committee on the subject of preparation of one or more instruments addressing new and emerging threats was held at ICAO Headquarters from 3 to 6 July 2007.[19] At this meeting, Australia submitted a proposal[20] to prohibit the intentional and unlawful transport by air of particularly dangerous goods and fugitives. In this paper, Australia quoted the Preamble[21] to the Chicago Convention and emphasized that ICAO was created to help ensure the safe and orderly growth of civil aviation and to encourage the operation of civil aircraft for peaceful purposes. It was also the view of Australia that there were gaps in the international legal framework with regard to the unlawful transport of biological, chemical and legal weapons and other dangerous material on board civil aircraft and that the international aviation community had a responsibility to address these lacunae and shortcomings, particularly when an opportunity such as the one presented through the ICAO meeting arose.

The Sub Committee meeting had the opportunity, through the Australian paper, to note other international legislation on the transportation of dangerous materials. For example, the 2005 Protocol to the Convention for the Suppression of Unlawful Acts Against the Safety of Maritime Navigation which underscores the extreme danger of use by unlawful activity of maritime transport of nuclear, chemical or biological weapons.[22] Additionally, there are other guidance material, such as those issued by the World Health Organization[23] which provide practical guidance to facilitate compliance with current international regulations for the transport of

---

[19] One of the terms of reference of the Sub Committee as agreed by the ICAO Council was: to prepare, in light of A33-1 and the guidance of the Council, one or more draft instruments addressing the new and emerging threats to civil aviation. See Special Sub Committee on the Preparation of One or More Instruments Addressing New and Emerging Threats, Introductory Note, *LC/SC-NET-WP/1*, 29/05/07 at p. 2.

[20] Proposal to Prohibit the International and Unlawful Transport by Air of Particularly Dangerous Goods and Fugitives, *LC/SC-NET-WP/3*, 5/07/07.

[21] The Preamble to the Chicago Convention recognizes that the future development of international civil aviation can greatly help to create and preserve friendship and understanding among the nations and peoples of the world, and yet its abuse can become a threat to the general security. It also states that it is desirable to avoid friction and to promote co-operation between nations and peoples upon which the peace of the world depends. In pursuance of these objectives, governments signed the Convention that contains certain principles and arrangements in order that international civil aviation may be developed in a safe and orderly manner and that international air transport services may be established on the basis of equality of opportunity and operated soundly and economically.

[22] United Nations (2009, p. 1–2).

[23] Guidance on Regulation for the Transport of Infectious Substances, World Health Organization, September 2005, WHO/CDS/CSR/LYO/2005.22.

infectious substances[24] and patient specimens by all modes of transport, both nationally and internationally, and include the changes that apply from 1 January 2005.[25] The WHO regulations categorically state that The *Technical Instructions for the Safe Transport of Dangerous Goods by Air* published by ICAO are the legally binding international regulations.[26] IATA *Dangerous Goods Regulations* (DGR) that incorporate the ICAO provisions and may add further restrictions (where necessary such restrictions are included in these guidelines). The ICAO rules apply on all international flights. For national flights, i.e. flights within one country, national civil aviation authorities apply national legislation. This is normally based on the ICAO provisions, but may incorporate variations. State and operator variations are published in the ICAO Technical Instructions and in the IATA Dangerous Goods Regulations. The WHO guidelines also contain detailed packing instructions regarding infectious substances.[27]

A Special Sub Committee of the Legal Committee of ICAO met in Montreal from 3 to 6 July 2007 to discuss the preparation of one or more instruments addressing new and emerging threats. One of the issues addressed at this meeting was the unlawful transport of biological, chemical, nuclear weapons and other dangerous substances on board aircraft.

Earlier, the Secretary General of ICAO, Dr. Taieb Cherif, addressing the China Civil Aviation Development Forum on 9 May 2007, stated that although the global air transport system remains as secure as ever, yet events such as the illegal terrorist plot in the United Kingdom in the Summer of 2006, potentially involving liquids

---

[24] For the purposes of transport, infectious substances are defined as substances which are known or are reasonably expected to contain pathogens. Pathogens are defined as microorganisms (including bacteria, viruses, rickettsia, parasites, fungi) and other agents such as prions, which can cause disease in humans or animals. The definition is applied to all specimens except those explicitly excluded in the WHO Guidance Material.

[25] The international regulations for the transport of infectious substances by any mode of transport are based upon the Recommendations made by the Committee of Experts on the Transport of Dangerous Goods (UNCETDG), a committee of the United Nations Economic and Social Council. The Recommendations are presented in the form of Model Regulations. The United Nations Model Regulations are reflected in international law through international modal agreements.

[26] Technical Instructions for the Safe Transport of Dangerous Goods by Air: Supplement 2009/10 Edition (Doc 9284).

[27] The system of packaging is recommended for use all infectious substances. It consists of three layers as follows: Primary receptacle – which is a primary watertight, leak-proof receptacle containing the specimen. The receptacle is packaged with enough absorbent material to absorb all fluid in case of breakage. Secondary packaging – which is a second durable, watertight, leak-proof packaging to enclose and protect the primary receptacle(s). Several cushioned primary receptacles may be placed in one secondary packaging, but sufficient additional absorbent material shall be used to absorb all fluid in case of breakage. Outer packaging – which are secondary packagings placed in outer shipping packagings with suitable cushioning material. Outer packagings protect their contents from outside influences, such as physical damage, while in transit. The smallest overall external dimension shall be 10 cm × 10 cm. Each completed package is normally required to be marked, labelled and accompanied with appropriate shipping documents (as applicable). Special Sub Committee on the Preparation of One or More Instruments Addressing New and Emerging Threats, Introductory Note, *LC/SC-NET-WP/1*, 29/05/07 at p. 6.

C. Emerging Threats 19

used as explosives, reminds us how vulnerable the system is. On another aviation platform, Giovanni Bisignani, Director General and CEO of the International Air Transport Association (IATA) stressed at its Annual General Meeting held in Vancouver from 3 to 5 June 2007 that the industry has changed tremendously in five years since 9/11. Bisignani stated that, six years after the tragic events of 2001, air travel was much more secure but there were unlimited ways to attack the aircraft integrity. He added that there was no perfect security system and terrorists change tactics and weapons. Bisignani rightly pointed out that terrorists are studying what measures the industry is adopting; and that all the air industry can do is make the system strong enough to constitute sufficient deterrent and make aircraft a harder target to hit.

VI. The AVSEC Panel

The Aviation Security Panel of ICAO met at its Twentieth Meeting in Montreal from 30 March to 3 April 2009. One of the key areas of discussion at this meeting concerned new and emerging threats to civil aviation. The Panel worked through the Working Group on New and Emerging Threats and noted that significant progress in efforts to proactively identify vulnerabilities and potential gaps in existing measures had been made, that would strengthen *Annex 17* (Aviation Security) to the Convention on International Civil Aviation (Chicago Convention).[28] At this meeting, the European Civil Aviation Conference (ECAC) stressed the importance of the challenge posed by cyber threats in light of the current lack of related provisions in Annex 17.

Consequently, the Panel considered the threat of cyber attacks, and some members stressed that this threat is significant. With reference to a proposal to include a Recommended Practice in Annex 17 to ensure that information and communication technology systems used for civil aviation purposes be protected from cyber attacks, the Panel agreed that, given the complexity of this issue, which involves air traffic management systems, aircraft design and operations, the matter requires further analysis by the Working Group on New and Emerging Threats prior to inclusion in Annex 17 or any guidance material. This analysis will be disseminated over the secure website by the end of June 2009 and, depending on the results of the analysis, the Working Group on Amendment 12 to Annex 17 will develop a proposal for amending the Annex, to be presented to the Panel at its 21st meeting.

The Panel also considered the merits of building unpredictability into the aviation security regime. While concern was expressed regarding the impact of

---

[28] Convention on International Civil Aviation signed at Chicago on 7 December 1944. See ICAO Doc 7300/9 Ninth Edition, 2006.

unpredictable security measures on passenger confidence in aviation security, many Panel members supported implementation of the concept because of its value as a deterrent. It was suggested that States adopt an approach providing for a baseline regime, but with the addition of unpredictable measures, thus achieving a balance between certainty and unpredictability. With regard to an amendment to Annex 17 in this regard, the need for introducing unpredictability into the aviation security regime was considered, and it was agreed that unpredictability should be promoted in principle but not prescribed. The Panel suggested that if an Annex 17 specification related to unpredictability were to be developed, it would be necessary to ensure that the introduction of this concept by States does not diminish the level of security or result in delays for passengers. Further, the Panel noted that appropriate guidance material may be required to address the potential negative impact of introducing the concept of unpredictability, and proposed the development of guidance material related to unpredictability prior to the introduction of an amendment to Annex 17.

A Conclusion of the Panel was, *inter alia*, that the threat of cyber attacks is real and cannot be ignored, and that further analysis by the Working Group on New and Emerging Threats would be appropriate. Another Conclusion was that the ICAO focal point of contact (PoC) Network is an important tool for sharing critical threat information and should be used more effectively, and that the Secretariat should consider the establishment of a web-based community page. Yet another was that the concept of building unpredictability into the aviation security regime is in principle a useful tool, however, concerns expressed regarding the possible impact on the level of security and the impact on passenger confidence should be resolved before its inclusion as a Recommended Practice in Annex 17.

The Recommendations of the Panel were that:

(a) The Working Group on New and Emerging Threats propose its new name, terms of reference and composition, including suggestions on how observers might participate in the Working Group, as well as details of its evolving collaboration with the G8 Group, at the 21st Panel meeting.
(b) The Working Group evaluate the threat of cyber attacks and disseminate the results of its analysis on the secure website by the end of June 2009 and that, depending on the results of this analysis, the Working Group on Amendment 12 to Annex 17 consider developing an amendment to Annex 17 for presentation at the 21st Panel meeting.
(c) The ICAO Secretariat issue an electronic bulletin reminding States of the importance of subscribing to the PoC Network and providing information on its usage.
(d) The concept of building unpredictability into the aviation security regime be further considered.[29]

---

[29] See Report of the Aviation Security (AVSEC) Panel, Twentieth Meeting, AVSECP/20 at 2.1.

VII. Bioterrorism

The recent recreation of the Spanish flu virus that killed 50 million people worldwide in 1918 proves that deadly viruses are being revisited and are undergoing genetic modification. This brings to bear the inevitable question as to whether there is enough security to stop them from falling into the wrong hands. There is also the ominous prospect – that transportation of these dangerous pathogens by air would leave aviation vulnerable. This article examines precautionary measures currently being taken and the legal and regulatory significance of such measures.

The leakage of dangerous pathogens[30] from laboratories presents an ominous analogy to the aviation sector in that the same could well occur in the carriage of such dangerous goods by air. Although past instances of escaping dangerous pathogens are small in number, nonetheless their occurrence and the threat posed to the wellbeing of humanity cannot be underestimated. In 2002 when Anthrax spores escaped from two military laboratories in the United States, the authorities agreed that the leakage was due to a security lapse.[31] In 2003 a string of such leakages occurred in Asia, this time of the SARS virus.[32]

It is now known that the laboratory is not the only place where security lapses could occur. Modern exigencies require samples of deadly pathogens to be transported regularly over vast distances to reach researchers across the world. This calls for a delicate balance between recognizing the compelling need for scientists to exchange and collectively use different strains in order to identify naturally occurring diseases and mutations on the one hand and ensuring that the transport of these infectious substances[33] are carried out according to United Nations Model Regulations[34] on the other. These model regulations are the base upon which specific

---

[30] Pathogens are micro-organisms (including bacteria, viruses, rickettsia, parasites, fungi) or recombinant micro-organisms (hybrid or mutant) that are known or are reasonably expected to cause infectious disease in humans or animals.

[31] An year earlier, a covert event occurred in October 2001 when anthrax spores were sent through the mail exposing persons in the eastern USA to contaminated mail resulting in deaths, illnesses and identified exposures to Anthrax. Overt, announced events, in which persons are warned that an exposure has occurred, have taken place in the United States, although most of these were determined to have been hoaxes, that is, there were no true exposures to infectious agents.

[32] The leakages occurred in China, Taiwan and Singapore. See Air-Tight Security, *Intersec*, June 2007, 33–35 at 34.

[33] Infectious substances are defined as substances known to contain, or reasonably expected to contain, pathogens.

[34] The United Nations has developed recommendations on model regulations for the transport of dangerous goods which recognize that various chemical combinations and mixtures have different requirements in packing for the purpose of transport. See Recommendations on the Transport of Dangerous Goods, Model Regulations, Volume 1, 14th Revised Edition: 2005, Chap. 2.6, p. 113–114. Furthermore, the United Nations Model regulations contain packing instructions for primary, secondary and outer packaging of hazardous goods. See Model Regulations *Id.* Volume 11, Instruction P620 at p. 70. Specimens (human, animal, food, environmental, etc.) known or reasonably expected to contain pathogens are now to be classified as infectious substances. When

provisions for the carriage by air are formulated in the packing of samples of infectious pathogens for transportation by air. The shipment of infectious agents or diagnostic specimens by air must comply with local, national and international regulations. International air transport regulations are contained in various documentation of the International Civil Aviation Organization and *Dangerous Goods Regulations* – an annual publication of the International Air Transport Association published in January, and usually revised on a yearly basis. These ICAO and IATA documents will be discussed in some detail later in this article.

Dangerous Goods Regulations are implicitly accompanied by the requirement that anyone requesting samples should provide the necessary evidence that they are registered with their government for the receipt of such substances and that they have the appropriate facilities, staff and security measures in place to carry out work on the samples received.

There are four diseases recognized as most likely to be associated with bioterrorism potential: anthrax; botulism; plague; and smallpox. Although these agents are considered to be the most likely to be used in bioterrorism they are not usually prioritized in any order of importance. There are other agents which offer potential to bioterrorism such as those causing tularemia, brucellosis, Q fever, viral hemorrhagic fevers, viral encephalitis, and a disease associated with staphylococcal enterotoxin B.

There are others which cause security experts concern as emergent threats to security through bio terrorism. These are Severe Acute Respiratory Syndrome (SARS), monkeypox and pandemic influenza. These are naturally occurring diseases,[35] which are of concern because they are new and/or epidemic.[36] Outbreaks of dangerous pathogens may occur naturally or as covert or overt events. An outbreak is suspected only upon recognition of unusual disease clusters or symptoms.[37] For example, SARS was recognized as a naturally occurring event initially from Southeast Asia in February 2003.

---

these specimens are transported/shipped for any purpose, including initial or confirmatory testing for the presence of pathogens, they are to be packaged and shipped as infectious substances.

[35] It is widely recognized that SARS is not a disease but a syndrome. See generally, Abeyratne (2002, pp. 53–80).

[36] Rapid response to a dangerous pathogen event requires prompt identification of its onset. Because of the rapid progression to illness and potential for dissemination of some of these agents, it may not be practical to await diagnostic laboratory confirmation. Instead, it is necessary to initiate a response based on the recognition of high-risk syndromes, i.e., typical combination of clinical features of the illness at presentation that might alert healthcare practitioners to the possibility of an outbreak. Examples of syndromes potentially resulting from infections with dangerous pathogens include: encephalitis/meningitis, hemorrhagic mediastinitis, severe pneumonia, papulopustular rash, hemorrhagic fever, descending paralysis and nausea/vomiting/diarrhoea.

[37] An outbreak is usually identified consequent to a rapidly increasing disease incidence (e.g., within hours or days) in a normally healthy population, such as unexplained death with fever in a non-trauma patient, or a botulism-like syndrome, meningitis or encephalitis in more than one patient.

C. Emerging Threats

A bioterrorism attack is the deliberate release of viruses, bacteria, or other germs (agents) used to cause illness or death in people, animals, or plants. These agents are typically found in nature, but it is possible that they could be changed to increase their ability to cause disease, make them resistant to current medicines, or to increase their ability to be spread into the environment. Biological agents can be spread through the air, through water, or in food. Terrorists may use biological agents because they can be extremely difficult to detect and do not cause illness for several hours to several days. While some bioterrorism agents, such as the smallpox virus, can be spread from person to person some agents such as anthrax are incapable of doing so.

There have been several noteworthy instances of bioterrorism in the past[38] as early as 1915,[39] which send an ominous message that it is a distinct possibility in the aviation context. Until recently in the United States of America, most biological defense strategies have been geared to protecting soldiers on the battlefield rather than looking after ordinary people in cities. In 1999, the University of Pittsburgh's Center for Biomedical Informatics deployed the first automated bioterrorism detection system, called RODS (Real-Time Outbreak Disease Surveillance). RODS is designed to draw collect data from many data sources and use them to perform signal detection, that is, to detect a possible bioterrorism event at the earliest possible moment. RODS, and other similar systems, collect data from sources including clinical data, laboratory data, and data from over-the-counter drug sales. In 2000, Michael Wagner, the co director of the RODS laboratory, and Ron Aryel, a subcontractor, conceived of the idea of obtaining live data feeds from "non-traditional" (non-health-care) data sources. The RODS laboratory's first efforts eventually led to the establishment of the National Retail Data Monitor, a system which collects data from 20,000 retail locations nation-wide.

On 5 February 2002, President Bush visited the RODS laboratory and used it as a model for a $300 million spending proposal to equip all 50 States with bio surveillance systems. In a speech delivered at the nearby Masonic temple, Bush compared the RODS system to a modern "DEW" line (referring to the Cold War ballistic missile early warning system).

---

[38] In 1984 followers of the Bhagwan Shree Rajneesh attempted to control a local election by incapacitating the local population by infecting salad bars in 11 restaurants, doorknobs, produce in grocery stores and other public domains with *Salmonella typhimurium* in the city of The Dalles, Oregon. The attack caused about 751 people to get sick (there were no fatalities). This incident was the first known bioterrorist attack in the United States in the twentieth century. In September and October of 2001, several cases of anthrax broke out in the United States which were reportedly caused deliberately. This was a well-publicized act of bioterrorism. It motivated efforts to define biodefense and biosecurity.

[39] In 1915 and 1916, Dr Anton Dilger, a German-American physician used cultures of anthrax and glanders with the intention of committing biological sabotage on behalf of the German government. Other German agents are known to have undertaken similar sabotage efforts during World War I in Norway, Spain, Romania and Argentina.

The principles and practices of bio surveillance, a new interdisciplinary science, were defined and described in a handbook published in 2006.[40] Data which potentially could assist in early detection of a bioterrorism event include many categories of information. Health-related data such as those collected from hospital computer systems, clinical laboratories, electronic health record systems, medical examiner record-keeping systems, 911 call center computers, and veterinary medical record systems could be of help in the fight against bioterrorism. Researchers are also considering the utility of data generated by ranching and feedlot operations, food processors, drinking water systems, school attendance recording, and physiologic monitors, among others. Intuitively, one would expect systems which collect more than one type of data to be more useful than systems which collect only one type of information (such as single-purpose laboratory or 911 call-center based systems), and be less prone to false alarms. This indeed appears to be the case.

The inherently uncontrollable nature of a dangerous pathogen makes bioterrorism unattractive as a warfare strategy. However, the potential power of genetic engineering cannot be marginalized or underestimated and the compelling need for continuing vigilance cannot be ignored.

VIII. Cyber-Terrorism

As far back as in March 1998, the web site of the National Aeronautics and Space Administration (NASA) of the United States received a 'denial of service' attack, calculated to affect Microsoft Windows NT and Windows 95 operating systems.[41] These attacks prevented servers from answering network connections; crashed computers, causing a blue screen to appear on the computers. The attacked systems were revived, but this attack was a follow up of one in February of the same year, when, for two weeks the US Defence Department had unclassified networks penetrated, where hackers accessed personnel and payroll information.

Cyber-terrorism has the advantage of anonymity, which enables the hacker to obviate checkpoints or any physical evidence being traceable to him or her. It is a low budget form of terrorism where the only costs entailed in interfering with the computer programs of an air transport system would be those pertaining to the right computer equipment.

Any interference with air transport, which would be inextricably linked to the purpose of international civil aviation as enunciated in the *Preamble* to the Chicago

---

[40] Wagner et al. (2006). Bio surveillance is the science of real-time disease outbreak detection. Its principles apply to both natural and man-made epidemics (bioterrorism). It is worthy of note that in addition to activity in this field in the United States, there is also work being done in Europe, where disease surveillance is beginning to be organized on a continent-wide scale needed to track a biological emergencies. The system not only monitors infected persons, but also attempts to discern the origin of the outbreak.

[41] http://mgrossmanlaw.com/articles/1999.

C. Emerging Threats

Convention which categorically states that the future development of international civil aviation can greatly help to create and preserve friendship and understanding among the nations and people of the world, yet, its abuse can become a threat to the general security.

The maintenance of international peace and security is an important objective of the United Nations,[42] which recognizes one of its purposes as being *inter alia*:

> To maintain international peace and security, and to that end: take effective collective measures for the prevention and removal of threats to the peace, and for the suppression of acts of aggression or other breaches of peace, and to bring about by peaceful means, and in conformity with the principles of justice and international law, adjustment or settlement of international disputes or situations which might lead to a breach of the peace.[43]

It is clear that the United Nations has recognized the application of the principles of international law as an integral part of maintaining international peace and security and avoiding situations which may lead to a breach of the peace.

Manuela Guill, a leading expert in threat assessment says:

> Cyber-terrorism can be used in many ways. In its simplest form, it can be used as a means of disinformation or psychological warfare by manipulating media attention regarding possible threats, thus causing disruption to airport and aircraft operations. This could result in the reluctance of persons to travel which, in turn, could affect the economies of nations dependent on the movement of air passengers. In its most serious form, cyber-terrorism could lead to fatalities, injuries and major damage at airports and to aircraft in flight.[44]

The particularity of cyber-terrorism is that the threat is enhanced by globalization and the ubiquity of the Internet. It is a global problem in search of a global solution.

IX. MANPADS

Since the events of 11 September 2001, there have been several attempts against the security of aircraft in flight through the misuse of Man Portable Air Defense Systems (MANPADS). The threat of MANPADS to aviation security is by far the most ominous and the international aviation community has made some efforts through ICAO. In recent years, Man Portable Air Defense Systems (MANPADS) have posed a serious threat to aviation security.

MANPADS are extremely effective weapons which are prolific in their availability worldwide. Introduced in the 1950s and originally meant to deter terror attacks from air to ground to be used by State authorities and other protection agencies, these weapons have got into the wrong hands and are being used against

---

[42]Charter of the United Nations and Statute of the International Court of Justice, Department of Public Information, United Nations, New York, DPI/511 – 40108 (3-90), 100M at 1.

[43]Charter of the United Nations and Statute of the International Court of Justice, Department of Public Information, United Nations, New York, DPI/511 – 40108 (3-90), 100M at 3.

[44]Guill (2000, p. 18).

civil and military aviation. The surface to air MANPAD is a light weapon which offers very little warning before impact, and is often destructive and lethal.[45] They are cheap, easily carried, handled and concealed. It is claimed that there are at least 100,000 and possibly in excess of 500,000 systems in inventories around the world and several thousands of these are vulnerable to theft from State authorities.[46] It is also claimed that there is a 70% chance that a civil aircraft will be destroyed if hit by a MANPAD.[47] A study conducted and published in early 2005 by the Rand Corporation concludes that, based on the effects of the attacks of 11 September 2001, it is plausible for air travel in the United States to fall by 15–20% after a successful MANPADS attack on a commercial airliner in the United States.[48] The international aviation community is aware that civil aircraft are particularly vulnerable to hand held ground to air missiles and that susceptibility avoidance techniques (calculated to avoid being hit) and vulnerability avoidance (survival after being hit) systems must be in place. This is particularly so since tracking the proliferation of MANPADS is difficult since any intelligence gathered on this particular threat is usually *ex post facto*, through the recovery of launchers or fragments from expended missiles. Contrary to popular belief, the MANPAD is considerably durable and can be used several years after inactivity, with recharged batteries.

The World's attention to the deadly threat posed by MANPADS was further drawn in November 2002 when there was an unsuccessful attempt to bring down a civilian aircraft leaving Mombasa, Kenya. Over the past 35 years, significant developments have taken place in dangerous weapons systems creating more opportunities for terrorists. The ready acceptance of new modern technologies by the international community and our growing dependence on them have created many targets, such as nuclear and civil aircraft in flight. Similarly, developments in electronics and microelectronics, and the trend towards miniaturization and simplification have resulted in a greater availability of tactical weapons with longer ranges and more accuracy that are also simpler to operate. One of the most effective developments in individual weaponry is portable, precision-guided munitions (PGMs), which are lightweight and easy to operate. They can usually be carried and operated by a single person. The United States-made Stinger, the British-made Blowpipe and the Russian-made SA-7 missiles are examples of these smaller weapons. These are shoulder-fired, anti-aircraft missiles that have infra-red, heat-seeking sensors in the projectile that guide it to the heat emitted from an aircraft engine. It is known that more than 60 States possess SA-7 missiles and there is no

---

[45]The lethality of the weapon can be reflected by the 340 MANPADS used by Afghan Mujahedeen rebels to successfully hit 269 soviet aircraft. See: http://www.janes.com/security/international_security/news/.

[46]MANPADS, *Ploughshares Monitor*, Autumn 2004, at 83.

[47]MANPADS, *Ploughshares Monitor*, Autumn 2004, at 83. The deadly accuracy and ease of handling of MANPADS were demonstrated when Somali gunmen shot down two US MH-60 Black Hawk helicopters in October 1993.

[48]Infrastructure Safety and the Environment, *Protecting Commercial Aviation against the Shoulder-Fired Missile Threat*, Rand Corporation, 2005, at 9.

doubt that most of them maintain strict security measures to prevent the outflow of the weapons. However, it has been alleged that some States, including Libya, have supplied PGMs to terrorist organizations. It is incontrovertible that in the hands of terrorists these missiles are not likely to be used against conventional targets such as tanks and military fighter aircraft. Of particular concern is the prospect of civilian airliners being shot at by SAMs and anti-tank rockets as they land at or take off from airports[49] Dr. Richard Clutterbuck subsumes the great threat of missile attacks:

Recent years have seen increasing use of expensive and sophisticated surface-to-surface and surface-to-air missiles (SSM and SAM) by terrorists, generally of Russian or East European origin and redirected by Arab Governments, notably Colonel Gadafi's. Continuing development of these weapons for use by regular armies will ensure that new and more efficient versions will become available for terrorists.[50]

With increased airport security, the possibility of placing explosive devices on civil aircraft is becoming more difficult, but now the same destructive result can be achieved far more easily by using modern missiles or rockets.

There are a few incontrovertible truths that drive the issue of the illegal carriage of infectious pathogens by air. The first is that, as recognized by WHO, the lead role in legislative and regulatory control of the issue lies well within ICAO. The second is that, one has to go back to the basics of the rule book and start with the Preamble to the Chicago Convention. The Preamble unequivocally links the future development of aviation to "general security" which essentially means that aviation should not only be concerned with persons and property directly involved with air transport but also with the rest of the world that might be adversely affected by the release of infectious pathogens through aviation.

The third home truth is that it is a pre-eminent responsibility of States to ensure security at laboratories in their territories as the illegal carriage of infectious substances by air is liked to the initial leakage from a laboratory. Therefore it is extremely important for States to strictly enforce their dangerous goods legislation. It is also important to treat this subject holistically in terms of the world at large and not restrictively by singling out only those involved in the flight concerned. Finally, States have to adopt a security culture that admits of an overall approach to the threat as a potential harm to the health of humanity. This should inevitably include strict adherence by States to the provisions of Annexes 17 and 18 and inclusion of new Standards in the Annexes as necessary, together with an abiding understanding that the illegal carriage of infectious pathogens by air portends a threat both to safety and security of aviation.

As for cyber-terrorism, the offences related to cyber-terrorism should be addressed on the basis that individuals have international duties which transcend the national obligations of obedience imposed by an individual State. By the same token, it must also mean that individual States owe their citizens and the world at

---

[49] Hanle (1989, p. 185), Ofri (1984, p. 49), Pierre (1975, p. 1256), Dorey (1983, p. 142).
[50] Clutterbuck (1991, p. 175).

large a responsibility for maintaining world security. The philosophy of these two premises has to be vigorously employed in bringing to fruition the above measures. It is only then that a substantial legal contribution could be made to the controlling of this offence.

Finally, with regard to MANPADS, the gathering of reliable intelligence remains the first line of defense. Although modern technologies clearly aid terrorists in terms of weapons and targets, technology can also be used against terrorists. Governments which are endowed with the necessary technology can keep track of terrorist organizations and their movements with the aid of computers. At the same time, electronic collection methods and signals intelligence afford the possibility of eavesdropping on and intercepting terrorist communications, leading to better predictions of their operations. One of the instances where intelligence gathering has worked well to prevent terrorism occurred in September 1984, when the Provisional IRA spent an estimated £1.5 million in the United States on a massive shipment of seven tons of arms. With the help of an informer about a forthcoming shipment of weapons, including rockets, to the Provisional IRA from the United States, the FBI informed British intelligence, who in turn contacted the Irish, and the ship carrying the arms was tracked by a US satellite orbiting 300 km above the earth. The satellite photographed the transfer of the arms to a trawler. Finally, two Irish Navy vessels intercepted the trawler and British security forces arrested the crew.[51] This incident shows that intelligence gathering with the help of high technology can cut off the transfer of missiles and other weapons to terrorists.

The installation of a sophisticated antimissile system similar to that employed on military aircraft to divert surface-to-air missiles is an effective deterrent. One good example is the measure taken by the British government which, immediately after the discovery of 20 SA-7s in the coaster *Eksund*, which was intercepted by French authorities off the coast of Brittany in November 1987 when bound for the IRA, fitted all British Army helicopters flying in Northern Ireland with electronic and other decoy systems to confuse the missile's heat-seeking guidance system. These included the US-made Saunders, AN/ALG 144. This system, when linked to the Tracor AN/ALE 40 chaff dispenser, works by jamming the missile's homing radar and sending infra-red flares and chaff to act as a decoy for the heat-seeking device.[52] The system is used by both the US and the Israeli Armies, which have been well-pleased with its performance. Until the British realised that the IRA might be in possession of SAMs, the Ministry of Defense hesitated to install such a system because of the high cost involved, and its decision to do so shows the seriousness of the threat. Another example of a good counter-measure is the response of El Al airlines to the threat of such an attack which included the installation of electronic countermeasure equipment similar to that employed on military aircraft to divert surface-to-air missiles.[53] However the problem is that

---

[51]The *Daily Telegraph*, 16 October 1984; *The Times*, 12 December 1984.

[52]The *Daily Telegraph*, 7 January 1988.

[53]Lewis and Kaplan (1990, p. 226); Crenshaw (1987, p. 126).

these countermeasures are not yet fully effective, although they could minimize the threat. Hence there is a need to proceed diligently with the development of systems that are guaranteed to be able to prevent this type of attack against civil aviation.

For a successful missile attack against aircraft, the firing position has to be located within range of the flight path. A missile's guidance system is such that the weapon has to be fired within a few degrees of the flight path if the infra-red guidance is to locate the target. Accordingly, a possible preventive measure would be to prevent terrorists from getting into a firing position with their missiles. However, it would be very difficult to cut off areas of up to 6 km wide that lie in the paths of aircraft as they land and take off. This measure is therefore impracticable if not impossible.[54] This difficulty can be overcome to an extent by patrolling the outer areas of airports in times of stringent security conditions might prevent such attacks. Even in times when no specific threat has been received, it is within the capacity of most States to monitor those strips of land from which a SAM could be launched and thus minimize the risk. At the same time, these security operations would deter terrorists from spending vital resources on buying SAMs given the limited possibilities for their use.

Finally, it must be noted that, whatever counter measures are used in responding to new and emerging threats in aviation, the thread which binds the fabric of anti-terrorism is risk management. As one commentator said, the achievement of safety in aviation is no longer an esoteric activity.[55] It has to be a structured series of measures which starts by identifying hazards, and evaluating potential scenarios. It should end in the implementation of a management process. What is therefore needed is systematic assessment and a global means of ranking risks based on their seriousness and significance.

X. *The Diverse Nature of Missile Attacks*

The use of SAMs and anti-tank rockets by terrorists goes back to 1973. On 5 September 1973 Italian police arrested five Middle-Eastern terrorists armed with SA-7s. The terrorists had rented an apartment under the flight path to Rome Fumicino Airport and were planning to shoot down an El Al airliner coming in to land at the airport.[56] This arrest proved a considerable embarrassment to Egypt because the SA-7s were later traced back to a batch supplied to it by the Russian Union. It was alleged that the Egyptian government was supplying some of the missiles to the Libyan army but inexplicably, the SA-7s had been directly rerouted to the terrorists. This incident also placed the Russian Union in an awkward position

---

[54]Dorey (1983, p. 142).
[55]Stewart (1993, p. 12).
[56]Dobson and Payne (1987, p. 366).

because of the possibility that its new missile and its policy of the proxy use of surrogate warfare against democratic states were revealed to the West.[57]

The plot of the missile attack on El Al derived from an appalling incident on 21 February 1973, when a Libyan B-727 was shot down over the Sinai desert by an Israeli fighter, killing the 108 innocent people on-board.[58] The Libyan people called for vengeance against Israel. Libya urged the other Arab States to send their warplanes against Israel's major cities and to destroy Israeli airliners wherever they could be found.[59]

On 5 January 1974, 220 soldiers and 200 police sealed off five square miles around Heathrow International airport in London after receiving reports that terrorists had smuggled SA-7s into Britain in the diplomatic pouches of Middle-Eastern embassies and were planning to shoot down an El Al airliner.[60]

Another significant incident occurred on 13 January 1975 when an attempt by terrorists to shoot down an El Al plane with a missile was believed to have brought civil aviation to the brink of disaster. Two terrorists drove their car onto the apron at Orly airport, where they set up a rocket launcher and fired at an El Al airliner which was about to take off for New York with 136 passengers. The first round missed the target thanks to the pilot's evasive action and hit the fuselage of a Yugoslav DC-9 aeroplane waiting nearby to embark passengers for Zagreb. The rocket failed to explode and no serious casualties were reported. After firing again and hitting an administration building, which caused some damage, the terrorists escaped by car. A phone call from an individual claiming responsibility for the attack was received at Reuters. The caller clearly implied that there would be another such operation, saying 'Next time we will hit the target'.

In fact, six days later another dramatic though unsuccessful attempt did occur at Orly airport. The French authorities traced the attack to the PFLP Venezuelan terrorist, and leader of the PFLP group in Europe, Carlos.[61] It is also known that once again an El Al airliner had been deliberately chosen as a target by Gadafi in an attempt to avenge the loss of the Libyan airliner shot down by Israel over the Sinai Desert.[62]

Despite there failures, on 25 January 1976 another abortive attempt was carried out by three PFLP terrorists, who were arrested by Kenyan police at Nairobi Airport – following a tip-off by Israeli intelligence to the Kenyan General Service Unit – before they had time to fire SA-7 missiles at an El Al aircraft carrying 100 passengers. In connection with this operation, two members of the German Baader-Meinhoff Faction, Thomas Reuter and Brigitte Schultz, were also arrested. After 10 days of interrogation, the terrorists were handed over to Israel by the Kenyan

---

[57]Dobson and Payne (1977, p. 134).

[58]Keesing's Contemporary Archives, 5–11 March 1973, p. 25757.

[59]Keesing's Contemporary Archives, 5–11 March 1973, p. 25757.

[60]Mickolus (1980, p. 428).

[61]Dobson and Payne (1977, p. 53).

[62]Dobson and Payne (1977, p. 53).

C. Emerging Threats

government. However, it was not until March 1977, 14 months after the arrests in Kenya, that the Israelis officially announced that they were holding the three Palestinian and two German terrorists. During this period an unsuccessful attempt to gain their release was undertaken by the PFLP in June 1976, when Palestinian terrorists hijacked an Air France aircraft to Entebbe. The names of the five being held in Israel were included on the list of prisoners whose release was demanded in exchange for the hostages. The three Palestinians were released by the Israeli government in 1985.[63]

There has been a marked increase in missile attacks since 1984. On 21 September 1984 Afghan counter-revolutionaries fired a surface-to-air missile and hit a DC-10 Ariana Airliner carrying 308 passengers. The explosion tore through the aircraft's left engine, damaging its hydraulic system and a wing containing a fuel tank. The captain of the aircraft, however, managed to land the aircraft safely at Kabul International Airport.[64] Another significant incident took place on 4 April 1985, when a member of Abu Nidal group fired an RPG rocket at an Alia airliner as it took off from Athens Airport. Although the rocket did not explode, it left a hole in the fuselage.[65]

Advanced missiles and rockets can be found in many terrorist and insurgent armouries. It is suspected that some terrorist organizations, including Iranian militia in Lebanon, the Provisional Irish Republican Army and various African and Latin American insurgents, possess the sophisticated Russian-made RPG-7 portable rocket launcher, but it is disturbing to note that some terrorist organizations, most notably Palestinian groups, have their own RPG-7-manufacturing facilities. In addition, more than a dozen other terrorist and insurgent groups are known to possess portable surface-to-air missiles, These groups include various Cuban surrogates, Colombian drug dealers, and a number of African, European and Palestinian terrorist organizations.[66]

The possibility of undeterred use of missiles may be encouraged by the rapid proliferation of such weaponry and the publicity to be gained by using such systems. The enhanced effectiveness of missiles against aircraft makes the threat of such attacks real.

---

[63]Mickolus (1980, p. 581); Al-Hadaf, Al-Hadaf mao AlBabtal al-Muharrarin: Al-Muo taqilun hawwalu Dhallam al-Asr ila Nidhal Mushriq (Al-Hadaf with the Liberated Heroes: The detainees Transformed the Gloom of Imprisonment Into a Shining Struggle), June 1985, pp. 35–41; *Associated Press*, 15 August 1979.

[64]U.S. Department of Transportation (FAA), *Worldwide Significant Acts Involving Civil Aviation*, 1984, p. 14.

[65]U.S. Department of Defence, *Terrorist Group Profiles* (Washington DC: U.S. GPO, 1989), p. 7.

[66]Adams (1990, pp. 60–61); Wilkinson (1986, pp. 39–40); Dobson and Payne (1982, p. 119).

XI. Installation of an Anti-missile System

The installation of a sophisticated antimissile system similar to that employed on military aircraft to divert surface-to-air missiles is an effective deterrent. One good example is the measure taken by the British government which, immediately after the discovery of 20 SA-7s in the coaster *Eksund*, which was intercepted by French authorities off the coast of Brittany in November 1987 when bound for the IRA, fitted all British Army helicopters flying in Northern Ireland with electronic and other decoy systems to confuse the missile's heat-seeking guidance system. These included the US-made Saunders, AN/ALG 144. This system, when linked to the Tracor AN/ALE 40 chaff dispenser, works by jamming the missile's homing radar and sending infra-red flares and chaff to act as a decoy for the heat-seeking device.[67] The system is used by both the US and the Israeli Armies, which have been well-pleased with its performance. Until the British realised that the IRA might be in possession of SAMs, the Ministry of Defense hesitated to install such a system because of the high cost involved, and its decision to do so shows the seriousness of the threat. Another example of a good counter-measure is the response of El Al airlines to the threat of such an attack which included the installation of electronic countermeasure equipment similar to that employed on military aircraft to divert surface-to-air missiles.[68] However the problem is that these countermeasures are not yet fully effective, although they could minimize the threat. Hence there is a need to proceed diligently with the development of systems that are guaranteed to be able to prevent this type of attack against civil aviation.

XII. The Perimeter Guard

For a successful missile attack against aircraft, the firing position has to be located within range of the flight path. A missile's guidance system is such that the weapon has to be fired within a few degrees of the flight path if the infra-red guidance is to locate the target. Accordingly, a possible preventive measure would be to prevent terrorists from getting into a firing position with their missiles. However, it would be very difficult to cut off areas of up to 6 km wide that lie in the paths of aircraft as they land and take off. This measure is therefore impracticable if not impossible.[69] This difficulty can be overcome to an extent by patrolling the outer areas of airports in times of stringent security conditions might prevent such attacks. Even in times when no specific threat has been received, it is within the capacity of most States to monitor those strips of land from which a SAM could be launched and thus minimize the risk. At the same time, these security operations would deter terrorists

---

[67] The *Daily Telegraph*, 7 January 1988.
[68] Lewis and Kaplan (1990, p. 226); Crenshaw (1987, p. 126).
[69] Dorey (1983, p. 142).

C. Emerging Threats33

from spending vital resources on buying SAMs given the limited possibilities for their use.

Although the success rate so far of Western States in preventing terrorist missile attacks against civil aviation is satisfactory, and security forces, with the help of good intelligence, have been successful in tracking down and capturing missiles before they could be used, it is not unlikely that there will be attempts to use surface-to-air missiles to attack civil aviation in the near future. As some targets are becoming more difficult for terrorists to attack it can be anticipated that they will make efforts to overcome the enhanced security systems as well as redirecting their efforts towards less secure targets. The displacement of the increasingly ineffective system of hijacking by missile attacks against civil aviation is a real threat.

XIII. International Accord

In April 1996 in Vienna, States representatives of the "New Forum" held a Plenary to confirm the "Wassenaar Arrangement",[70] earlier agreed upon in the city of Wassenaar, the Netherlands, that addresses risks to regional and international security related to the spread of conventional weapons and dual-use goods and technologies while preventing destabilizing accumulations of weapons such as MANPADS. The *Wassenaar Arrangement* complements and reinforces, without duplication, the existing control regimes for weapons of mass destruction and their delivery systems, as well as other internationally recognized measures designed to promote transparency and greater responsibility, by focusing on the threats to international and regional peace and security which may arise from transfers of armaments and sensitive dual-use goods and technologies where the risks are judged greatest. It is also calculated to enhance co-operation in order to prevent the acquisition of armaments and sensitive dual-use items for military end-uses, if the situation in a region or the conduct of a state is, or becomes, a cause for serious concern to the Participating States. It is not the intent and purpose of the Arrangement to be directed against any state or group of states, nor will it impede bona fide civil transactions. Furthermore it will not interfere with the rights of states to acquire legitimate means with which to defend themselves pursuant to Article 51

---

[70] Wassenaar Arrangement on Export Controls for Conventional Arms and Dual-Use Goods and Technologies, Elements for Export Controls of MANPADS and the Inter-American Convention Against the Illicit Manufacturing of and Trafficking in Firearms, Ammunition, Explosives, and other Related Material. The participating States were Argentina, Australia, Austria, Belgium, Bulgaria, Canada, Croatia, Czech Republic, Denmark, Estonia, Finland, France, Germany, Greece, Hungary, Ireland, Italy, Japan, Latvia, Lithuania, Luxembourg, Malta, Netherlands, New Zealand, Norway, Poland, Portugal, Republic of Korea, Romania, Russian Federation, Slovakia, Slovenia, Spain, Sweden, Switzerland, Turkey, Ukraine, United Kingdom and United States.

of the Charter of the United Nations.[71] The Arrangement allows for Participating States to control all items set forth in a list of dual-use goods and technologies and in the munitions list, with the objective of preventing unauthorized transfers or re-transfers of those items. Participating States also agree to exchange general information on risks associated with transfers of conventional arms and dual-use goods and technologies in order to consider, where necessary, the scope for co-ordinating national control policies to combat the risks involved. At the tenth Plenary meeting of the *Wassenaar Arrangement*, held in Vienna on 8–9 December 2004, participating States reaffirmed their intent and resolve to prevent the acquisition by unauthorized persons of conventional arms and dual-use goods and technologies, in particular by terrorist groups and Organizations. States also exchanged information on various national measures adopted to implement the provisions of the Arrangement.

The Wassenaar Arrangement is the first global multilateral arrangement on export controls concerning conventional weapons and sensitive dual-use goods and technologies. It has not been designated the conventional term "convention" or "agreement" but nonetheless carries the agreement of participating States to collaborate in complementing, without duplication existing regimes on the non-proliferation of weapons of mass destruction and their delivery systems. The *Wassenaar Arrangement* is not a treaty in the sense of Article 102 of the United Nations Charter,[72] nor is it a treaty as defined by the *Vienna Convention on the Law of Treaties* of 1969, Article 2 of which defines a treaty *inter alia* as an international agreement concluded between States in written form and governed by international law. However, it remains an agreement between sovereign States concerning the implementation of municipal law of each participating State. This does not, however, mean that the *Wassenaar Arrangement* cannot be considered an international agreement or that it is invalid. It merely means that the Arrangement does not come within the purview of the Vienna Convention. It is worthy of note that Article 3 of the Convention explicitly recognizes that international agreements between States do not lose their validity merely because they do not come within the ambit of the Convention.[73]

As mentioned earlier, the Assembly of ICAO,[74] at its 36th Session (Montreal, 18–28 September 2007), adopted Resolution A36-2, wherein the Assembly

---

[71]Article 51 of the United Nations Charter provides, *inter alia*, that nothing in the Charter will impair the inherent right of individual or collective self-defense if an armed attack occurs against a member of the United Nations, until the Security Council has taken measures necessary to maintain international peace and security.

[72]Infrastructure Safety and the Environment Article 102 of the UN Charter stipulates that every treaty and every international agreement entered into by any member of the United Nations after the Charter comes into force shall be registered with the UN Secretariat as soon as possible.

[73]Shaw (2003, p. 812).

[74]The ICAO triennial Assembly, where its 190 member States gather to evaluate policy and make new policy as necessary through its Resolutions, is the supreme governing body of the Organization.

C. Emerging Threats

expressed its deep concern regarding the global threat posed to civil aviation by terrorist acts, in particular the threat posed by MANPADS, other surface-to-air missiles systems, light weapons and rocket propelled grenades.

The Assembly noted that the United Nations General Assembly, on 8 September 2006, adopted Counter-Terrorism Strategy, which is a unique global instrument that will enhance national, regional and international efforts to counter terrorism. The Strategy emphasizes the need to combat illicit arms trade, in particular small arms and light weapons, including MANPADS. Member States have agreed to a common strategic approach to fight terrorism, not only by sending a clear message that terrorism is unacceptable but also resolving to take practical steps individually and collectively to prevent and combat it. These steps include a wide range of measures ranging from strengthening State capacity to counter terrorist threats, to better coordinating United Nations System's counter-terrorism activities.

The Assembly recalled *United* Nations General Assembly resolutions 61/66 on the illicit trade in small arms and light weapons in all its aspects, 60/77 on prevention of the illicit transfer and unauthorized access to and use of man-portable air defence systems, 61/71 on assistance to States for curbing the illicit traffic in small arms and light weapons and collecting them and 60/288 on the United Nations Global Counter Terrorism Strategy. It also noted that the International Instrument to Enable States to Identify and Trace, in a Timely and Reliable Manner, Illicit Small Arms and Light Weapons (A/60/88) and the Wassenaar Arrangement on Export Controls for Conventional Arms and Dual-Use Goods and Technologies, Elements for Export Controls of MANPADS, and the Inter-American Convention Against the Illicit Manufacturing of and Trafficking in Firearms, Ammunition, Explosives, and other Related Material.

Noting with satisfaction the ongoing efforts of other international and regional organizations aimed at developing a more comprehensive and coherent response to the threat to civil aviation posed by MANPADS; and recognized that the specific threat posed by MANPADS requires a comprehensive approach and responsible policies on the part of States.

The Assembly urged all Contracting States to take the necessary measures to exercise strict and effective controls on the import, export, transfer or retransfer and stockpile management of MANPADS and associated training and technologies, as well as limiting the transfer of MANPADS production capabilities. It called upon all Contracting States to cooperate at the international, regional and sub-regional levels with a view to enhancing and coordinating international efforts aimed at implementing countermeasures carefully chosen with regard to their effectiveness and cost, and combating the threat posed by MANPADS. Furthermore, the Assembly called upon all Contracting States to take the necessary measures to ensure the destruction of non-authorized MANPADS in their territory, as soon as possible, while urging all Contracting States to implement the International instruments to enable States to identify and trace, in a timely and reliable manner, illicit small arms and light weapons as referred to in the United Nations General Assembly Resolution 61/66 on the illicit trade in small arms and light weapons in all its aspects. All Contracting States were urged to apply the principles defined in the Elements for

Export Controls of MANPADS of the Wassenaar Arrangement. Finally, the Assembly directed the ICAO Council to request the Secretary General to monitor on an on-going basis the threat to civil aviation posed by MANPADS and to continuously develop appropriate countermeasures to this threat and periodically request Contracting States to inform the Organization regarding the status of implementation of the resolution and the measures taken to fulfil its requirements.

XIV. Other Current Threats

Security restrictions on the carriage of liquids, aerosols and gels (LAGs) in hand baggage were introduced on 10 August 2006, in response to the foiling of an alleged terrorist plot in the United Kingdom against aviation using improvised explosive devices containing homemade liquid explosives. An initial ban on the carriage of all hand baggage on flights leaving the United Kingdom was subsequently modified to a restriction on the amounts of liquids, aerosols and gels (LAGs) which were permitted to be carried by passengers through screening points. These restrictions were adopted elsewhere in Europe and in North America. They were subsequently harmonised within the European Union by an amendment to the European Commission regulations which came into effect on 6 November 2006.

As a global follow up to these measures, ICAO recommended their universal adoption (not later than 1 March 2007) in a State Letter. ICAO also reacted to the new threat with urgency and efficiency in calling a special meeting of the Council on 17 August 2006 to explore ways of countering the new threat. As international civil aviation industry attaches great importance to the security screening of liquids, many countries have made a lot of efforts on the study of liquids detect methods. At present and in the near future, the most effective and safest security way is combination of regular measures, such as X-ray screening, visual examination, inspection by removing the bottle lids, restriction on carrying liquids, etc. ICAO temporary security control guidelines provide a uniform operation mode of liquids screening, which is helpful to the unification of international civil aviation security standard.

References

Abeyratne R (2000) Funding an international financial facility for aviation safety. The Journal of World Investment 1(2):383–407
Abeyratne R (2002) International responsibility in preventing the spread of communicable diseases through air carriage: the SARS crisis. Transportation Law Journal 30(1):53–80
Abeyratne R (2005) Emergent trends in aviation war risk insurance. Air and Space Law 30(2):117–129

References

Adams J (1990) Trading in death: weapons, warfare and the modern arms race. Hutchinson, London
Bobbitt P (2008) Terror and consent: the wars for the twenty first century. Knopf, New York
Clutterbuck R (1991) Living with terrorism. Butterworths, London
Crenshaw WA (1987) Terrorism and the threat to the civil aviation. Phd dissertation, University of Miami
Dershowitz A (2006) Pre-emption – a knife that cuts both ways. Norton, New York
Dobson C, Payne R (1977) The Carlos complex: a pattern of violence. Hodder and Stoughton, London
Dobson C, Payne R (1982) The terrorists: their weapons, leaders and tactics. Facts on File, New York
Dobson C, Payne R (1987) War without end. The terrorists: an intelligence dossier. Sphere Books, London, Appendix B
Dorey FC (1983) Aviation security. Granada, London
Ferguson N (2008) The ascent of money. Penguin, New York
Guill M (2000) Cyber-terrorism poses newest and perhaps elusive threat to civil aviation. ICAO Journal 55:18
Hanle DJ (1989) Terrorism: the newest face of warfare. Pergamon-Brassey's, New York
Keith KJ (2005) Clear and present danger: responses to terrorism. ICLQ 54:85–196
Lewis A, Kaplan M (eds) (1990) Terror in the skies: aviation security. In: Proceedings of the first international seminar on aviation security, Jerusalem, Israel, ISAS
Mickolus EF (1980) Transnational terrorism: a chronology of events, 1969–1979. Aldwych, London
Ofri A (1984) Intelligence and counterterrorism. ORBIS 28:41–52
Pierre AJ (1975) The politics of international terrorism. ORBIS 19:1256–1269
Shaw MN (2003) International law, 5th edn. Cambridge University Press, Cambridge
Stewart JP (1993) System approach to risk management focuses resources on most serious hazards. Aviation Safety 1993(November):12
United Nations (2009) Handbook on criminal justice responses to terrorism. Criminal Justice Handbook Series. United Nations, New York
Wagner M, Moore A, Aryel R (eds) (2006) Handbook of bio surveillance. Elsevier, New York
Wilkinson P (1986) Terrorism: international dimensions. In: Gutteridge W (ed) The new terrorism. Institute for the Study of Conflict and Terrorism, London

Chapter 2
Principles of Responsibility

A. State Responsibility

There are various offences that can be perpetrated by private individuals or groups of individuals against civil aviation, the earliest common species of which was hijacking of aircraft. Hijacking, in the late 1960s started an irreversible trend which was dramatised by such incidents as the skyjacking by Shiite terrorists of the TWA flight 847 in June 1985. The skyjacking of Egypt Air flight 648 in November the same year and the skyjacking of a Kuwait Airways Airbus in 1984[75] are other early examples of this offence. Aviation sabotage, where explosions on the ground or in mid air destroy whole aircraft, their passengers and crew, is also a threat coming through the past decades. The Air India flight 182 over the Irish Sea in June 1985, PAN AM flight 103 over Lockerbie, Scotland in 1988, and the UTA explosion over Niger in 1989 are examples. Missile attacks,[76] where aircraft are destroyed by surface to air missiles (SAM) have also occurred as early as in the 1970s. The destructions of the two Viscount aircraft of Air Rhodesia in late 1978 / early 1979 are examples of this offence. A re-emerging threat, namely armed attacks at airports, shows early occurrence in instances where terrorists opened fire in congested areas in the airport terminals. Examples of this type of terrorism are: The June 1972 attack by the Seikigunha (Japanese Red Army) at Ben Gurion Airport, Tel Aviv; The August 1973 attack by Arab gunmen on Athens Airport; and the 1985 attacks on Rome and Vienna Airports; Finally, the illegal carriage by air of narcotics and other psychotropic substances and crimes related thereto such as the seizure of or damage to aircraft, persons and property is also a threat that cannot be ignored in the present context, although, like other examples cited, it has been a perennial issue.

The most recent emerging threat to aviation security was reported by the United Kingdom authorities on 10 August 2006. It concerned, an alleged terrorist plot involving components of liquid explosives to be carried on board civil aircraft

---

[75] Abeyratne (1985, p. 120).

[76] Abeyratne (2005a) (on file with author).

flying across the North Atlantic, and it emphasized the vulnerability of the global air transport system. This plot revealed a new modus operandi, calling for immediate response. Accordingly, the ICAO Council convened a special session and directed the Aviation Security Panel to consider the wider implications for aviation security. Since technologies are not currently deployable to detect certain liquid explosives, the Council adopted security control guidelines for screening liquids, gels and aerosols, known as LAGs, and these were conveyed to States in December 2006, with an effective date of 1 March 2007. The guidelines recommended that all LAGs should be carried in containers with a capacity not greater than 100 mL each and should be placed in a transparent re-sealable plastic bag of maximum capacity not exceeding 1 L, each passenger being permitted to carry only one such bag. Exceptions are allowed for medications, baby food and special dietary or other medical requirements.

The issue of State responsibility for private acts of unlawful interference against civil aviation was not a contentious issue until the paradigm shift of 11 September 2001, when terrorists engaged in hijacking aircraft with a view to using them as weapons of mass destruction, causing damage to civilians on the ground.[77] The incidents of that day brought to bear serious implications for the continuity of air transport operations worldwide, particularly in the area of insurance of aircraft,[78] the cost of which the airlines had to bear. This prompted the International Air Transport Association (IATA) to raise the issue of State responsibility for the security of aviation within their territories at the 35th Session of the ICAO Assembly which was held in Montreal from 28 September to 8 October 2004. IATA drew the attention of the Assembly to the fact that the aviation underwriting community had announced formally its intention to exclude all hull, spares, passenger and third party liability claims resulting in damage caused by the hostile use of dirty bombs, electromagnetic pulse devices or biochemical materials.[79] IATA contended that such exclusions would place the airlines at risk of breaching state regulations as well as being destitute of access to adequate coverage, which in turn would compel airlines to cease their operations of air services. Recognizing the need for effectively precluding such market failure, IATA urged states to extend government guarantees that would ensure coverage for the categories mentioned above that would be affected by losses arising from state targeted act of terrorism. Furthermore, IATA requested the Assembly to consider the need to establish a legislative structure pertaining to limitation of liability for war and terrorism losses.

The main thrust of IATA's argument in seeking state coverage against losses in this context was that terrorists carried out their inimical acts against States and airlines and the air transport infrastructure were mere pawns or a proxy. As such it

---

[77] See Abeyratne (2002a, pp. 406–420).
[78] Abeyratne (2002b, pp. 521–533; 2005b, pp. 117–129).
[79] War Risk Exclusions, A35-WP/97, LE/8, 17/08/04, p. 1. See also, Abeyratne (2007, pp. 689–704).

A. State Responsibility

was the responsibility of governments to ensure indemnity of carriers and infrastructure.[80]

Although *ex facie* the claim of IATA – that governments indemnify air carriers and infrastructure against fiscal liability is based on the fact that terrorist attacks are aimed at governments and therefore the governments concerned should make good the losses to airlines and infrastructure – does not directly establish that governments are solely responsible for making aviation secure, it carries a general presumption of State responsibility. However, it must be noted that governments have the difficult and unenviable task of balancing the need for maintaining and encouraging anti-terrorists vigilance while, at the same time, putting in place workable security measures that do not compromise the commercial basis of the air transport sector. This problem is compounded by the fact that huge economic damage can be suffered even when terrorist plots are foiled. There also can be well-founded human rights legislation that might impact negatively across a host of air security issues, from the need for legal statutes, identification and the criminality clearance of airport workers, to legal rulings hindering the deportation of aircraft hijackers. These and other issues must be addressed in order to reconcile security with the efficiency and sustainability of the industry.

In spite of the delicate balance between enforcing security in a transport system which has as its main advantage speed and expediency of carriage of passengers and freight and coping with new and emerging threats that would compromise that advantage, States have, to their credit, taken initiatives that demonstrate their responsibility. A good example is Europe where over the last 40 years, States have followed the philosophy that if compensation does not come from the perpetrators, States would step in to assume that responsibility.[81] A similar approach was taken by the United States consequent to the events of 11 September 2001. The European Union imposed Council Directive 12(2) which, by 1 July 2005 at the latest, all 25 EU Member States should have complied with and required EU member States to ensure that their national rules provide for the existence of a scheme on compensation to victims of violent international crimes committed in their respective territories, which guarantees fair and appropriate compensation to victims.

A leading commentator Harold Caplan states that the idea that States have a responsibility to ensure that victims of crime are compensated is not confined to Europe. He observes that the US Department of Justice has long had an Office for Victims of Crime [OVC] which oversees the schemes in individual States,[82] and in collaboration with the State Department, has compiled and updated a Directory of

---

[80] War Risk Exclusions, A35-WP/97, LE/8, 17/08/04, p. 3.

[81] See 1983 Strasbourg "Convention on the Compensation of Victims of Violent Crime"; and EU Council Directive 2004/80/EC (29 April 2004) relating to compensation to crime victims.

[82] See Crime Victim Compensation Programs Directory 2002 http://www.ojp.usdoj.gov/ovc/publications.

schemes in 35 countries principally for the information of US citizens who travel or reside overseas.[83]

Finally, Caplan states that:

> Disregarding the 9/11 Victim Compensation Fund, none of the known State crime compensation schemes around the world can be said to provide lavish compensation. What is important is that they exist and they demonstrate unmistakable evidence of a widely-accepted principle of State responsibility.[84]

I. Principles of State Responsibility

The fundamental issue in the context of State responsibility for the purposes of this article is to consider whether a State should be considered responsible for its own failure or non-feasance to prevent a private act of terrorism against civil aviation or whether the conduct of the State itself can be impugned by identifying a nexus between the perpetrator's conduct and the State. One view is that an agency paradigm, which may in some circumstances impute to a state reprehensibility on the ground that a principal–agent relationship between the State and the perpetrator existed, can obfuscate the issue and preclude one from conducting a meaningful legal study of the State's conduct.[85]

II. The Theory of Complicity

At the core of the principal–agent dilemma is the theory of complicity, which attributes liability to a State that was complicit in a private act. Hugo Grotius (1583–1645), founder of the modern natural law theory, first formulated this theory based on State responsibility that was not absolute. Grotius' theory was that although a State did not have absolute responsibility for a private offence, it could be considered complicit through the notion of *Patienta* or *receptus*.[86] While the concept of *Patienta* refers to a State's inability to prevent a wrongdoing, *receptus* pertains to the refusal to punish the offender.

The eighteenth century philosopher Emerich de Vattel was of similar view as Grotius, holding that responsibility could only be attributed to the State if a sovereign refuses to repair the evil done by its subjects or punish an offender or

---

[83]Directory of International Crime Victim Compensation Programs 2004–2005.

[84]Harold Caplan, Damage to third parties on the ground caused by aircraft, Some basic issues of policy which re-merit examination in the context of modernization of the 1952 Rome Convention, unpublished Aide Memoire.

[85]Caron (1998, pp. 109, 153–154) cited in Becker (2006, p. 155).

[86]Grotius (1646, pp. 523–526).

A. State Responsibility   43

deliver him to justice whether by subjecting him to local justice or by extraditing him.[87] This view was to be followed and extended by the British jurist Blackstone a few years later who went on to say that a sovereign who failed to punish an offender could be considered as abetting the offence or of being an accomplice.[88]

A different view was put forward in an instance of adjudication involving a seminal instance where the Theory of Complicity and the responsibility of States for private acts of violence was tested in 1925. The case[89] involved the Mexico–United States General Claims Commission which considered the claim of the United States on behalf of the family of a United States national who was killed in a Mexican mining company where the deceased was working. The United States argued that the Mexican authorities had failed to exercise due care and diligence in apprehending and prosecuting the offender. The decision handed down by the Commission distinguished between complicity and the responsibility to punish and the Commission was of the view that Mexico could not be considered an accomplice in this case.

The Complicity Theory, particularly from a Vattellian and Blackstonian point of view is merely assumptive unless put to the test through a judicial process of extradition. In this Context it becomes relevant to address the issue through a discussion of the remedy.

III. Mechanisms for Extradition of Offenders: The Lockerbie Case

At present, the issue of extradition could be settled through the United Nations and its Organs such as the Security Council[90] and the International Court of Justice (ICJ).[91] Of noteworthy practical relevance with regard to the complicity theory, particularly on the issue of extradition and whether one State can demand the

---

[87]De Vattel (1916, p. 72).

[88]Blackstone (2001, p. 68).

[89]Laura M.B. Janes (USA) v. United Mexican States (1925) 4 R Intl Arb Awards 82.

[90]The Security Council is the branch of the United Nations charged with the maintenance of international peace and security. Its powers, outlined in the Charter of the United Nations, include the establishment of peacekeeping operations, the establishment of international sanctions, and the authorization for military action. The Security Council's power are exercised through its Resolutions. The Permanent members of the Security Council are the United States of America, United Kingdom, France, the Russian Federation and the Republic of China.

[91]The International Court of Justice (ICJ) is the principal judicial organ of the United Nations (UN). It was established in June 1945 by the Charter of the United Nations and began work in April 1946. The Court's role is to settle, in accordance with international law, legal disputes submitted to it by States and to give advisory opinions on legal questions referred to it by authorized United Nations organs and specialized agencies. The Court is composed of 15 judges, who are elected for terms of office of 9 years by the United Nations General Assembly and the Security Council. It is assisted by a Registry, its administrative organ. Its official languages are English and French.

extradition of offenders in another State is the opinion given by the ICJ[92] on the explosion over Lockerbie, Scotland on 21 December 1988 of PAN AM Flight 103. The explosion is believed to have been caused by the detonation of a plastic explosive concealed in a portable cassette player/radio. The ICJ was encumbered with the discussion as to whether the Court had jurisdiction over a United Nations Security Council Resolution on the issue.

The United States, in its submission to the ICJ in the Lockerbie case, contended that the Security Council was actively seized of the situation which was the subject of the issue of the offenders from Libya to the United States that therefore the Court should not indicate provisional measures as requested.[93] Judge Oda, Acting President of the Court, observing that Libya instituted proceedings against the United States in respect of the interpretation and application of the Montreal Convention,[94] noted that it was a general principle of international law that no State could be compelled to extradite its nationals and that the State concerned held the prerogative of trying the accused of a crime in its own territory. Judge Oda seems to have recognized two principles in his opinion:

(1) While any State can request extradition, no State can coerce extradition of nationals of another State.
(2) Whether or not a State can compel extradition is a matter for resolution by the general principles of international law and not necessarily those stipulated in the Montreal Convention.

It appears that the question in Judge Oda's mind was therefore whether the Security Council, by its Resolution 748 (1992) which required Libya to extradite its

---

[92] I.C.J. Reports 1980, 116.

[93] I.C.J. Reports 1980, 122. By letter of 2 April 1992, a copy of which was transmitted to Libya by the Registrar, the Agent of the United States drew the Court's attention to the adoption of Security Council Resolution 748 (1992) the text of which he enclosed. In that letter the Agent for the United States stated:
That resolution, adopted pursuant to Chapter V11 of the United Nations Charter, "decides that the Libyan Government must now comply without any further delay with paragraph 3 of resolution 731 (1992) of 21 January 1992 regarding the requests contained in documents S/23306, S/23308 and S/23309." It will be recalled that the referenced requests include the request that Libya surrender the two Libyan suspects in the bombing of Pan Am flight 103 to the United States or to the United Kingdom. For this additional reason, the United States maintains its submission of 28 March 1992 that the request of the Government of the Great Socialist Peoples' Libyan Arab Jamahiriya for the indication of provisional measures for protection should be denied, and that no such measures should be indicated. See I.C.J. Reports 1980, 125.

[94] *Convention for the Suppression of Unlawful Acts Against the Safety of Civil Aviation*, signed at Montreal on 23 September 1971, ICAO, Doc 8966. Article 8 of the Montreal Convention stipulates that the offences under the Convention shall be deemed to be included as extraditable offences in any extradition treaty existing between the Contracting States and that States undertake to include these offences as extraditable in any extradition treaty to be concluded by them. Article 11 of the same Convention requires Contracting States to afford one another the greatest measures of assistance in connection with criminal proceedings brought in respect of the offences, stating further that in such instances the law of the State requested shall apply at all times.

A. State Responsibility

nationals either to the United States or to the United Kingdom, had the authority to override an established principle of international law. The answer to this question was, in Judge Oda's view, in the affirmative when he opined:

> I do not deny that under the positive law of the United Nations Charter a resolution of the Security Council may have binding force, irrespective of the question whether it is consonant with international law derived from other sources. There is certainly nothing to oblige the Security Council, acting within its terms of reference, to carry out a full evaluation of the possibly relevant rules and circumstances before proceeding to the decisions it deems necessary. The Council appears, in fact, to have been acting within its competence when it discerned a threat against international peace and security in Libya's refusal to deliver up the two Libyan accused. Since, as I understand the matter, a decision of the Security Council, properly taken in the exercise of its competence, cannot be summarily reopened and since it is apparent that resolution 748 (1992) embodies such a decision, the Court has at present no choice but to acknowledge the pre-eminence of that resolution.[95]

Judge Oda was emphatic that the Security Council Resolution had overriding effect over any principle of international law. He observed however, that if the Court appeared to have *prima facie* jurisdiction over a legal issue that was the subject of its consideration, the Court was not precluded from indicating provisional measures applied for, merely because of the absolute preemptive powers of the Security Council Resolution. The learned judge concluded that, in this case, the application of the Libyan Government would have been rejected by the Court in any case, as the application was based on the Montreal Convention and not on the general customary international law principle *aut didere aut judicare*. Judge Oda was also unequivocal in his view that the Security Council Resolution would prevail over any established rule of international law.

Judge Ni on the other hand, observed that the Security Council and the International Court of Justice could simultaneously exercise their respective functions without being excluded by each other. Citing the arbitration that came up before the ICJ in respect of the United States diplomatic consular staff in Teheran, Judge Ni quoted from that judgment:

> ...it is for the Court, the principal judicial organ of the United Nations, to resolve any legal questions that may be in issue between parties to a dispute; and the resolution of such legal questions by the Court may be an important and sometimes decisive factor in promoting the peaceful settlement of the dispute.

This is indeed recognized by Article 36 of the Charter, paragraph 3 of which specifically provides that:

> In making recommendations under this Article the Security Council should also take into consideration that legal disputes should as a general rule be referred by the parties to the International Court of Justice in accordance with the provisions of the Statute of the Court.[96]

---

[95] I.C.J. Reports 1980, 129.

[96] I.C.J. Reports 1980, paragraph 40.

Judge Ni also analyzed Article 24 of the United Nations Charter which provides:

> In order to ensure prompt and effective action by the United Nations, its members confer on the Security Council primary responsibility for the maintenance of international peace and security....

The learned judge reasoned the Charter did not confer exclusive responsibility upon the Security Council and observed that the Council has functions of a political nature assigned to it whereas the Curt exercised purely judicial functions. According to Judge Ne therefore, both organs could perform their separate but complementary functions with respect to the same events. On the above reasoning, Judge Ne concluded that since the Court held independent jurisdiction, it could interpret the applicable law, which in this case was Article 14(1) of the Montreal Convention.[97] The ICJ could therefore, according to Judge Ni, by no means be pre-empted by a Security Council resolution, in its exercise of jurisdiction and application of the principles of international law.

Judges Evensen, Tarassov, Guillaume, and Aguila Mawdsley expressed their collective opinion that prior to the adoption by the Security Council of Resolution 748 (1992), The United States and the United Kingdom, although having the right to request extradition, could only take measures towards ensuring such extradition that were consistent with the principles of international law. With the Resolution in force however, the judges concluded that the Court was precluded from indicating provisional measures against the United States.

Judge Lachs, although in a separate opinion declared that the ICJ was bound to respect the binding decisions of the Security Council, seems to have recognized the co-existence of the two institutions, and the right of the Court to render its opinion irrespective of the application of Security Council resolutions. Judge Lachs said:

> The framers of the Charter, in providing for the existence of several main organs, did not effect a complete separation of powers, nor indeed is one to suppose that such was their aim. Although each organ has been allotted its own Chapter or Chapters, the functions of two of them, namely the General Assembly and the Security Council, also pervade other Chapters other than their own. Even the International Court of Justice receives, outside its own Chapter, a number of mentions which tend to confirm its role as the general guardian of legality within the system. In fact the Court is the Guardian of legality for the international community as a whole, both within and without the United Nations. One may therefore legitimately suppose that the intention of the founders was not to encourage a blinkered parallelism of functions but a fruitful interaction.

Two main organs of the United Nations have the delivery of binding decisions explicitly included in their powers under the Charter: the Security Council and the International Court of Justice. There is no doubt that the Court's task is "to ensure

---

[97] Article 14(1) of the Montreal Convention requires any dispute between two or more Contracting States concerning the interpretation or application of the Convention which cannot be settled through negotiation to be, at the request of one of them, submitted to arbitration. The provision goes on to say that if within six months from the date of the request for arbitration the Parties are unable to agree on the organization of the arbitration, any one of those parties may refer the dispute to the International Court of Justice by request in conformity with the Statute of the Court.

A. State Responsibility

respect for international law...."[98] It is its principal guardian. Now, it has become clear that the dividing line between political and legal disputes is blurred, as law becomes ever more frequently an integral element of international controversies. The Court, for reasons well known so frequently shunned in the past, is thus called upon to play an even greater role. Hence it is important for the purposes and principles of the United Nations that the two main organs with specific powers of binding decision act in harmony – though not, of course, in concert – and that each should perform its functions with respect to a situation or dispute, different aspects of which appear on the agenda of each, without prejudicing the exercise of the other's powers. In the present case the Court was faced with a new situation which allowed neither room for further analysis nor the indication of effective interim measures. The order made should not, therefore, be seen as an abdication of the Court's powers; it is rather a reflection of the system within which the Court is called upon to render justice.[99]

Judge Shahabuddeen, recognizing that there is no superior authority to that of the Security Council, added in his opinion that treaty obligations can be overridden by a decision of the Security Council's sanctions. Addressing the critical question whether a decision of the Security Council may override the legal rights of States, Judge Shahabuddeen did not attempt an answer, but merely concluded that such a decision may stand in the way of the legal rights of a State or its subjects being judicially scrutinized. Judge Bedjaoui in his opinion added to the opinion of Judge Shahabuddeen, saying that as a rule, the International Court of Justice does not exercise appellate jurisdiction over the Security Council.[100] The learned judge however, strongly dissented from the views of his colleagues which recognised that a Security Council resolution completely pre empted the jurisdiction of the International Court of Justice and effectively precluded the latter from performing its judicial functions. Judge Bedjaoui stringently maintained that there were two aspects to the problem between Libya and the United States on this issue – political and judicial. In his view, although it was not possible for the Court to override the Security Council Resolution, the Court was by no means precluded from declaring provisional measures, as applied for by Libya, even if such a declaration by the Court was rendered destitute of effect by the Security Council Resolution.[101]

Judge Weeramantry, concurring with Judge Bedjaoui, conceded that although Article 25 of the United Nations Charter required member States of the United Nations to accept and carry out decisions of the Security Council, the Court was not deprived of its jurisdiction in issuing provisional measures as applied for by Libya. Adding that the International Court of Justice and the Security Council were created by the same Charter to fulfil the common purposes and principles of the United Nations, Judge Weeramantry concluded that the two agencies are complementary

---

[98] I.C.J. Reports 1949, 35.
[99] I.C.J. Reports 1980, 138–139.
[100] I.C.J. Reports 1980, 140.
[101] I.C.J. Reports 1980, 155–156.

to each other, each performing a special role assigned to it. The Court was however, unlike most courts that were vested with domestic jurisdictions, not enabled to sit in review of the executive (which in this context was the Security Council). The dichotomy arose, in Judge Weeramantry's mind, when the principal judicial organ of the United Nations was restrained by decisions of its executive arm when deciding, according to the principles of international law, disputes that are submitted to it. The conclusions reached by judge Weeramantry were based on the Kelsenian observation that the Security Council and the General Assembly were only quasi-judicial organs of the United Nations and that the Security Council was by no means a judicial organ since its members were not independent;[102] and the Court ought to collaborate (emphasis added) with the Security Council if the circumstances so require.[103] Judge Weeramantry therefore emphasised that the Court must at all times preserve its independence, particularly in view of the fact that Article 24(2) of the Charter provides that the Security Council Shall act in accordance with the purposes and principles of the United Nations, which are set out in Article 1(1) of the Charter as being those aims to settle international disputes and situations that might lead to breaches of the peace, according to the principles of justice and international law.

The essence of these views of the learned judges of the ICJ is that the complimentary roles played by the United Nations Security Council and the ICJ would devolve responsibility on States to respect both these organs on the subject of extradition of private offenders who unlawfully interfere with civil aviation.

IV. The Condonation Theory

The emergence of the Condonation Theory was almost concurrent with the *Jane* case[104] decided in 1925 which emerged through the opinions of scholars who belonged to a school of thought that believed that States became responsible for private acts of violence not through complicity as such but more so because their refusal or failure to bring offenders to justice, which was tantamount to ratification of the acts in question or their condonation.[105] The theory was based on the fact that it is not illogical or arbitrary to suggest that a State must be held liable for its failure to take appropriate steps to punish persons who cause injury or harm to others for the reason that such States can be considered guilty of condoning the criminal acts and therefore become responsible for them.[106] Another reason attributed by

---

[102]Kelsen (1951, pp. 476–477). See I.C.J. Reports 1957 *supra* note 14, 167.

[103]I.C.J. Reports 1959, *Id.*, 168.

[104]Laura M.B. Janes (USA) v. United Mexican States (1925) 4 R Intl Arb Awards 82.

[105]*Black's Law Dictionary* defines condonation as "pardon of offense, voluntary overlooking implied forgiveness by treating offender as if offense had not been committed".

[106]Laura M.B. Janes (USA) v. United Mexican States (1925) 4 R Intl Arb Awards 82, at 92.

A. State Responsibility

scholars in support of the theory is that during that time, arbitral tribunals were ordering States to award pecuniary damages to claimants harmed by private offenders, on the basis that the States were being considered responsible for the offences.[107]

The responsibility of governments in acting against offences committed by private individuals may sometimes involve condonation or ineptitude in taking effective action against terrorist acts, in particular with regard to the financing of terrorist acts. The United Nations General Assembly, on 9 December 1999, adopted the International Convention for the Suppression of the Financing of Terrorism,[108] aimed at enhancing international co-operation among States in devising and adopting effective measures for the prevention of the financing of terrorism, as well as for its suppression through the prosecution and punishment of its perpetrators.

The Convention, in its Article 2 recognizes that any person who by any means directly or indirectly, unlawfully or, provides or collects funds with the intention that they should be used or in the knowledge that they are to be used, in full or in part, in order to carry out any act which constitutes an offence under certain named treaties, commits an offence. One of the treaties cited by the Convention is the International Convention for the Suppression of Terrorist Bombings, adopted by the General Assembly of the United Nations on 15 December 1997.[109]

The Convention for the Suppression of the Financing of Terrorism also provides that, over and above the acts mentioned, providing or collecting funds toward any other act intended to cause death or serious bodily injury to a civilian, or to any other person not taking an active part in the hostilities in the situation of armed conflict, when the purpose of such act, by its nature or context, is to intimidate a population, or to compel a government or an international organization to do or to abstain from doing any act, would be deemed an offence under the Convention.

The United Nations has given effect to this principle in 1970 when it proclaimed that:

> Every State has the duty to refrain from organizing or encouraging the organization of irregular forces or armed bands, including mercenaries, for incursion into the territory of another State. Every State has the duty to refrain from organizing, instigating, assisting or participating in acts of civil strife or terrorist acts in another State or acquiescing in organized activities within its territory directed towards the commission of such acts, when the acts referred to in the present paragraph involve a threat or use of force.[110]

Here, the words *encouraging* and *acquiescing in organized activities within its territory directed towards the commission of such acts* have a direct bearing on the

---

[107]Hyde (1928, pp. 140–142).

[108]International Convention for the Suppression of the Financing of Terrorism, adopted by the General Assembly of the United Nations in resolution 54/109 of 9 December 1999.

[109]A/52/653, 25 November 1997.

[110]Declaration on Principles of International Law Concerning Friendly Relations and Co-operation Among States in Accordance with the Charter of the United Nations, UN General Assembly Resolution 2625 (XXV) 24 October 1970.

concept of condonation and would call for a discussion about how States could overtly or covertly encourage the commission of such acts. One commentator[111] identifies three categories of such support: *Category I* support entails protection, logistics, training, intelligence, or equipment provided terrorists as a part of national policy or strategy; *Category II* support is not backing terrorism as an element of national policy but is the toleration of it; *Category III* support provides some terrorists a hospitable environment, growing from the presence of legal protections on privacy and freedom of movement, limits on internal surveillance and security organizations, well-developed infrastructure, and émigré communities

Another commentator[112] discusses what he calls the *separate delict theory'* in State responsibility, whereby the only direct responsibility of the State is when it is responsible for its own wrongful conduct in the context of private acts, and not for the private acts themselves. He also contends that indirect State responsibility is occasioned by the State's own wrong-doing in reference to the private terrorist conduct. The State is not held responsible for the act of terrorism itself, but rather for its failure to prevent and/or punish such acts, or for its active support for or acquiescence in terrorism.[113] Arguably the most provocative and plausible feature in this approach is the introduction by the commentator of the desirability of determining State liability on the theory of causation. He emphasizes that:

The principal benefit of the causality based approach is that it avoids the automatic rejection of direct State responsibility merely because of the absence of an agency relationship. As a result, it potentially exposes the wrong-doing State to a greater range and intensity of remedies, as well as a higher degree of international attention and opprobrium for its contribution to the private terrorist activity.[114]

The causality principle is tied in with the rules of State Responsibility enunciated by the International Law Commission and Article 51 of the United Nations Charter which states that nothing in the Charter will impair the inherent right of individual or collective self-defense if an armed attack occurs against a Member of the United Nations, until the Security Council has taken measures necessary to maintain international peace and security. The provision goes on to say that measures taken by Members in the exercise of this right of self-defense will be immediately reported to the Security Council and will not in any way affect the authority and responsibility of the Security Council under the present Charter to take at any time such action as it deems necessary in order to maintain or restore international peace and security.

The International Law Commission has established that a crime against the peace and security of mankind entails individual responsibility, and is a crime of aggression.[115] A further link drawing civil aviation to the realm of international

---

[111]Metz (2002).

[112]Becker (2006).

[113]Becker (2006, Chap. 2, p. 67).

[114]Becker (2006, p. 335).

[115]Draft Code of Crimes Against the Peace and Security of Mankind, International Law Commission Report, 1996, Chapter II Article 2.

peace and security lies in the Rome Statute of the International Criminal court, which defines a war crime, *inter alia*, as intentionally directing attacks against civilian objects; attacking or bombarding, by whatever means, towns, villages, dwellings or buildings which are undefended and which are not military objects; employing weapons, projectiles, and material and methods of warfare that cause injury.[116] The Statute also defines as a war crime, any act which is intentionally directed at buildings, material, medical units and transport, and personnel using the distinctive emblems of the Geneva Conventions in conformity with international law.[117]

V. *The Role of Knowledge*

Another method of determining State responsibility lies in the determination whether a State had actual or presumed knowledge of acts of its instrumentalities, agents or private parties which could have alerted the State to take preventive action. International responsibility of a State cannot be denied merely on the strength of the claim of that State to sovereignty. Although the Chicago Convention in Article 1 stipulates that the contracting States recognize that every State has complete and exclusive sovereignty over the airspace above its territory, the effect of this provision cannot be extended to apply to State immunity from responsibility to other States. Professor Huber in the *Island of Palmas* case[118] was of the view:

> Sovereignty in the relations between States signifies independence. Independence in regard to a portion of the globe is the right to exercise therein, to the exclusion of any other State, the functions of a State...Territorial sovereignty...involves the exclusive right to display the activities of a State.[119]

Professor Huber's definition, which is a simple statement of a State's rights, has been qualified by Starke as the residuum of power which a State possesses within the confines of international law.[120] Responsibility would devolve upon a State in whose territory an act of unlawful interference against civil aviation might occur, to other States that are threatened by such acts. The International Court of Justice (ICJ) recognised in the *Corfu Channel* Case:

> Every State's obligation not to allow knowingly its territory to be used for acts contrary to the rights of other States.[121]

---

[116] Rome Statute of the International Criminal Court, Article 8.2 (b) (ii), (V) and (XX).
[117] Rome Statute of the International Criminal Court, Article 8.2 (b) (XXIV).
[118] The *Island of Palmas* Case (1928) 11 U.N.R. I.A.A. at 829.
[119] The *Island of Palmas* Case (1928) 11 U.N.R. I.A.A. at 829.
[120] Starke (1989, p. 3).
[121] (1949) *I.C.J.R.* 1, 22.

In the famous *Corfu Channel* case, the International Court of Justice applied the subjective test and applied the fault theory. The Court was of the view that:

> It cannot be concluded from the mere fact of the control exercised by a State over its territory and waters that the State necessarily knew, or ought to have known, of any unlawful act perpetrated therein, nor yet that it necessarily knew, or should have known the authors. This fact, by itself and apart from other circumstances, neither involves prima facie responsibility nor shifts the burden of proof.[122]

The Court, however, pointed out that exclusive control of its territory by a State had a bearing upon the methods of proof available to establish the involvement or knowledge of that State as to the events in question.

Apart from the direct attribution of responsibility to a State, particularly in instances where a State might be guilty of a breach of treaty provisions, or violate the territorial sovereignty of another State, there are instances where an act could be imputed to a State.[123] Imputability or attribution depends upon the link that exists between the State and the legal person or persons actually responsible for the act in question. The legal possibility of imposing liability upon a State wherever an official could be linked to that State encourages a State to be more cautious of its responsibility in controlling those responsible for carrying out tasks for which the State could be ultimately held responsible. In the same context, the responsibility of placing mines was attributed to Albania in the *Corfu Channel* case since the court attributed to Albania the responsibility, since Albania was known to have knowledge of the placement of mines although it did not know who exactly carried out the act. It is arguable that, in view of the responsibility imposed upon a State by the Chicago Convention on the provision of air navigation services, the principles of immutability in State responsibility could be applied to an instance of an act or omission of a public or private official providing air navigation services.

The sense of international responsibility that the United Nations ascribed to itself had reached a heady stage at this point, where the role of international law in international human conduct was perceived to be primary and above the authority of States. In its Report to the General Assembly, the International Law Commission recommended a draft provision which required:

---

[122]The *Corfu Channel* Case, ICJ Reports, 1949, p. 4.

[123]There are some examples of imputability, for example the incident in 1955 when an Israeli civil aircraft belonging to the national carrier El Al was shot down by Bulgarian fighter planes, and the consequent acceptance of liability by the USSR for death and injury caused which resulted in the payment of compensation to the victims and their families. See 91 *ILR* 287. Another example concerns the finding of the International Court of Justice that responsibility could have been be imputed to the United States in the *Nicaragua* case, where mines were laid in Nicaraguan waters and attacks were perpetrated on Nicaraguan ports, oil installations and a naval base by persons identified as agents of the United States. See *Nicaragua* v. *the United States*, ICJ Reports 1986, 14. Also, 76 *ILR* 349. There was also the instance when the Secretary General of the United Nations mediated a settlement in which a sum, *inter alia* of $7 million was awarded to New Zealand for the violation of its sovereignty when a New Zealand vessel was destroyed by French agents in New Zealand. See the *Rainbow Warrior* case, 81 *AJIL*, 1987 at 325. Also in 74 *ILR* at 241.

A. State Responsibility

Every State has the duty to conduct its relations with other States in accordance with international law and with the principle that the sovereignty of each State is subject to the supremacy of international law.[124]

This principle, which forms a cornerstone of international conduct by States, provides the basis for strengthening international comity and regulating the conduct of States both internally – within their territories – and externally, towards other States. States are effectively precluded by this principle of pursuing their own interests untrammelled and with disregard to principles established by international law.

The United Nations General Assembly, in its Resolution 56/83,[125] adopted as its Annex the International Law Commission's *Responsibility of States for Internationally Wrongful Acts* which recognizes that every internationally wrongful act of a State entails the international responsibility of that State[126] and that there is an internationally wrongful act of a State when conduct consisting of an action or omission is attributable to the State under international law and constitutes a breach of an international obligation of the State.[127] Article 5 of the ILC document provides that the conduct of a person or entity which is not an organ of State but which is empowered by the law of that State to exercise elements of the governmental authority shall be considered an act of State under international law, provided the person or entity is acting in that capacity in the particular instance.

In the *Pan Am* case, where an aircraft was destroyed over Lockerbie, which has been referred to earlier in this article, the British allegation against Libya's involvement in the act of terrorism was that the accused individuals (Libyan nationals) had acted as part of a conspiracy to further the purposes of the Libyan Intelligence Services using criminal means that amounted to terrorism. The United Kingdom appeared to stress the point in the UN Security Council that Libya had failed to respond to the request for extradition of the implicated Libyan nationals, and arguably as a consequence, the Security Council adopted Resolution 731 on 21 January 1992 which expressed concerns over certain investigations which imputed reprehensibility to officials of the Libyan Government.[128]

The above discussion leads one to conclude that the responsibility of a State for private acts of individuals which unlawfully interfere with civil aviation is determined by the quantum of proof available that could establish intent or negligence of the State, which in turn would establish complicity or condonation on the part of the State concerned. One way to determine complicity or condonation is to establish the extent to which the State adhered to the obligation imposed upon it by international law and whether it breached its duty to others. In order to exculpate itself, the State concerned will have to demonstrate that either it did not tolerate the offence or that

---

[124]Report of the International Law Commission to the General Assembly on the Work of the 1st Session, A/CN.4/13, June 9 1949, at 21.

[125]A/RES/56/83, 56th Session, 28 January 2002.

[126]A/RES/56/83, Article 1.

[127]A/RES/56/83, Article 2.

[128]For a discussion on this point see Jorgensen (2000, pp. 249–254).

it ensured the punishment of the offender. *Brownlie* is of the view that proof of such breach would lie in the causal connection between the private offender and the State.[129] In this context, the act or omission on the part of a State is a critical determinant particularly if there is no specific intent.[130] Generally, it is not the intent of the offender that is the determinant but the failure of a State to perform its legal duty in either preventing the offence (if such was within the purview of the State) or in taking necessary action with regard to punitive action or redress.[131]

Finally, there are a few principles that have to be taken into account when determining State responsibility for private acts of individuals that unlawfully interfere with civil aviation. Firstly, there has to be either intent on the part of the State towards complicit or negligence reflected by act or omission. Secondly, where condonation is concerned, there has to be evidence of inaction on the part of the State in prosecuting the offender. Thirdly, since the State as an abstract entity cannot perform an act in itself, the imputability or attribution of State responsibility for acts of its agents has to be established through a causal nexus that points the finger at the State as being responsible. For example, The International Law Commission, in Article 4 of its Articles of State Responsibility states that the conduct of any State organ which exercises judicial, legislative or executive functions could be considered an act of State and as such the acts of such organ or instrumentality can be construed as being imputable to the State. This principle was endorsed in 1999 by the ICJ which said that according to well established principles of international law, the conduct of any organ of a state must be regarded as an act of State.[132]

The law of State responsibility for private acts of individuals has evolved through the years, from being a straightforward determination of liability of the State and its agents to a rapidly widening gap between the State and non State parties. In today's world private entities and persons could wield power similar to that of a State, bringing to bear the compelling significance and modern relevance of the agency nexus between the State and such parties. This must indeed make States more aware of their own susceptibility.

VI. *Profiling of Passengers*

It is an incontrovertible fact that profiling is a useful tool in the pursuit of the science of criminology. Profiling is also a key instrument in a sociological context and therefore remains a sustained social science constructed through a contrived

---

[129]Brownlie (1983, p. 39).

[130]Report of the International Law Commission to the United Nations General Assembly, UNGOAR 56th Session, Supp. No. 10, *UN DOC A/56/10*, 2001 at 73.

[131]de Arechaga (1968, p. 535).

[132]Differences Relating to Immunity from Legal Process of a Special Rapporteur, ICJ Reports 1999, 62 at 87.

process of accumulation of single assumptions and propositions that flow to an eventual empirical conclusion. However, profiling raises well reasoned latent fears when based on a racial platform. Jonathan Turley, Professor of Constitutional Law at George Washington University, in his testimony before a United States House of Representatives Committee on Airport Security regarding the use of racial profiling to identify potentially dangerous persons observed:

> [R]acial profiling is to the science of profiling as forced confessions are to the art of interrogation. Like forced confessions, racial profiling achieves only the appearance of effective police work. Racial profiling uses the concept of profiling to shield or obscure a racist and unscientific bias against a particular class or group. It is the antithesis of profiling in that it elevates stereotypes over statistics in law enforcement.

Notwithstanding this telling analogy, and the apprehensions one might have against racial profiling, it would be imprudent to conclude that racial profiling is *per se* undesirable and unduly discriminatory, particularly in relation to profiling at airports which should essentially include some considerations of ethnic and national criteria. This article will examine the necessary elements that would go into effective and expedient airport profiling of potential undesirable passengers. It will also discuss legal issues concerned with the rights of the individual with regard to customs and immigration procedures. The rights of such persons are increasingly relevant from the perspective of ensuring air transport security and refusing carriage to embarking passengers who might show profiles of criminality and unruly persons on board.

VII. Airport Profiling

A legitimate profiling process should be based on statistically established indicators of criminality which are identified through a contrived aggregation of reliable factors. The application of this criterion to airport profiling would immediately bring to bear the need to apply nationality and ethnic factors to passenger profiles. Although one might validly argue that racial profiling would entail considerable social and political costs for any nation, while at the same time establishing and entrenching criminal stereotypes in a society, such an argument would be destitute of effect when applied to airport security which integrally involves trans boundary travel of persons of disparate ethnic and national origins. This by no means implies that racial profiling is a desirable practice. On the contrary, it is a demeaning experience to the person subjected to the process and a *de facto* travel restriction and barrier. It is also a drain on law enforcement resources that effectively preclude the use of proven and conventional uses of enforcement.

The sensitive conflict of interests between racial profiling *per se,* which at best is undesirable in a socio-political context, and airport profiling, raises interesting legal and practical distinctions between the two. Among these the most important distinction is that airport profiling is very serious business that may concern lives of hundreds if not thousands in any given instance or event. Profiling should therefore be considered justifiable if all its aspects are used in screening passengers at airports. Nationality and ethnicity are valid baseline indicators of suspect persons together with other indicators which may raise a 'flag' such as the type of ticket a passenger holds (one way instead of return) and a passenger who travels without any luggage.

Racial profiling, if used at airports, must not be assumptive or subjective. It must be used in an objective and non discriminatory manner alongside random examinations of non-targeted passengers. All aspects of profiling, including racial and criminal profiling, should as a matter of course be included in the Computer Assisted Passenger Screening System (CAPS) without isolating one from the other. In this context the now popular system of compliance examination (COMPEX) is a non threatening, non discriminatory process which transcends the threshold debate on "profiling" by ensuring a balanced and proper use of profiling in all its aspects by examining "non targeted" passengers as well as on a random basis.

Another critical distinction to be drawn between discriminatory and subjective racial profiling on the one hand and prudent airport profiling on the other is the blatant difference between racism and racial profiling. The former is built upon the notion that there is a causal link between inherent physical traits and certain traits of personality, intellect or culture and, combined with it, the idea that some races are inherently superior to others. The latter is the use of statistics and scientific reasoning that identify a set of characteristics based on historical and empirical data. This brings to bear the clear difference between "hard profiling", which uses race as the only factor in assessing criminal suspiciousness and "soft profiling" which uses race as just one factor among others in gauging criminal suspiciousness.

Article 13 of the *Convention on International Civil Aviation* of 1944 (Chicago Convention), provides that the laws and regulations of a Contracting State as to the admission to and departure from its territory of passengers, crew or cargo of aircraft, such as regulations relating to entry, clearance, immigration, passports, customs and quarantine shall be complied with by or on behalf of such passengers, crew or cargo upon entrance into or departure from, or while within the territory of that State. This provision ensures that a Contracting State has the right to prescribe its own internal laws with regard to passenger clearance and leaves room for a State to enact laws, rules and regulations to ensure the security of that State and its people at the airport. However, this absolute right is qualified to preclude unfettered and arbitrary power of a State, by Article 22 which makes each Contracting State agree to adopt all practicable measures, through the issuance of special regulations or otherwise, to facilitate and expedite navigation of aircraft between the territories of Contracting States, and to prevent unnecessary delays to aircraft, crews, passengers and cargo, especially in the administration of the laws relating to immigration, quarantine, customs and clearance. Article 23 follows this trend to its conclusion by providing that each Contracting State undertakes, to the extend practicable, to establish customs and immigration procedures affecting air navigation in accordance with the practices which may be established or recommended from time to time pursuant to the Convention.

Annex 9 to the Chicago Convention (Facilitation) in Standard 3.2, recognizes that, in developing procedures aimed at the efficient application of border controls on passengers and crew, Contracting States shall take into account the application of aviation security, border integrity, narcotics control and immigration control measures, where appropriate. This Standard gives States the flexibility of enacting procedures, rules and regulations to ensure the security and integrity of their borders and view passengers with necessary caution.

The aims and objectives of the International Civil Aviation Organization (ICAO), as contained in Article 44 of the Chicago Convention are, *inter alia*, to ensure the safe and orderly growth of international civil aviation; meet the needs of the peoples of the world for safe, regular, efficient and economical air transport; and avoid discrimination between Contracting States. Annex 17 – *Security* to the Chicago Convention, in paragraph 2.1.1, identifies the main aim of aviation security as being the safeguarding of international civil aviation operations against acts of unlawful interference. These aims have been established by the international community with a view to ensuring that security and safety of civil aviation apply not only in the context of individual States, but also of their peoples, in accordance with the distinction made in the Preamble of the Chicago Convention.

The above-mentioned principles and aims make it abundantly clear that the role of ICAO and the international community is to consider the offence of unlawful interference with civil aviation in its entirety, as a generic term encompassing a wide range of offensive activity on the part of the perpetrators. This does not admit of the offence being restricted to a species or other category of offensive activity. It is for the above reason that the ICAO Assembly, at its 33rd Session, held subsequent to the events of 11 September 2001, adopted Resolution A33-2, strongly condemning all acts of unlawful interference against civil aviation wherever and by whomsoever and for whatever reason they are perpetrated. This all-encompassing approach to the offence of unlawful interference with civil aviation effectively precludes parochial assumptions that the offence would be recognizable as such only if it is perpetrated pursuant to or as an act which can only or mostly be perpetrated by a certain type of individual of a certain race, nationality or religious persuasion.

The first issue, that of encompassing aviation security under the umbrella of threats or considering aviation security as being affected only by certain acts, is therefore a critical consideration if a balanced and focussed approach to remedial action on aviation security were to be seriously addressed. Broadly considered, unlawful interference with civil aviation is an offence, which is any wrong that affects the security or well-being of civil aviation, in the context of persons and property, and creates an interest in the public in its suppression. The offence itself should therefore not be confused with a type of conduct or activity.

The second issue, within the purview of the security guidelines of the Chicago Convention, pertains to stereotyping and racial profiling of individuals in the pursuit of ensuring aviation security. In essence, racial profiling is intersectional in nature and may consist of multiple grounds of institutionalized discrimination such as nationality, race, age, gender, socio-economic status, disability, health status, descent, language, class, culture and religion. At the World Conference against Racism, Racial Discrimination, Xenophobia and Related Intolerance, held in Durban, South Africa from 31 August to 7 September 2001, the Conference, referring to the International Convention on the Elimination of All Forms of Racial Discrimination, urged States to implement or strengthen legislation and administrative measures prohibiting racial discrimination and related intolerance.

In its Declaration, the Conference urged States "to design, implement and enforce effective measures to eliminate the phenomenon popularly known as 'racial profiling' and comprising the practice of police and other law enforcement officers relying, to any degree, on race, colour, descent or national or ethnic origin as the basis for subjecting persons to investigatory activities or for determining whether an individual is engaged in criminal activity".

It becomes important to draw attention to practices of racial profiling involving discrimination during civil aviation operations, at airports at departure and arrival, that have re-emerged, especially after the events of 11 September 2001. Such unacceptable practices, which are diametrically opposed to international tenets and norms of human rights, only succeed in causing insult and injury to individuals discriminated upon, creating rancour, and being totally inconsistent with the fundamental principles of civil aviation as enunciated in the Chicago Convention.

VIII. *Profiling and the Right of Privacy*

Profiling of airline passengers could be carried out primarily through examination of the passenger's passport. At its 33rd Session, held in September/October 2001, the ICAO Assembly, while acknowledging that new measures should be taken to enhance security, observed that such measures should not impede ICAO's ongoing work in improving border control systems at airports and ensuring the smooth flow of passengers and cargo. Consequently, the Assembly stressed that ICAO's work on these issues should continue on an urgent basis. The machine readable travel document was among specific areas mentioned by the Assembly as requiring urgent continuing work, in keeping with UN Security Council Resolution 1373 of 23 September 2001, which re-affirmed the need for continuing work to ensure the integrity and security of passports and other travel documents. In this context, the Assembly agreed that all contracting States should be urgently encouraged to issue their travel documents in machine readable format and enhance their security in accordance with ICAO specifications, while introducing automated travel document reading systems at their international airports.

These measures of the ICAO Assembly bring to bear the essential link between aviation security and facilitation and the fact that one cannot be ignored while the other is given some prominence, as is the case with aviation security at the present time.

The data subject, like any other person, has an inherent right to his privacy. The subject of privacy has been identified as an intriguing and emotive one. The right to privacy is inherent in the right to liberty, and is the most comprehensive of rights and the right most valued by civilized man. This right is susceptible to being eroded, as modern technology is capable of easily recording and storing dossiers on every man, woman and child in the world. The data subject's right to privacy, when applied to the context of the machine readable travel document (MRTD) is brought into focus by Alan Westin who says:

> Privacy is the claim of individuals, groups or institutions to determine for themselves when, how, and to what extent information is communicated to others.

There are three rights of privacy relating to the storage and use of personal data:

(1) The right of an individual to determine what information about oneself to share with others, and to control the disclosure of personal data.
(2) The right of an individual to know what data is disclosed, and what data is collected and where such is stored when the data in question pertains to that individual; the right to dispute incomplete or inaccurate data.
(3) The right of people who have a legitimate right to know in order to maintain the health and safety of society and to monitor and evaluate the activities of government.

It is incontrovertible therefore that the data subject has a right to decide what information about oneself to share with others and more importantly, to know what data is collected about him. This right is balanced by the right of a society to collect data about individuals that belong to it so that the orderly running of government is ensured.

The role played by technology in modern day commercial transactions has affected a large number of activities pertaining to human interaction. The emergence of the information superhighway and the concomitant evolution of automation have inevitably transformed the social and personal life styles and value systems of individuals, created unexpected business opportunities, reduced operating costs, accelerated transaction times, facilitated accessibility to communications, shortened distances, and removed bureaucratic formalities. Progress notwithstanding, technology has bestowed on humanity its corollaries in the nature of automated mechanisms, devices, features, and procedures which intrude into personal lives of individuals. For instance, when a credit card is used, it is possible to track purchases, discovering numerous aspects about that particular individual, including, food inclination, leisure activities, and consumer credit behaviour. In similar vein, computer records of an air carrier's reservation system may give out details of the passenger's travel preferences, *inter alia*, seat selection, destination fondness, ticket purchasing dossier, lodging keenness, temporary address and telephone contacts, attendance at theatres and sport activities, and whether the passenger travels alone or with someone else. This scheme of things may well give the outward perception of surveillance attributable to computer devices monitoring individuals' most intimate activities and preferences, leading to the formation of a genuine "traceable society".

A good airport profiling system must originate from a repository of research based on the characteristics of a person evoking criminal suspiciousness. These characteristics must match the automated passenger information of airlines including the

API technology generated prior to a flight. In addition, they must also be consistent with information on travel documents used by passengers. One way of accomplishing this objective is to use the profiles of known or suspected criminals and terrorist categories. Additionally, a diligent and energetic State instrumentality must be established for the purpose of constantly monitoring and ensuring that airport profiling does not discriminate between categories of persons on a subjective basis and that a balanced system of compliance examination is in place.

In the ultimate analysis, the socio-legal relevance of non discrimination in the profiling process lies in the importance of respecting and safeguarding human rights and the rights of certain identified ethnic groups which may form a minority in a particular jurisdiction. Due recognition and active protection of a minority's rights pertaining to racial, cultural and religious issues is the first characteristic of a prudent and balanced profiling process. This feature guarantees freedom from discrimination based on race, language, nationality and national origin or religion. Western democracies, particularly after World War II and the Nuremberg trials which ensured punishment for those responsible for the organized murder of thousands of innocent persons by the commission of atrocities during the war, have been particularly sensitive to the need to ensure human rights. This has led to a gradual evolution where focus on collective rights of national minorities has replaced earlier emphasis on individual rights.

The protection of human rights is the most significant and important task for a modern State, particularly since multi ethnic States are the norm in today's world. Globalization and increased migration across borders is gradually putting an end to the concept of the nation State, although resistance to reality can be still seen in instances where majority or dominant cultures impose their identity and interests on groups with whom they share a territory. In such instances, minorities frequently intensify their efforts to preserve and protect their identity, in order to avoid marginalization. Polarization between the opposite forces of assimilation on the one hand and protection of minority identity on the other inevitably causes increased intolerance and eventual armed ethnic conflict. In such a scenario, the first duty of governance is to ensure that the rights of a minority society are protected.

Racial profiling is an issue related to minority rights and must not be ignored in essence, racial profiling is intersectional in nature and may consist of multiple grounds of institutionalized discrimination such as nationality, race, age, gender, socio-economic status, disability, health status, descent, language, class, culture and religion. At the World Conference against Racism, Racial Discrimination, Xenophobia and Related Intolerance, held in Durban, South Africa from 31 August to 7 September 2001, the Conference, referring to the International Convention on the Elimination of All Forms of Racial Discrimination, urged States to implement or strengthen legislation and administrative measures prohibiting racial discrimination and related intolerance.

Truth and justice are unhappily mutually exclusive. While in legal terms, legislative parameters will define acts and qualitize their reprehensibility, in truth, speech and conduct that ingratiate themselves to a society have to be addressed politically. This is the dilemma that legislators will face in dealing with racial discrimination. Racial discrimination primarily erode ethical boundaries and convey an unequivocal message of contempt and degradation. The operative question then becomes ethical, as to whether societal mores would abnegate their vigil and tolerate some members of society inciting their fellow citizens to degrade, demean and cause indignity to other members of the very same society, with the ultimate aim of harming them. Conversely, the question arises as to whether there is any obligation on a society to actively protect all its members from indignity and physical harm caused by hatred. The answer to both these questions lies in the fundamental issue of restrictions on racial hatred and discrimination and the indignity that one would suffer in living in a society that might tolerate such discrimination. Obviously, a society committed to protecting principles of social and political equality cannot look by and passively endorse such atrocities, and much would depend on the efficacy of a State's coercive mechanisms. These mechanisms must not only be punitive, but should also be sufficiently compelling to ensure that members of a society not only respect a particular law but also internalize the effects of their proscribed acts.

The intrinsic value to a society and perhaps to the whole world, of eschewing racial and national discrimination is portrayed in the aftermath of the Holocaust – the defining event of this century. Human rights in our lifetime cannot be comprehended without touching our own conceptual proximity to this and other recent events which marred the dignity of human civilization. The result of the Holocaust was the adoption by the United Nations of the Universal Declaration of Human Rights which has now stood its ground over the past 50 plus years. The Universal Declaration, which has flourished both internationally and nationally, has been supplemented by the International Covenant on Economic, Social and Cultural Rights adopted in 1966. Both the Universal Declaration and the International Covenant have committees established to oversee their implementation. The Universal Declaration of Human Rights is composed of 30 articles which asserts a human being's just rights to civilized and dignified living. In this context, we are traversing a thin line between genuine, acceptable processes of profiling possible criminal elements and the danger of racial and national discrimination.

B. Other Aspects of Responsibility

I. Prelude to the Rome Convention of 1952

Even prior to ICAO's coming into being in 1944, there existed a Rome Convention of 1933[133] which provided that damage caused by an aircraft in flight to persons or property on the surface gave rise to a right to compensation on proof only that

---

[133]Convention for the Unification of Certain Rules Relating to Damage Caused by Aircraft to Third Parties on the Surface, signed at Rome on 29 May 1933. Weishaupt (1979, p. 223).

damage exists and that it is attributable to the aircraft.[134] This included damage caused by an object of any kind falling from the aircraft, including in the event of the proper discharge of ballast or of jettison made in case of necessity and instances where damage was caused to any person on board the aircraft. Exceptions were made in the case of an act unconnected with the management of the aircraft which was committed intentionally by a person other than a crew member and where inability of the operator, his servants or agents to prevent such an act was evident. For purposes of the Convention, the aircraft was deemed to be 'in flight' from the beginning of the operations of departure until the end of the operations of arrival.[135] The operator, on whom liability devolved, was considered to be any person who had the aircraft at his disposal and who made use of the aircraft for his own account. The Convention, although not based on principles of fault liability, mitigated damages if the person injured was found to have contributed to the damage by his own negligence.

Article 12 of the Convention required every aircraft registered in the territory of a high contracting party to obtain insurance coverage relating to flight over the territory of a high contracting party, determined upon liability limits set in Article 8 which made the operator liable for each occurrence up to an amount determined at the rate of 250 francs for each kilogramme of the weight of the aircraft[136] to a limit not less than 600,000 francs, nor greater than 2,000,000 francs.

A mandatory requirement in the 1933 Convention for insurance coverage gave rise to the need to specify provisions regarding legalities. A Protocol was concluded in Brussels in September 1938[137] which linked Article 12 of the 1933 Rome Convention to principles of insurance, giving the insurer the right to invoke a defence (in addition to the defence of contributing negligence) in the event the insurance coverage ceased to have effect; the damage occurred outside territorial limits prescribed in the contract in instances not resulting from *force majeure,* justifiable deviation of the aircraft; negligence in piloting; or if the damage was the direct consequences of international armed conflict or civil disorder.

At the 23rd Meeting of the ICAO Legal Committee, held on 21 January 1950, where the Committee was considering a draft Convention to replace the 1933

---

[134]Convention for the Unification of Certain Rules Relating to Damage Caused by Aircraft to Third Parties on the Surface, signed at Rome on 29 May 1933, Article 2.

[135]The meaning imputed to the words "beginning of the operations of departure until the end of the operations of arrival" is debatable. It is interesting that an earlier treaty, the Convention for the Unification of Certain Rules Relating to International Carriage by Air (Warsaw Convention) signed at Warsaw on 12 October 1929 applies liability for accidents taking place on board the aircraft or any of the operations of embarking or disembarking. The word "on board" has been interpreted judicially in different circumstances. See Abeyratne (2001, pp. 197–198).

[136]The weight of the aircraft was the weight with total maximum load as indicated in the certificate of airworthiness or any other official document.

[137]Protocol Supplementing the Convention for the Unification of Certain Rules Relating to Damage Caused by Aircraft to Third Parties on the Surface (signed at Rome on 29 May 1933) concluded at Brussels, on 30 September 1938. Article 2 of the Protocol provides that the Protocol forms an integral part of the 1933 Convention.

B. Other Aspects of Responsibility

Convention, the Committee observed that the 1933 Convention not only applied to damage caused through contact but also to damage caused through fire or explosion or any person or things falling from the aircraft.[138] The Committee placed on record its view that the 1933 Rome Convention's definition of "in flight" caused considerable difficulty, along with its concern with regard to the double system enforced by the Convention of a ceiling of 2,000,000 francs on liability which was made unlimited for gross negligence or wilful misconduct on the part of the operator or his servants or agents or if the operator has not furnished adequate security in accordance with the Convention to cover his liability.[139] The Legal Committee suggested that a new Convention should raise the liability of the operator to 6,000,000 francs.

The ICAO Air Transport Committee (ATC), at the 14th Session of the Council, in December 1951, considered a draft Convention developed by the Legal Committee at its 7th Session in Mexico City. The ATC noted that the Mexico City Draft Convention, like the original 1933 Rome Convention, attempted to regulate the liability of aircraft operators to persons on the surface who sustained injury, death or property damage as a result of aircraft accidents involving foreign aircraft. The ATC noted that the Legal Committee wished States to balance legitimate interests of aircraft operators engaged in international air transport against those of the general public who may suffer as third parties in accident involving foreign aircraft.[140]

It was recognized that the operator needed protection against the risk of catastrophic loss and the draft Convention (1952) accorded him this protection by providing that in no one accident shall his liability to third on the surface exceed a certain maximum figure. On the other hand, it was also noted that the third party on the surface needed the assurance that in accidents in which he suffered loss, he would be able to recover, with a minimum of litigation, the full amount of his damages. The courts gave him this assurance by:

(a) Allowing him to sue where the damage occurred
(b) In certain cases a right of direct action against the insurer
(c) Taking from the operator the "no negligence" defence and ensuring adequate recompense[141]

Representatives of the aviation insurers expressed the view that the policy suggested in the Draft Convention, of making aircraft operators absolutely liable for damage to third parties on the surface, would likely have resulted in a substantial increase in claims and settlements[142] and therefore the Legal Committee had

---

[138] Minutes of the 23rd Meeting of the Legal Committee, Annex VIII Appendix D, p. 355.
[139] Minutes of the 23rd Meeting of the Legal Committee, Annex VIII Appendix D, p. 357.
[140] AT-WP/247, 7/12/51, p. 3.
[141] AT-WP/247, 7/12/51, p. 4.
[142] AT-WP/224, 10 October 51 at 10.

definitely decided on recommending for the new Convention unconditional liability except in four cases:

- When an aircraft was made use of without the consent of the owner
- When damage was a direct consequence of armed conflict or civil disturbance
- When the operator was deprived of the use of the aircraft by public authority
- When the injured person was himself responsible for injury due to negligence or other wrongful act[143]

The Committee believed that the limits should not be set so high as to cause the cost of third party insurance to become an excessive burden to the development of air navigation. Furthermore, it was thought that the limits should be set high enough to cover compensation to third parties in all but extremely rare catastrophic circumstances.[144]

The basic issues regarding compensation (in chapter 3 of the Draft Convention) were whether the Convention should include provisions indicating that Contracting States would accept certain specified proof that aircraft wishing to fly over their territory were adequately insured according to the terms of the Convention, proof to the effect that:

(a) The operator was insured for the aircraft in question against his liability under the Convention.
(b) The insurer was financially sound and able to meet his commitments.
(c) Necessary foreign exchange would be made available so that the compensation could be paid in the currency of the third party.[145]

Following the recommendations of the Legal Committee, the ATC presented to the Council some draft comments – which included, *inter alia*, special reduced limits for gliders, since they did not consume fuel, had no weight, making damage caused by gliders minimal.[146]

It was also recommended to the Council the definition of "in flight" as existing in the 1933 Rome Convention was inappropriate in the case of helicopters as they did not have a landing run.[147]

It is worthy of note that the Chicago Conference of 1944 did not make any reference to the Rome Convention of 1933 although the Conference encouraged States to give consideration to the early calling of an international conference on private air law for the purpose of adopting a convention dealing with transfer of title to aircraft and to ratify or adhere to a Convention for the Unification of Certain

---

[143] AT-WP/224, 10 October 51 at 10.

[144] At that time it was statistically shown that aircraft accidents involving large third party claims occurred infrequently. See AT-WP/247, 10 October 1951, p. 6.

[145] AT-WP/247, 7/11/51, p. 11.

[146] Economic Aspects of the Mexico City Draft Convention on Damage Caused by Foreign Aircraft to Third Parties on the Surface. AT-WP/248 7/12/51, p. 2.

[147] C-WP/1077, 10/12/51, p. 2.

B. Other Aspects of Responsibility

Rules Relating to the Precautionary Attachment of Aircraft, also done in Rome in 1933. Following the Chicago Conference, the Interim Assembly of PICAO in 1946 also made no mention of the Rome Convention of 1933 relating to damage caused by aircraft to third parties on the surface.

At its first session of the ICAO Assembly in 1947, certain delegates pointed out that their Governments had been advised not to ratify that Convention in view of the changing conditions of air transport. Consequently, task of revising that Convention was placed on a sub-Committee, and thereafter by a Commission and the Plenary of the Assembly itself, which examined the work programme of the Legal Committee which had just been established by that Assembly. A high priority was given to that revision; discussions having followed immediately after the questions of recognition of rights in aircraft, the Warsaw Convention and definitions.

At its first session held in Brussels in September 1947 the Legal Committee appointed a Sub-Committee on the Revision of the Rome Convention.

From that moment the revision of the Rome Convention was continuously under study, firstly by a sub-committee which held three sessions and thereafter by a rapporteur entrusted with the preparation of a draft text, leading finally, during three sessions, to its examination by the Committee itself and the Legal Commission of the Assembly. *Ad hoc* sub-committees were also established from time to time for the consideration of special problems. All these bodies together held a total of 160 meetings, averaging approximately three hours each.[148]

During this phase of the work, the Organization maintained close contact with the Governments and the international organizations concerned. The Governments were consulted six times and not less than 20 States sent detailed answers. The international organizations, and in particular the International Air Transport Association and the International Union of Aviation Insurers, were represented at almost all the meetings. The Committee, the Commission and the various sub-committees had before them numerous studies prepared by the Secretariat.[149]

The result of that work was the adoption by the Legal Committee at its seventh session (Mexico City, January 1951) of a "final draft" which was transmitted to the Council together with a report by the Chairman. The Committee recommended that the Council circulate the draft "with such comments as it deems appropriate".

In accordance with that recommendation and a suggestion of the Chairman of the Legal Committee, the Council, on 6 April 1951, referred the draft Convention to the ATC for consideration and report on the desirability, on economic and policy grounds, of the retention or modification of Chapter III (Security for Operator's Liability) and Article II (Limits of Liability) and on such other economic aspects as the Committee deemed appropriate for comment. The Council also decided not to request Contracting States to provide material to assist the ATC in its study of the two specific questions referred to it; however, on 20 June 1951, a questionnaire was

---

[148] See Vol. II, page 13 for the details of the meetings.

[149] These studies are listed in Vol. II, page 14.

circulated to States in order to obtain further factual information bearing on the economic aspects of the convention.

The ATC studied the questions referred to it at its session of October–December 1951 during 11 meetings, and reported to the Council, which on 12 December 1951 approved comments on the economic aspects of the draft convention for transmission to States. The Council reserved the possibility of further study of the expression "in flight" and of comment on other points.[150]

II. The Rome Convention of 1952

Even prior to ICAO's coming into being in 1944, there existed a Rome Convention of 1933[151] which provided that damage caused by an aircraft in flight to persons or property on the surface gave rise to a right to compensation on proof only that damage exists and that it is attributable to the aircraft.[152] This included damage caused by an object of any kind falling from the aircraft, including in the event of the proper discharge of ballast or of jettison made in case of necessity and instances where damage was caused to any person on board the aircraft. Exceptions were made in the case of an act unconnected with the management of the aircraft which was committed intentionally by a person other than a crew member and where inability of the operator, his servants or agents to prevent such an act was evident. For purposes of the Convention, the aircraft was deemed to be 'in flight' from the beginning of the operations of departure until the end of the operations of arrival.[153] The operator, on whom liability devolved, was considered to be any person who had the aircraft at his disposal and who made use of the aircraft for his own account. The Convention, although not based on principles of fault liability, mitigated damages if the person injured was found to have contributed to the damage by his own negligence.

Article 12 of the Convention required every aircraft registered in the territory of a high contracting party to obtain insurance coverage relating to flight over the territory of a high contracting party, determined upon liability limits set in Article 8 which made the operator liable for each occurrence up to an amount determined at the rate

---

[150]No supplementary comment was formulated by the Council.

[151]Convention for the Unification of Certain Rules Relating to Damage Caused by Aircraft to Third Parties on the Surface, signed at Rome on 29 May 1933. Weishaupt (1979, p. 223).

[152]Convention for the Unification of Certain Rules Relating to Damage Caused by Aircraft to Third Parties on the Surface, signed at Rome on 29 May 1933, Article 2.

[153]The meaning imputed to the words "beginning of the operations of departure until the end of the operations of arrival" is debatable. It is interesting that an earlier treaty, the Convention for the Unification of Certain Rules Relating to International Carriage by Air (Warsaw Convention) signed at Warsaw on 12 October 1929 applies liability for accidents taking place on board the aircraft or any of the operations of embarking or disembarking. The word "on board" has been interpreted judicially in different circumstances. See Abeyratne (2001, pp. 197–198).

B. Other Aspects of Responsibility

of 250 francs for each kilogramme of the weight of the aircraft[154] to a limit not less than 600,000 francs, nor greater than 2,000,000 francs.

A mandatory requirement in the 1933 Convention for insurance coverage gave rise to the need to specify provisions regarding legalities. A Protocol was concluded in Brussels in September 1938[155] which linked Article 12 of the 1933 Rome Convention to principles of insurance, giving the insurer the right to invoke a defence (in addition to the defence of contributing negligence) in the event the insurance coverage ceased to have effect; the damage occurred outside territorial limits prescribed in the contract in instances not resulting from *force majeure*, justifiable deviation of the aircraft; negligence in piloting; or if the damage was the direct consequences of international armed conflict or civil disorder.

At the 23rd Meeting of the ICAO Legal Committee, held on 21 January 1950, where the Committee was considering a draft Convention to replace the 1933 Convention, the Committee observed that the 1933 Convention not only applied to damage caused through contact but also to damage caused through fire or explosion or any person or things falling from the aircraft.[156] The Committee placed on record its view that the 1933 Rome Convention's definition of "in flight" caused considerable difficulty, along with its concern with regard to the double system enforced by the Convention of a ceiling of 2,000,000 francs on liability which was made unlimited for gross negligence or wilful misconduct on the part of the operator or his servants or agents or if the operator has not furnished adequate security in accordance with the Convention to cover his liability.[157] The Legal Committee suggested that a new Convention should raise the liability of the operator to 6,000,000 francs.

The ICAO Air Transport Committee (ATC), at the 14th Session of the Council, in December 1951, considered a draft Convention developed by the Legal Committee at its 7th Session in Mexico City. The ATC noted that the Mexico City Draft Convention, like the original 1933 Rome Convention, attempted to regulate the liability of aircraft operators to persons on the surface who sustained injury, death or property damage as a result of aircraft accidents involving foreign aircraft. The ATC noted that the Legal Committee wished States to balance legitimate interests of aircraft operators engaged in international air transport against those of the general public who may suffer as third parties in accident involving foreign aircraft.[158]

---

[154] The weight of the aircraft was the weight with total maximum load as indicated in the certificate of airworthiness or any other official document.

[155] Protocol Supplementing the Convention for the Unification of Certain Rules Relating to Damage Caused by Aircraft to Third Parties on the Surface (signed at Rome on 29 May 1933) concluded at Brussels, on 30 September 1938. Article 2 of the Protocol provides that the Protocol forms an integral part of the 1933 Convention.

[156] Minutes of the 23rd Meeting of the Legal Committee, Annex VIII Appendix D, p. 355.

[157] Minutes of the 23rd Meeting of the Legal Committee, Annex VIII Appendix D, p. 357.

[158] AT-WP/247, 7/12/51, p. 3.

It was recognized that the operator needed protection against the risk of catastrophic loss and the draft Convention (1952) accorded him this protection by providing that in no one accident shall his liability to third on the surface exceed a certain maximum figure. On the other hand, it was also noted that the third party on the surface needed the assurance that in accidents in which he suffered loss, he would be able to recover, with a minimum of litigation, the full amount of his damages. The courts gave him this assurance by:

(a) Allowing him to sue where the damage occurred
(b) In certain cases a right of direct action against the insurer
(c) Taking from the operator the "no negligence" defence and ensuring adequate recompense[159]

Representatives of the aviation insurers expressed the view that the policy suggested in the Draft Convention, of making aircraft operators absolutely liable for damage to third parties on the surface, would likely have resulted in a substantial increase in claims and settlements[160] and therefore the Legal Committee had definitely decided on recommending for the new Convention unconditional liability except in four cases:

– When an aircraft was made use of without the consent of the owner
– When damage was a direct consequence of armed conflict or civil disturbance
– When the operator was deprived of the use of the aircraft by public authority
– When the injured person was himself responsible for injury due to negligence or other wrongful act[161]

The Committee believed that the limits should not be set so high as to cause the cost of third party insurance to become an excessive burden to the development of air navigation. Furthermore, it was thought that the limits should be set high enough to cover compensation to third parties in all but extremely rare catastrophic circumstances.[162]

The basic issues regarding compensation (in chapter 3 of the Draft Convention) were whether the Convention should include provisions indicating that Contracting States would accept certain specified proof that aircraft wishing to fly over their territory were adequately insured according to the terms of the Convention, proof to the effect that:

(a) The operator was insured for the aircraft in question against his liability under the Convention.
(b) The insurer was financially sound and able to meet his commitments.

---

[159] AT-WP/247, 7/12/51, p. 4.

[160] AT-WP/224, 10 October 51 at 10.

[161] AT-WP/224, 10 October 51 at 10.

[162] At that time it was statistically shown that aircraft accidents involving large third party claims occurred infrequently. See AT-WP/247, 10 October 1951, p. 6.

B. Other Aspects of Responsibility

(c) Necessary foreign exchange would be made available so that the compensation could be paid in the currency of the third party.[163]

Following the recommendations of the Legal Committee, the ATC presented to the Council some draft comments – which included, *inter alia*, special reduced limits for gliders, since they did not consume fuel, had no weight, making damage caused by gliders minimal.[164]

It was also recommended to the Council the definition of "in flight" as existing in the 1933 Rome Convention was inappropriate in the case of helicopters as they did not have a landing run.[165]

It is worthy of note that the Chicago Conference of 1944 did not make any reference to the Rome Convention of 1933 although the Conference encouraged States to give consideration to the early calling of an international conference on private air law for the purpose of adopting a convention dealing with transfer of title to aircraft and to ratify or adhere to a Convention for the Unification of Certain Rules Relating to the Precautionary Attachment of Aircraft, also done in Rome in 1933. Following the Chicago Conference, the Interim Assembly of PICAO in 1946 also made no mention of the Rome Convention of 1933 relating to damage caused by aircraft to third parties on the surface.

At its first session of the ICAO Assembly in 1947, certain delegates pointed out that their Governments had been advised not to ratify that Convention in view of the changing conditions of air transport. Consequently, task of revising that Convention was placed on a sub-Committee, and thereafter by a Commission and the Plenary of the Assembly itself, which examined the work programme of the Legal Committee which had just been established by that Assembly. A high priority was given to that revision; discussions having followed immediately after the questions of recognition of rights in aircraft, the Warsaw Convention and definitions.

At its first session held in Brussels in September 1947 the Legal Committee appointed a Sub-Committee on the Revision of the Rome Convention.

From that moment the revision of the Rome Convention was continuously under study, firstly by a sub-committee which held three sessions and thereafter by a rapporteur entrusted with the preparation of a draft text, leading finally, during three sessions, to its examination by the Committee itself and the Legal Commission of the Assembly. *Ad hoc* sub-committees were also established from time to time for the consideration of special problems. All these bodies together held a total of 160 meetings, averaging approximately three hours each.[166]

During this phase of the work, the Organization maintained close contact with the Governments and the international organizations concerned. The Governments were consulted six times and not less than 20 States sent detailed answers.

---

[163] AT-WP/247, 7/11/51, p. 11.

[164] Economic Aspects of the Mexico City Draft Convention on Damage Caused by Foreign Aircraft to Third Parties on the Surface. AT-WP/248 7/12/51, p. 2.

[165] C-WP/1077, 10/12/51, p. 2.

[166] See Vol. II, page 13 for the details of the meetings.

The international organizations, and in particular the International Air Transport Association and the International Union of Aviation Insurers, were represented at almost all the meetings. The Committee, the Commission and the various sub-committees had before them numerous studies prepared by the Secretariat.[167]

The result of that work was the adoption by the Legal Committee at its seventh session (Mexico City, January 1951) of a "final draft" which was transmitted to the Council together with a report by the Chairman. The Committee recommended that the Council circulate the draft "with such comments as it deems appropriate".

In accordance with that recommendation and a suggestion of the Chairman of the Legal Committee, the Council, on 6 April 1951, referred the draft Convention to the ATC for consideration and report on the desirability, on economic and policy grounds, of the retention or modification of Chapter III (Security for Operator's Liability) and Article II (Limits of Liability) and on such other economic aspects as the Committee deemed appropriate for comment. The Council also decided not to request Contracting States to provide material to assist the ATC in its study of the two specific questions referred to it; however, on 20 June 1951, a questionnaire was circulated to States in order to obtain further factual information bearing on the economic aspects of the convention.

The ATC studied the questions referred to it at its session of October–December 1951 during 11 meetings, and reported to the Council, which on 12 December 1951 approved comments on the economic aspects of the draft convention for transmission to States. The Council reserved the possibility of further study of the expression "in flight" and of comment on other points.[168]

C. The Rome Convention of 1952

I. Background

The Rome Convention of 1952[169] entered into force in February 1958 and was ratified by only 46 States Parties, a fact which largely brings to bear its irrelevance to modern day exigencies of liability in air transport. As was stated earlier, the principles of the Convention were conceived by the Legal Committee of ICAO, at its 7th Session in Mexico City, which completed a final draft of the Convention containing principles of liability for damage caused to third parties on the surface by foreign aircraft. The text of the completed draft convention was presented to the ICAO Council for comments.[170] In particular, the Legal Committee requested the

---

[167] These studies are listed in Vol. II, page 14.

[168] No supplementary comment was formulated by the Council.

[169] Convention on Damage Caused by Foreign Aircraft to Third Parties on the Surface, signed at Rome on 7 October 1952. See ICAO Doc 7364.

[170] Volume 7, Minutes and Documents of the ICAO Legal Committee, p. 337.

C. The Rome Convention of 1952

Council's views on limits of liability contained in Article II and provisions regarding the security for the operator's liability appearing in Chapter III. Although the Legal Committee considered its Mexico City draft as being ripe for submission either to the ICAO Assembly or a diplomatic conference for finalization and opening of signature, the Committee was not entirely convinced that some issues of an economic and policy nature incorporated in the draft showed the considered final without their being examined by the Council.[171]

The Council recognized *in limine* that the Mexico City draft was similar to the Rome Convention of 1933 in that both instruments attempted to regulate and establish uniformity with respect to the liability of aircraft operators to persons on the surface who sustain injury, death or property damage as a result of aircraft accidents involving foreign aircraft. The main thrust of the Council's reasoning behind the recognition of the draft convention's relevance was the need to balance the legitimate interests of aircraft operators engaged in international air transport against those of the general public on the ground who may suffer from accidents involving foreign aircraft.

II. *Insurance*

The Council recognized that the operator needed protection against the risk of catastrophic loss, which the draft convention afforded him by limiting his liability to a maximum amount in any one instance, regardless of the damage caused except in the case of intentional damage caused by the operator or in an instance where the operator had wrongfully taken possession of an aircraft without the consent of the owner. Furthermore, the instrument afforded additional protection to the operator by placing a special limitation on the amount of his liability for personal injury or death, to any one person. With regard to protection offered to a third party on the surface, the Council noted that in the event of loss or damage, he will be able to recover the full amount of damages with the minimum of litigation. The Convention ensured the abovementioned protection by allowing the person so injured to bring an action in courts of the place where the damage occurred and by denying the operator's defence of "no negligence" which many jurisdictions afforded and, most importantly, by identifying maximum and minimum limits of liability which would assure adequate compensation to the injured.

The Council carefully examined detailed statistics provided to it by States and others with regard to the cost of insurance together with past records of settlement of third party claims in accidents involving aircraft.

In framing comments on the liability limits in the draft convention, two trends of opinion emerged in Council, the one holding that the limits should be substantially increased, the other that they should be retained at approximately the level in the

---

[171] Volume 7, Minutes and Documents of the ICAO Legal Committee, p. 379.

draft. There was, however, general agreement that the two chief factors to be taken into account in considering the level at which such limits should be set were:

(a) The limits should not be set so high as to cause the cost of third party insurance to become an excessive burden on international civil aviation.
(b) The limits should be set high enough to cover compensation to third parties in all but extremely rare catastrophic accidents.

There was broad agreement also that the influence of the first of these two factors was not strong up to levels of liability limits considerably higher than those under discussion. The disagreement lay chiefly in the evaluation and application of the second factor.

It was noted that the available statistical data indicated clearly that, under existing conditions, the cost of third party insurance under the provisions of the Mexico City draft would generally be a small proportion of total insurance costs for an aircraft operator and an almost negligible part of his total operating costs. Of the rates reported for third party insurance for commercial aircraft in several different parts of the world, none represented more than 0.06 cents (U.S.) per ton-mile available, even for twice the amount of those in the Mexico City draft. Rates for private operators were relatively higher owing to the comparatively small figure of their total operational costs and the low utilization of their aircraft; rates reported to Council for third party insurance costs for private operators varied between 2% and 5% of estimated operating costs for liability limits such as those in the Mexico City draft.

The insurers emphasized that future premium rates for aircraft third party insurance could not be predicted with certainty and that the low rates then current might be increased if a series of accidents occurred involving large payment to third parties. It was impossible to predict the effects on insurance costs of the provisions in the Convention relating to absolute liability, the granting of jurisdiction to the State where the damage occurs, and the right of direct action against the insurers in certain cases. Some insurers believed, however, that these provisions would cause substantial increases in the cost of third party insurance both by increasing the costs of litigation and by tending to raise the amounts of compensation claimed and awarded.

There were, however, other factors tending towards the reduction of the cost of third party insurance for aircraft. In the first place the number of aircraft accidents in relation to the amount of flying done constantly tended to decrease as the quality and efficiency of aircraft construction, maintenance and operation improved. The Council was aware that any decrease in the number of accidents per aircraft would ultimately have produced a decrease in insurance premiums for third party insurance as well as for other forms of aviation insurance. In the second place the growing experience of third party risks gained by the insurers as the volume of flying increases should tend to produce a stabilization of the market and hence to reduce third party insurance rates. Some operators considered the limits proposed in the Mexico City draft as being considerably below the limits of their present third party insurance limits and a premium reduction might result to these operators on this account.

C. The Rome Convention of 1952

Making due allowance for the effect of these various factors, the Council felt that, it was fairly certain that although the cost of third party insurance under the provisions of the Mexico City draft Convention could have been increased, it still would not have imposed an undue burden on aircraft operators, and that, at least for commercial operators, the liability limits in the Convention might be substantially raised without this part of their operating costs becoming excessive. The cost of third party insurance to private operators of aircraft, however, was a considerably higher proportion of operating costs and an increase in liability limits might impose burdens of cost on this section of aviation that would seriously impede its development.

It was suggested that an increase in the liability limits substantially above those proposed in Article II would have produced overall limits for the larger aircraft so high as to strain the insurance market. It seemed, however, that the insurers did not believe that this would occur as long as existing conditions prevail.

The information in the Appendices submitted to the Council clearly showed that aircraft accidents involving large third party claims occurred infrequently. At that time, out of over 2,000 accidents in the British Royal Air Force, only 124 had caused third party damage or casualties and in the vast majority of cases the damage done was minor. It was also noted that, in 118 of these accidents only property damage was done, and in the remaining six where injury was caused to persons, the casualties were one dead and eight injured. Reports from States concerning 23 accidents that caused substantial third party damage included only two instances in which the claims paid and outstanding exceeded the limits in the medico City draft for the aircraft involved.

This information did not, however, point directly to any exact conclusion as to where the liability limits should be set so as to cover all but rare catastrophic accidents. In the first place, it was possible to disagree as to what constituted a catastrophic accident; in the second place past experience as to the frequency of accidents causing large third party damage was inadequate to predict their incidence in the future. It was clear that at whatever level the liability limits were established, the possibility would still have existed that accidents might have occurred where legitimate compensation for third party damage would have been greater than those limits, that is to say, where the third parties concerned would not have been able to obtain full compensation. It was also clear that the higher the limits were placed the smaller that possibility will become and the more nearly the compensation paid in such cases would approach to the full compensation level. The Legal Committee had raised the limits from those proposed in the original Rome draft of the Convention to those now in the Mexico City draft. The divergence of opinion was as to whether they should be further raised or not. The Council believed that it would have been of assistance to Contracting States to have had a brief analysis of the arguments that cause this divergence of opinion.

Those who favoured increasing the liability limits pointed out that the cost of third party insurance, at least to commercial operators, was very small and would still have been small with much higher limits than those in the Mexico City draft. They believed that the limits should be substantially increased and could have been so increased without placing an unreasonable insurance burden on aircraft operators or an excessive strain on the insurance market. They pointed out that accidents causing third party damage greater than the limits in the Mexico City draft had occurred in the past; in the case of the two most serious third party accidents which had been brought to the Council's attention, the Mexico City limits, if applicable, would have resulted in grossly inadequate compensation to the damaged third parties, amounting in one case to approximately one-fifth of their losses and in the other case to approximately one-eighth of their losses. They believed that it was only reasonable to assume that such accidents would occur again in the future. They pointed to the rapid growth of large industrial installations that might be destroyed by fire caused by an aircraft accident; to the possibility of an aircraft crashing into a large public audience or other large collection of people; and to the growing recognition of the value of human life as reflected in increasing compensation awarded in cases of death or permanent injury. They felt that States would be mindful of the legitimate demands for the protection of the general public, and that they will not surrender the rights which the citizens of most States had to claim full compensation for losses caused by foreign aircraft, unless a very strong case could have been made that it would have been unfair to ask the operators of those aircraft to pay the necessary insurance premiums to cover full compensation. They believed that aviation had become an accepted medium of transport and that its further development depended less on special privileges, than on its ability to maintain the confidence of the public. They held that it was not in the best interest of international civil aviation to accord to it privileges which could not be justified by sound technical and economic analysis.

The Council found that, on the basis of data available to the Council, the limits proposed in the Mexico City draft were justifiably low and should be increased for all aircraft, except those in the smallest weight class. The Council noted that the Legal Committee had recognized that the smaller types of aircraft could cause personal injury and death, as distinct from property damage, disproportionate to their weights. For this reason they recommended higher per kilogramme limits for the smaller aircraft, and recommended successively decreasing limits per kilogramme for the successively larger weights of aircraft. They proposed that the increase of liability with weight should commence at a lower weight limit (1,000 kg) and thus operate for all aircraft except those of the smallest weight class, i.e., generally the two-seater private aircraft. Recognizing that the burden of insurance costs was heavier for small privately owned aircraft in the smallest weight class, they proposed no increase in the limit of liability applicable to such aircraft.

The Council could see no justification for fixing an absolute upper limit to the liability limits at 10 million francs, a limit which abruptly ceased to bear a fixed relation to the weight of aircraft. Aircraft were being constructed and others would be built during the period in which the Convention was effective which would considerably have exceeded in weight, and therefore in potential destructiveness, would have probably tended to diminish as the weight of aircraft increases beyond that of the largest types then in general use, and it was for this reason that they

C. The Rome Convention of 1952

recommended a lower rate of increase in the limit of liability per kilogramme for aircraft weighing in excess of 50,000 kg.

The Council appreciated the fact that those who favoured retaining liability limits approximating those in the Mexico City draft considered the proposed Convention as primarily having been designed to establish a fair relationship between operator and third party in given circumstances. Accordingly, cost to the operator was not the first consideration. The nature of the relationship established was the first consideration. In this connection they attached importance to the other provisions of the Convention which affected the conditions in which the liability would be liquidated, such as absolute liability, jurisdiction in the country where the damage occurs, limited defence to the operator, and direct access to the insurers in certain cases. They considered that the limits of liability should not be set unnecessarily high, but at a level which experience and judgment indicated to be adequate to meet all normal cases.

Considerable importance was attached to the information concerning past experience, which in the Council's view, demonstrated the rarity of accidents affecting third parties; that in such accidents it was property, and not persons, which was damaged in the overwhelming majority of instances; and that moreover in all cases of which information was available, save two in North America, the proposed limits would have been more than adequate. They felt also that account should be taken of varying cost levels in different parts of the world. They noted that it was only in North America that there was any evidence of a case in which the proposed limits would not suffice, and in this respect they considered that an equilibrium must be set between the high cost and the low cost areas of the world.

The limits presented an acceptable compromise between the views of various States. They pointed out that the decision taken to raise the limits from those in the original Rome Convention to those in the Mexico City draft had not been unanimous and that some States favoured lower limits than those now in the Mexico City draft. They considered that the economic evidence brought before Council subsequent to the last meeting of the Legal Committee did not justify any modification of the decision reached at Mexico City.

The relating of the liability limits for different types of aircraft to the weight of the aircraft concerned, as proposed in the Mexico City draft, was generally agreed to accord approximately with the potential of each type of aircraft to cause damage to third parties on the surface if an accident occurs. It was recognized, however, that this general rule was subject to certain exceptions. Small and medium aircraft, for example, may cause injury and death, as distinct from property damage, in somewhat greater proportion to their weights. Taking account of this fact, the Council agreed that the proportion of weight to liability limit may vary for different classes of aircraft as it does in the scale of limits proposed in the Mexico City draft Convention, but felt it desirable that the limits should increase without abrupt changes throughout the scale of weights. The introduction of the fixed limit for aircraft weighing more than 2,000 but not exceeding 6,000 kg in paragraph (1) (b) of Article II caused an undesirable discontinuity at the 2,000 kg point. Aircraft just below that weight would have a liability limit of 500,000 francs, aircraft just above that weight would have a liability limit of 1,500,000 francs although the difference in the ability to cause damage between the two types of aircraft might be small. The Council agreed that this discontinuity should be removed.

In the course of examining the liability limits in the draft Convention, the Council considered a number of specific proposals for modifications of these limits. The following proposals were recommended to States as warranting further study since they illustrated the two trends of opinion mentioned above:

Proposal A

(a) This proposal aimed to retain the general level of limits in the Mexico City draft and merely to eliminate the discontinuity at the 2,000 kg level. It would be achieved by substituting the following for sub-paragraph (b) in paragraph (1) of Article II.
(b) 500,000 francs[172] plus 250 francs per kilogramme over 2,000 kg for aircraft weighing more than 2,000, but not exceeding 6,000 kg.

Proposal B This proposal aimed to eliminate the discontinuity at the 2,000 kg level, to increase the limits substantially for all aircraft except those of less thank 1,000 kg and to permit the limits to rise continuously with weight for the larger aircraft. It would be achieved by substituting for sub-paragraphs (a), (b) and (c) of paragraph (1) of Article II, the following sub-paragraphs:

(a) 500,000 francs for aircraft weighing 1,000 kg or less
(b) 500,000 francs plus 400 francs per kilogramme over 1,000 kg for aircraft weighing more than 1,000, but not exceeding 6,000 kg
(c) 2,500,000 francs plus 250 francs per kilogramme over 6,000 kg for aircraft weighing more than 6,000, but not exceeding 20,000 kg
(d) 6,000,000 francs plus 150 francs per kilogramme over 20,000 kg for aircraft weighing more than 20,000 but not exceeding 50,000 kg
(e) 10,500,000 francs plus 100 francs per kilogramme over 50,000 kg for aircraft weighing more than 50,000 kg

Proposals had been made that the sub-limit of 300,000 francs per person killed or injured in paragraph (2) of Article II should be deleted. The Council recognized the importance of this question but felt that the issues raised by the individual limit of 300,000 francs per person killed or injured were largely legal in their implications and that the Council was not in possession of any information on this point not available to the Legal Committee. The Council therefore decided that it was not in a position to give advice to States as to this limit.

---

[172]Article I. The franc used in the Convention, and defined in Article II (4) thereof, equalled US $0.66335, as indicated by the International Monetary Fund.

III. Provisions of the Convention

The 1952 Rome Convention provides that, any person who suffers damage on the surface shall, upon proof only that the damage was caused by an aircraft in flight or by any person or thing falling therefrom, be entitled to compensation as provided by the Convention. Nevertheless, there shall be no right to compensation if the damage is not a direct consequence of the incident giving rise thereto, or if the damage results from the mere fact of passage of the aircraft through the airspace in conformity with existing air traffic regulations.

Article 1 of the Convention provides that, for the purpose of the Convention, an aircraft is considered to be in flight from the moment when power is applied for the purpose of actual take-off, until the moment when the landing run ends. In the case of an aircraft lighter than air, the expression "in flight" relates to the period from the moment when it becomes detached from the surface until it becomes again attached thereto.

Article 2 provides that, the liability for compensation contemplated by Article 1 attaches to the operator of the aircraft and that, for the purposes of the Convention, the term "operator" shall mean the person who was making use of the aircraft at the time the damage was caused, provided that if control of the navigation of the aircraft was retained by the person whom the right to make use of the aircraft was derived, whether directly or indirectly, that person shall be considered the operator. A person shall be considered to be making use of an aircraft when he is using it personally or when his servants or agents are using the aircraft in the course of their employment, whether or not within the scope of their authority. The registered owner of the aircraft is presumed to be the operator and is liable as such unless, in the proceedings for the determination of his liability, he proves that some other person was the operator and, in so far as legal procedures permit, takes appropriate measures to make that other person a party in the proceedings.

If the person who was the operator at the time the damage was caused had not the exclusive right to use the aircraft for a period of more than 14 days, dating from the moment when the right to use commenced, the person from whom such right was derived is deemed to be liable jointly and severally with the operator, each of them being bound under the provisions and within the limits of liability of this Convention.[173] If a person makes use of an aircraft without the consent of the person entitled to its navigational control, the latter, unless he proves that he has exercised due care to prevent such use, is jointly and severally liable with the unlawful user for damage giving a right to compensation under Article 1, each of them being bound under the provisions and within the limits of liability of the Convention.[174]

Article 5 provides that, any person who would otherwise be liable under the provisions of the Convention is not deemed liable if the damage is the direct

---

[173] Article 3.
[174] Article 4.

consequence of armed conflict or civil disturbance, or if such person is deprived of the use of the aircraft by act of public authority.

Article 6 of the Convention lays down principles of fault liability as applicable to the person claiming compensation. It provides that, any person who would otherwise be liable under the provisions of the Convention shall not be liable for damage if he proves that the damage was caused solely through the negligence of other wrongful act or omission of the person who suffers the damage or of the latter's servants or agents. If the person liable proves that the damage was contributed to by the negligence or other wrongful act or omission of the person who suffers the damage, or of his servants or agents, the compensation is reduced to the extent to which such negligence or wrongful act or omission contributed to the damage. Nevertheless, there shall be no such exoneration or reduction if, in the case of the negligence or other wrongful act or omission of a servant or agent, the person who suffers the damage proves that his servant or agent was acting outside the scope of his authority.

When an action is brought by one person to recover damages arising from the death or injury of another person, the negligence or other wrongful act or omission of such other person, or of his servants or agents, shall also have the effect provided in the preceding paragraph.

In terms of liability limits, the Rome Convention stipulates that, subject to Article 12[175] the liability for damage giving a right to compensation under Article 1, for each aircraft and incident in respect of all persons liable under this Convention, shall not exceed:

(a) 500,000 francs for aircraft weighing 1,000 kg or less
(b) 500,000 francs plus 400 francs per kilogramme over 1,000 kg for aircraft weighing more than 1,000 but not exceeding 6,000 kg
(c) 2,500,000 francs plus 250 francs per kilogramme over 6,000 kg for aircraft weighing more than 6,000 but not exceeding 20,000 kg
(d) 6,000,000 francs plus 150 francs per kilogramme over 20,000 kg for aircraft weighing more than 20,000 but not exceeding 50,000 kg
(e) 10,500,000 francs plus 100 francs per kilogramme over 50,000 kg for aircraft weighing more than 50,000 kg

The liability in respect of loss of life or personal injury shall not exceed 500,000 francs per person killed or injured. "Weight" means the maximum weight of the aircraft authorized by the certificate or airworthiness for take-off, excluding the effect of lifting gas when used. The sums mentioned in francs in this Article refer to a currency unit consisting of 65.5 mg of gold of millesimal fineness 900.

---

[175] Article 12 states that if the person who suffers damage proves that it was caused by a deliberate act or omission of the operator, his servants or agents, done with intent to cause damage, the liability of the operator shall be unlimited, provided that, in the case of such act or omission of such servant or agent, it is also proved that he was acting in the course of his employment and within the scope of his authority. If a person wrongfully takes and makes use of an aircraft without the consent of the person entitled to use it, his liability shall be unlimited.

C. The Rome Convention of 1952

These sums may be converted into national currencies in round figures. Conversion of the sums into national currencies other than gold shall, in case of judicial proceedings, be made according to the gold value of such currencies at the date of the judgment, or, in cases covered by Article 14, at the date of the allocation.

A Contracting State is given the option of requiring the operator of an aircraft registered in another Contracting State to obtain insurance with respect to his liability for damage sustained in its territory for which a right to compensation exists under Article 1.

The Convention prescribes, in Article 15 that such insurance shall be accepted as satisfactory if it conforms to the provisions of the Convention and has been effected by an insurer authorized to effect such insurance under the laws of the State where the aircraft is registered, or of the State where the insurer has his residence or principal place of business, and whose financial responsibility has been verified by either of those States. If insurance has been required by any State under paragraph 1 of Article 15 and a final judgment in that State is not satisfied by payment in the currency of that State, any Contracting State may refuse to accept the insurer as financially responsible until such payment, if demand has been made.[176]

Notwithstanding the above, the State overflown may refuse to accept as satisfactory insurance effected by an insurer who is not authorized for that purpose in a Contracting State. The Convention also provides that an appropriate cash deposit, a bank guarantee or a guarantee given by a Contracting State may suffice instead of insurance, for the purpose of Article 15. The State overflown may also require that the aircraft carrying the certificate of insurance issued by the insurer certifying that insurance has been affected in accordance with the Convention. Article 20 prescribes that actions under the Convention may be brought only before the courts of the Contracting State where the damage occurred. Exceptionally, however, the parties to an action under the Convention may consensually agree to bring the action in a court of another jurisdiction, provided such option does not impugn or jeopardise the right of the plaintiff to bring the action in a jurisdiction in which the accident occurred.

Perhaps the most significant feature of the Rome Convention of 1952, which currently impacts the modernization process it is going through, is the nature of liability. It will be recalled that at the Rome Conference, the United States made a strong case for basing liability of the operator, on fault liability as accepted by common law to be rebutted by the operator in the absence of fault and not an absolute liability. The United States delegation contended that air transportation, whether commercial or private, served a great public purpose, both national and international. Therefore, it was proper to take every reasonable step to encourage its development bearing in mind that the development of aviation included a proper relation between the responsibility of the operator and the third parties who might be damaged as a consequence of an aviation accident. The opinion of the United States was that there was no necessity of imposing on aviation a heavier burden than

---

[176] Article 15.

that imposed on other means of transportation. The draft Convention presented to the Conference made the operator liable even if it were evident that he had committed no fault. This principle was in contradiction with the basic principles of common law and seemed to be incompatible with the general principle applicable in civil law countries. The draft Convention provided that once an operator had put an aircraft in the air, if the aircraft crashed on the surface, the operator was liable up to the limits which might be included in the Convention, no matter what the cause of the accident, and that liability would exist no matter how far beyond the control of the operator might be the force which actually caused the accident. In this latter case the operator could not defend himself by proving that he was without fault.

The Delegation of the United States was of the opinion that this system imposed an unwarranted burden on aviation. This burden was not necessary for the proper protection of the public and was not consistent with the status of aviation which was playing an ever-growing part in the life of everyone, and which should be treated on the same basis as other activities of the same nature. Harold Caplan[177] makes the point, referring to modernization of the 1952 Rome Convention, that incredibly the ICAO Working Group is following the analogy of a service to the public (such as a restaurant) with regard to aviation, which is expected to bear its own losses and also pay for the consequences in the event of a terrorist attack.

IV. The Montreal Protocol of 1978

ICAO convened, from 6 to 23 September 1978, an international Conference[178] on private air law at is headquarters in Montreal which resulted in a Protocol to amend the Rome Convention of 1952. The Conference was the direct outcome of a request by the ICAO Council, made in 1964 to the Legal Committee, to study the Convention which showed a marked lack of acceptance. Consequent to several sessions of the Legal sub-Committee in 1965 and 1966, the Legal Committee, in 1967 examined issues arising from the sub-Committee's meetings, particularly regarding the then contentious issue of the sonic boom, and other areas such as nuclear damage and liability. Following the Legal Committee's work, the International Conference on Air Law in September 1978[179] held eight plenary meetings. The main discussions of the Conference ranged from increasing the limits of the Rome Convention; making a clearer pronouncement in the Convention on sonic boom; the single

---

[177]Caplan (2004, p. 5).

[178]The Conference was attended by delegates from 58 States and observers from four organizations.

[179]See ICAO Doc. 9527. The Protocol opened for signature on 23 September 1978.

forum for litigation of claims; effect of increasing the limits on the cost of insurance; and achieving specificity in definitions.[180]

According to FitzGerald, the International Conference on Air Law of 1978 demonstrated the serious difficulties faced when one attempts to revise existing liability Conventions. He attributed the failure of these difficulties to the inherent differences between States on economic issues.[181]

V. Modernizing the Rome Convention

During the 31st Session of the ICAO Legal Committee in September 2000, a formal proposal made by Sweden calling for the modernization of the 1952 Rome Convention under the aegis of ICAO received the endorsement of the Committee. Inspiration for initiating the modernization process was drawn from the adoption of the Montreal Convention of 1999,[182] which the 30th Session of the Legal Committee in 1997 had initiated and which entered into force on 4 November 2003. The Legal Committee, at its 31st Session had recognized that the Montreal Convention[183] enhanced the rights of claimants in respect of death or bodily injury of passengers, and that such rights should also be given formal recognition through treaty with regard to damage to third parties on the surface. Subsequently, in 2002, the Council considered a Secretariat study on the subject and agreed to establish a study group to assist the Secretariat in future work. The Secretariat developed a draft Convention with the assistance of this Study Group.

The draft Convention is similar in scope to the 1952 Convention and 1978 Protocol, and attributes liability to the State of registration of a foreign aircraft if it causes damage over the Exclusive Economic Zone of a State or over the high seas. A prominent feature of the text is that there are two operative systems of liability, one which introduces a two tier liability structure imposing liability for damages not exceeding 100,000 special drawing rights where liability is absolute and the carrier is expressly precluded from denying or limiting his liability, and one, in Chapter III which stipulates various layers of liability based on the weight of the aircraft causing the damage, in case of acts of unlawful interference. This dichotomy prompted one delegation to the 32nd Session of the Legal Committee (Montreal, 15–21 March 2004) to raise the question as to whether it was not preferable to have one basic system for determining all forms of compensation.[184] It is also

---

[180]For a detailed discussion of these issues, see Gerald (1979, pp. 29–74).

[181]Gerald (1979, p. 72).

[182]Report of the 31st Session of the Legal Committee, ICAO Doc 9765.

[183]*Convention for the Unification of Certain Rules for International Carriage by Air*, 28 May 1999, ICAO Doc 9740. The text of the Montreal Convention is also contained in *Annals of Air and Space Law*, vol. 24, p. 425 (1999).

[184]Legal Committee 32nd Session, Montreal, 15–21 March 2004, Report, Doc 9832-LC/192 at 3-1.

noteworthy that, linked to the absolute liability of the operator regarding the 100,000 SDR limit, is a proviso to the effect that, in instances where liability exceeds the limit, the operator shall not be liable for damages if he proves that it was free from fault (that the damage was not due to its negligence or other wrongful act or omission of that of its servants) or that the damage was solely due to the negligence or other wrongful act or omission of another person.[185] This provision is identical to Article 21 of the Montreal Convention of 1999. Another similarity between the two Conventions is found in Article 6 of the new draft Convention and Article 20 of the Montreal Convention, both of which provide that if the operator proves contributory negligence or other wrongful act or omission of the person claiming compensation or another deriving rights therefrom, as having caused the damage, the operator could exonerate himself wholly or partly from liability to the extent that such actions or omissions caused the damage. There is a similar provision for death or injury of passengers in both the Conventions. This symbiosis between the modernized Rome Convention and the Montreal Convention brings to bear the need to clearly identify the scope of liability of the two regimes. Simply put, there is absolute or strict liability of the operator up to a limit of SDR 100,000. Thereafter, over and about this amount of liability, if the carrier proves no fault on his part, or a wrongful act or omission or contributory negligence on the part of another, he could exculpate himself. In other words, for large sum claims, the onus of proving that the operator is not liable is on the operator based on a no fault theory. The entire theory of liability for claims over SDR 100,000 revolves round the word "negligence" or lack thereof applying equally for both plain negligence and contributory negligence. These are essentially tort law concepts but the essential feature in this instance is that there is seemingly a presumption of negligence on the part of the carrier to prove his innocence. It is an ineluctable principle of tort law that tortuous liability exists primarily to compensate the victim by compelling the wrongdoer to pay for the damage he has done.[186] Theoretical bases of tort liability in air law have repeatedly been aligned to presume fault on the part of the carrier until he rebuts the presumption of liability. An earlier instance in 1929 at the Warsaw Conference leading up to the Warsaw Convention on private liability of the carrier also adopted a similar approach where one of the fundamental deviations from the fault liability principle in the context of the Warsaw Conference was that, instead of retaining the basic premise that the person who alleges injury must prove

---

[185] Article 3.3(a) and (b). In the context of private air carrier liability under the Warsaw system there are two analogies that are worthy of note. In *Haddad v. Cie Air France* (1982) 36 RFDA 355, where an airline had to accept suspicious passengers who later perpetrated a hijacking, the court held that the airline could not deny boarding to the passengers who later proved to be hijackers. In that instance the airline had found it impossible to take all necessary precautions and was considered sound in defence under Article 20 (1). A similar approach was taken in the case of *Barboni v. Cie Air-France* (1982) 36 RFDA 358, where the court held that when an airline receives a bomb threat whilst in flight and performs an emergency evacuation, a passenger who is injured by evacuation through the escape chute cannot claim liability of the airline since it would have been impossible for the airline to take any other measure.

[186] Fleming (1983, p. 1).

C. The Rome Convention of 1952

that the injury was caused by the alleged wrong does, the Conference recognized the obligation of the carrier to assume the burden of proof. This was done seemingly to obviate the inherent difficulties which are posed in situations of air carriage where it would be difficult, if not impossible, to determine fault from evidence which is reduced to debris and wreckage after an aircraft accident.

The Conference succinctly subsumed its views on liability through the words of its Rapporteur:

> These rules sprang from the fault theory of the liability of the carrier toward passengers and goods, and from the obligation of the carrier to assume the burden of proof. The presumption of fault on the shoulders of the carrier was, however, limited by the nature itself of the carriage in question, carriage whose risks are known by the passenger and consignor. The Conference had agreed that the carrier would be absolved from all liability when he had taken reasonable and ordinary measures to avoid the damage ... one restriction on this liability had been agreed upon. If for commercial transactions one could concede the liability of the carrier, it did not seem logical to maintain this liability for the navigational errors of his servants, if he proves that he himself took proper measures to avoid a damage.[187]

Negligence is grounded on the notion of duty of care, and the *love thy neighbor* principle enunciated in the 1932 decision of *M'Alister (or Donoughue)* v. *Stevenson*[188] where Lord Atkin stated:

> The rule you are to love your neighbor becomes in law you must not injure your neighbor; and the lawyer's question, who is my neighbor, receives a restricted reply. You must take reasonable care to avoid acts or omissions which you can reasonably foresee would be, likely to injure your neighbor? The answer seems to be: persons who are so closely and directly affected by my act that I ought reasonably to have them in contemplation as being so affected when I am directing my mind to the acts or omissions which are called in question.[189]

This fundamental postulate needs cautious appraisal in the context of the Rome Convention, which calls for a duty of care on the part of the operator toward everyone on the surface, imputing to the scope of air transport an untenably broad focus of foreseeability. Is the operator to foresee damage to every citizen of a State flown over in the event of an act of unlawful interference? If so, how is the operator to exercise due diligence and care to prevent such damage? Since the *Donoughue* decision was handed down, further clarification was provided on the issue of foreseeability in the 1977 House of Lords decision of *Anns v. Merton London Borough Council*[190] where Lord Wilberforce said:

> First, one has to ask whether, as between the alleged wrongdoer and the person who has suffered damage there is a sufficient relationship of proximity or neighborhood such that, in

---

[187] Second International Conference on Private International Law, 4–12 October 1929, Warsaw, Minutes, (translated by Robert C. Herner and Didier Legrez), Fred B. Rottman & Co., New Jersey, 1975, at 21.

[188] [1932] A.C. 562 (H.L.) Hereafter *Donoughue*.

[189] [1932] A.C. 562 (H.L.), 580.

[190] [1978] A.C. 728 (H.L.).

the reasonable contemplation of the former, carelessness on his part may be likely to cause damage to the latter – in which case a prima facie duty of care arises. Secondly, if the first question is answered affirmatively, it is necessary to consider whether there are any considerations which ought to negate, or to reduce or limit the scope of duty or the class of person to whom it is owed or the damages to which a breach may give rise.[191]

The foreseeability requirement would particularly be relevant under the Rome Convention when it comes to items dropped from the aircraft while in flight which damage people or property not specifically foreseen as a vulnerable category by the operator, in view of the remoteness of such damage. The case of *Palsgraf* v. *Long Island Railway Co.*[192] created relevant precedent to the aeronautical context. In *Palsgraf*, a person carrying a parcel of fireworks was being assisted to a train by a porter, when the parcel accidently escaped from the defendant's custody and landed on the railroad tracks beneath, exploding and injuring a passerby. The court held that the defendant owed no duty of care to the plaintiff since the plaintiff was beyond the range of the defendant's foreseeability.

There is also the issue of proximity in the law of negligence which would apply in an instance of damage by foreign aircraft caused to third parties on the surface. Proximity involves three elements: legal closeness; factual closeness; and broad policy factors. Legal closeness relates to the extent to which the proposed duty is related to the concept of duty of care in conventional negligence. In tort law, scholars and legislators have called for a more rigorous evaluation of proximity between plaintiff and defendant, aligned to enforcing liability on a more liberal basis than is practiced at present.[193] Factual closeness relates to the nature of the relationship between the plaintiff and defendant. Policy considerations, often referred to as residual policy factors, are macro-considerations concerning the overall needs and interests of a community which is at risk of damage. A modernized Rome Convention would therefore have to involve and consider all three factors: foreseeability; proximity; and policy in the principles of negligence that are envisioned under the scope of the operator's liability.

The exemption given to the operator in instances of the plaintiff's contributory negligence affords the operator an opportunity to prove that the plaintiff failed to take reasonable care for his own safety. Contributory negligence can arise in three scenarios. Firstly, the plaintiff could be the cause of the accident and the damage suffered must be directly linked to the negligence which contributes to such damage. In *Cork* v. *Kirby Maclean Ltd.*,[194] the court held that a person suffering from epilepsy could not claim that his employer was liable for his fall from his workstation after suffering from an epileptic attack as the employer was not advised by the plaintiff of his illness. The worker was presumed to have known, or ought to

---

[191][1978] A.C. 728 (H.L.), 751–752. See also the decision of *Neilson* v. *Kamloops (City of)*, [1984] 2.S.C.R. 2 handed down by the Supreme Court of Canada which adopted the principle enunciated by Lord Wilberforce and adopted it consistently from 1984 to 2001.

[192]162 N.E. 99 (N.Y. 1928).

[193]Osborne (2003, p. 68).

[194][1952] 2 All.E.R. 402 (C.A.).

C. The Rome Convention of 1952

have known the consequences of his working in a high place. Secondly, a person cannot allege liability of a person successfully if he willingly and knowingly puts himself in a position of foreseeable harm. For instance, a person cannot enforce liability on an inebriated driver who offers her a lift, if she knows prior to accepting the ride of the driver's condition. Thirdly, a person cannot avoid being found responsible for contributory negligence if she does not take protective measures in the face of foreseeable danger, such as not harnessing the seat belt in a fast moving vehicle.[195] Of particular relevance to an accident whereby an aircraft injures persons on the surface is the "agony of the moment" principle which allows some leeway for a person's actions which might not be rational due to the stress of the moment. In the 1969 case of *Walls v Mussens Ltd.*,[196] The court refused to recognize the conduct of the defendant as being guilty of contributory negligence even though he had not used a conventional fire extinguisher which he had ready access, to douse a fire but chose to throw heaps of snow at a fuel ignited fire in his business premises.

Article 5 of the draft Convention provides that, in the instance of disruption to insurance coverage of operators caused by acts of unlawful interference,[197] The ICAO Council may recommend to States Parties to suspend their obligations under the convention. This provision, which was the result of a recommendation by the Council[198] brought in some discussion in the Legal Committee with apprehension being voiced by several delegations that such a provision will throw back responsibilities regarding insurance to national legislatures.[199] Concerns were also raised that there was undue focus only on terrorism oriented issues and that victims should not have to face different regimes depending on nationality. The general view of the Committee on this point seemed to be to prefer a suitable substitute for the Rome Convention to address insurance matters.[200]

---

[195] Under the Warsaw regime there are analogies in contributory negligence. For example, in *Goldman v. Thai Airways International Ltd* (1983) 3 All E.R. 693, it was held that a passenger is not guilty of contributory negligence if he keeps his seatbelt unfastened through the flight and suffers injury when there is no sign given by the aircraft control panel to keep the seat belt on. However, if a passenger removes a bandage or braces that he is required to keep on for an existing injury and he suffers injury in flight due to the removal of the support he would be found to have contributed to the negligence resulting in his injuries.

[196] (1969) 2 N.B.R. (2d) 131 (S.C.A.D).

[197] Following the events of 11 September 2001, where civil aircraft were used as weapons of destruction, aviation insurers gave seven days' notice on 17 September that war risk third party liability coverage according to policy terms applying to the write back coverage for war, hijacking and other perils would be withdrawn. The most compelling reason for the cancellations was the emergence of an exposure in terms of third party bodily injury and property damage that was unquantifiable.

[198] See Report of the Special Group on the Modernization of the Rome Convention of 1952, Montreal, 10–14 January 2005, SG-MR/1 at 1&2-4.

[199] See Report of the Special Group on the Modernization of the Rome Convention of 1952, Montreal, 10–14 January 2005, SG-MR/1 at 1&2-4.

[200] See Report of the Special Group on the Modernization of the Rome Convention of 1952, Montreal, 10–14 January 2005, SG-MR/1 at 1&2-5.

With regard to insurance, the new draft Convention contains two significant provisions, the first being Article 5 which has been already discussed, and provides that States rights and obligations may be suspended, at the request of the Council where acts of terrorism or unlawful interference may severely disrupt the availability of insurance. The second is Article 13 which is consistent with the requirement of Article 50 of the Montreal Convention of 1999, providing that States Parties require their Operators to maintain adequate insurance or guarantees in order to cover their liability under the Convention. As regards compensation, in keeping with Article 28 of the Montreal Convention, the new draft Convention, in Article 23, provides for up-front or advance payments to be made by the Operator in the event of death or injury caused to passengers and immovable property which is left uninsured, provided in both instances the national laws provide for such payments. The new draft Convention offers only one jurisdiction for adjudication, which is the territory of a State in which the damage occurred.

Article 25 provides that the Convention shall not apply to damage caused by State aircraft and that aircraft used in military, customs and police services are deemed to be State aircraft. The inclusionary text is similar to that contained in Article 3(b) of the Chicago Convention. The determination as to whether an aircraft is "used" in military, customs or police services is largely determinant upon a multiplicity of factors such as the nature of the cargo carried and whether such cargo comprises military, customs or police equipment; the ownership of the aircraft in terms of private or public/State ownership; persons carried; registration markings of the aircraft; and the purpose of and publicity given to the flight *inter alia*.[201] The draft text in Article 25 implies that no country which has accepted the Convention in a manner consistent with international recognition of a treaty would use civil aircraft for purposes that would erode the aim of the Convention – which is to provide adequate compensation to those injured on the surface by aircraft. By this provision, therefore, the modernized Convention would serve well to protect both the integrity of civil aviation as well as the interests of those who might be injured. Based on this fundamental postulate, a commercial civil airline could object to its aircraft being used for military purposes not only on grounds of safety, but also on the ground that such use would adversely affect the economic interests of the airline concerned.[202]

The draft Convention has shed any connotation of damage caused to persons and property on the surface from its title owing to the fact that Article 9, a new provision, admits of recovery for third parties suffering damage on board aircraft in a mid-air collision from the operator whose aircraft collided with the aircraft in which the aggrieved person was on board. It would be interesting to examine the liability implications and insurance coverage that would address this situation, as

---

[201] In 1993, the ICAO Secretariat undertook a study on civil/State aircraft with a view to advising the Council on the various determinants that go to differentiate between the two types of aircraft. The results of that study can be found in C-WP/9835, 22/9/93. 3.

[202] See generally, Abeyratne (1997, pp. 1–2).

C. The Rome Convention of 1952

both the Warsaw Convention and Montreal Convention allow an aggrieved person to claim compensation from the operator of the aircraft in which he was travelling. The introduction of another avenue of claim would, at least theoretically, double the opportunity of recovery. In 1999, the U.S. Supreme Court in *El Al Israel Airlines, Ltd. v. Tseng*[203] ruled that the Warsaw Convention preempts all state law personal injury actions arising from international flights. Therefore, in the United States, Courts have recognized the exclusive application of the Warsaw Convention, relying on the United States Supreme Court's decision in the *Tseng* case, that the Convention preempts statutory discrimination claims as well as common law claims.[204]

At its 174th Session, the ICAO Council was presented with the findings of the meeting of the Legal Committee held in January 2005, whereby certain general principles agreed upon by the Legal Committee were placed before the Council. The Committee started by observing that protection of the victim in terms of compensation for damage should be comparable at least to the extent of being as good as compensation offered under the Montreal Convention of 1999 and that the main aim of the Convention ought to be to address incidents with an international element, although it was not ruled out that provision should not be made in the Convention to cover domestic incidents. A valid consideration in this regard was that damage envisioned under the Rome Convention was not only to the person concerned but also to his property which might offer him shelter and in some instances his livelihood as well. It was also contended that instances of "catastrophic losses" ought to be well considered and provisions for compensation therefore be well thought out. Any compensation must not endanger the financial well being of the operator, which might in turn jeopardize the sustainability of the air transport system. One of the compelling convictions of the Committee was that the system of victim compensation must be sufficiently durable so as to survive catastrophic events that would threaten the viability and continuity of the air transport system worldwide. The Committee also suggested that due consideration be given to the establishment of a supplementary funding mechanism for compensation that would

---

[203] 525 U.S. 155 (1999). In *Gibbs v. American Airlines, Inc.*, the Court rejected the plaintiff's argument that his statutory claims under Section 1981 of the Civil Rights Act were not preempted by the Warsaw Convention because they are based on a federal statute and Congress did not intend for the Convention to impede civil rights statutes. In rejecting the plaintiff's argument, the Court relied on the United States Supreme Court's decision in El *Al Israel Airlines, Ltd. v. Tseng* and the decisions of several other district courts that have held that the Convention preempts statutory discrimination claims as well as common law claims. See also Speiser and Krause (1978, Sect. 10.2) for a chapter on mid air collisions in the United States involving domestic air transport.

[204] In *Buchbinder v. American Airlines*, a passenger asserted state law claims against the air carrier and its catering company after becoming ill from a meal consumed on board the aircraft. The catering company filed a motion for summary judgment on the grounds that the plaintiffs' state law claims were preempted. The Court, relying on Tseng decision, held that the Warsaw Convention provided plaintiff's exclusive remedy and dismissed plaintiff's state law claims against the catering company.

succeed in bridging the gap between limits of compensation offered and ensuring protection for the civil aviation sector while maintaining a durable system.[205]

In considering issues relating to the justification for modernizing the Rome Convention, no single issue stands out as much as environmental damage caused to persons and property on the surface by foreign aircraft. The formal proposal made by Sweden at the 31st Session of the Legal Committee with regard to international rules contained in the Rome Convention of 1952, drew the attention of the Legal Committee to the fact that the concept of damage to the environment, including preventive measures and measures of reinstatement, was a factor to be considered and that environmental questions are of great importance. Environmental awareness of carriers and their responsibilities in terms of minimally affecting the environment by their operations was considered a critical issue. The proposal also called for recognition that measures should be in place for the repair and reinstatement of the environment in case of damage. *A fortiori*, the Committee was advised that such measures were already being contemplated in other fora, necessitating a revision of existing principles of liability under the Rome Convention.

At its 35th Assembly, held in September/October 2004, ICAO Contracting States adopted Resolution A35-5 containing a consolidated statement regarding environmental protection which recognized *in limine* that many of the adverse environmental effects of civil aviation activity can be reduced by the application of integrated measures embracing technological improvements, appropriate operating procedures, proper organization of air traffic and the appropriate use of airport planning, land-use planning and management and market-based measures. The resolution also recognizes that other international organizations are becoming involved in activities relating to environmental policies affecting air transport and in fulfilling its role and that ICAO strives to achieve a balance between the benefit accruing to the world community through civil aviation and the harm caused to the environment in certain areas through the progressive advancement of civil aviation.

ICAO has adopted certain goals toward ensuring optimal environmental protection with regard to air transport in the context of the adverse environmental impacts that may be related to civil aviation activity and its responsibility and that of its Contracting States. At the Assembly session, ICAO Contracting States recognized the need to achieve maximum compatibility between the safe and orderly development of civil aviation and the quality of the environment. Therefore, in carrying out its responsibilities, ICAO's goals will be to limit or reduce the number of people affected by significant aircraft noise; limit or reduce the impact of aviation emissions on local air quality; and limit or reduce the impact of aviation greenhouse gas emissions on the global climate.

The Resolution also emphasizes the importance of ICAO's leadership role in all civil aviation matters related to the environment and requests the Council to maintain the initiative in developing policy guidance on these matters, and not leave such initiatives to other organizations; to regularly assess the present and

---

[205] C-WP/12391, 11/02/05, Modernizing the Rome Convention of 1952, Appendix A.

C. The Rome Convention of 1952

future impact of aircraft noise and aircraft engine emissions and to continue to develop tools for this purpose; and to disseminate information on the present and future impact of aircraft noise and aircraft engine emissions and on ICAO policy and guidance material in the environmental field, in an appropriate manner, such as through regular reporting and workshops.

States are invited by Resolution A35-5 to continue their active support for ICAO's environment-related activities on all appropriate occasions and to provide, together with international organizations the necessary scientific information to enable ICAO to substantiate its work in this field. They are urged to refrain from unilateral environmental measures that would adversely affect the orderly development of international civil aviation. The Resolution also encourages the Council to pursue co-operative arrangements with the United Nations Environment Programme for the execution of environmental projects financed by the United Nations Environment Fund if and when it deems this desirable.

With regard to aircraft noise the Resolution contains new guidance material on a balanced approach to noise management and on land-use planning and management. On the subject of aircraft engine emissions, there are new guidance material on operational opportunities to minimize fuel use and reduce emissions and to take account of further work undertaken on emission-related levies and emissions trading.

In the context of seeking justification for the modification of the Rome Convention as amended in 1978, the main consideration must be environmental protection and reparation for damage caused by aircraft operations on the environment. The 35th Session of the ICAO Assembly, reviewed the progress that had been made over the years in ICAO's work in the environmental field, on both aircraft noise and the impact of aircraft engine emissions. The Assembly noted that, regarding noise, ICAO had developed new guidance for States on a balanced approach to aircraft noise management consisting of four principal elements: reduction at source; land-use planning and management; noise abatement operational procedures; and operating restrictions. With regard to aircraft engine emissions the Assembly noted that there were concerns about the impact of emissions at the local level, on local air quality in the vicinity of some of the world's airports. There were also concerns about the impact at the global level, on climate change. ICAO's work on emissions is based on three main approaches:

(a) Improving technology and tightening standards so that aircraft produce less emissions. There is a particular focus on oxides of nitrogen (NOx). It is worthy of note that the ICAO Council has reduced the permitted levels of NOx twice, and is currently considering proposals for a further reduction.
(b) Identifying operational measures that will reduce fuel consumption, which results in less emissions. ICAO has recently published guidance to States on this subject.[206]

---

[206]Circular 303, *Operational Opportunities to Minimise Fuel Use and Reduce Emissions.*

(c) Exploring the possible use of market-based measures. These include voluntary measures, emissions trading and emissions-related levies, which are charges and taxes.[207]

Studies are being undertaken at ICAO to determine whether greenhouse gas charges would be an appropriate approach to market based measures. One of the drawbacks to this approach is that there are a considerable number of outstanding issues on which States' views differ widely. For example, many States question the cost-effectiveness of charges. The Assembly therefore agreed that studies should continue in this area of work. It was understood that States will not introduce greenhouse gas emissions charges internationally during the next three years and the matter will be discussed again at the 36th Session of Assembly in 2007. However, the agreement leaves room for some countries to introduce market based measures amongst themselves under certain circumstances if they wish. It must be noted that some States in Europe have introduced charges at the local level to address problems associated with local air quality in the vicinity of airports. This matter will also be studied more closely by ICAO over the next three years.[208]

In view of its dexterity and flexibility in synthesizing elements of the existing 1952 Rome Convention, its 1978 Protocol and the Montreal Convention of 1999, while adding on new provisions and approaches to liability for damage, a new revised Convention on damage caused to persons and property by foreign aircraft bears promise in providing the much needed balance between economic theory and social justice as well as interests of the public and those of the air transport industry. The advantages offered by an international and globally applicable instrument are clearly evident, particularly in the context of new and emergent threats arising from nuclear devices and chemical and biological weapons, as well as conventionally used terrorist tools, as such would provide an umbrella of regulation particularly to States which are inadequately legislated and carriers who might be under-insured and destitute of liquid assets. A back to back regime provided by the Montreal Convention and the new Convention could supplement each other well in providing coverage in instances of passenger and third party liability, while at the same time looking after the interests of the carrier.

---

[207]Market based measures are targeted through voluntary measures and emissions trading. With regard to the former, ICAO has developed a Template Agreement – Memorandum of Understanding that States and other parties concerned could use as a basis for voluntary measures. On emissions trading, the Assembly endorsed the further development of an open emissions trading system for international aviation.

[208]See the following working papers for background information: A35-WP/56: Civil Aviation and the Environment (which provides an overview); A35-WP/76: Market-based Measures regarding Aircraft Engine Emissions; A35-WP/77: Updating of Resolution A33-7; and A35-WP/352: Report on Item 15 for some details relating to the outcome of Assembly.

References

Abeyratne RIR (1985) Hijacking and the Tehran incident – a world in crisis. Air Law 10:120
Abeyratne RIR (1997) The use of civil aircraft and crew for military purposes. Annals of Air and Space Law 22(2):1–23
Abeyratne RIR (2001) Aviation trends in the new millennium. Ashgate, London
Abeyratne RIR (2002a) The events of 11 September 2001 – ICAO's responses to the security and insurance crises. Air and Space Law 27(6):406–420
Abeyratne RIR (2002b) Investing in insurance of air transport: some perspectives. The Journal of World Investment 3(3):521–533
Abeyratne RIR (2005a) Has the Rome Convention adequately responded to security concerns? What of MANPADS? In: Proceedings of the American Bar Association Forum on Air & Space Law, Annual Meeting and Conference, Sept. 2005
Abeyratne RIR (2005b) Emergent trends in aviation war risk insurance. Air and Space Law 30(2):117–129
Abeyratne RIR (2007) The safe carriage of dangerous pathogens by air: legal and regulatory issues. European Transport Law 42(6):689–704
Becker T (2006) Terrorism and the state: rethinking the rules of state responsibility. Hart Monographs in transnational and international law. Hart, Portland, OR
Blackstone W (2001) Commentaries on the Laws of England (1765–1769), vol 4 (editor: Morrison W). Cavendish, London
Brownlie I (1983) System of the law of nations: state responsibility, Part 1. Clarendon, Oxford
Caplan H (2004) War and terrorism insurance: how to promote long-term international stability and affordability. Air and Space Law 29(1):3–28
Caron DD (1998) The basis of responsibility: attribution and other trans-substantive rules. In: Lillich RB, Magraw DB (eds) The Iran-United States Claims Tribunal: its conclusions to state responsibility. Transnational, Irvington-on-Hudson, NY
de Arechaga EJ (1968) International responsibility. In: Sorenson M (ed) Manual of public international law. New York, St. Martin's
De Vattel E (1916) The law of nations or, the principles of natural law: applied to the conduct and to the affairs of nations and sovereigns, vol 2 (trans: Fenwick CG). Legal Classics Library, New York
Fleming JG (1983) The law of torts, 6th edn. The Law Book Company, Sydney
Gerald F (1979) The protocol to amend the convention on damage caused by foreign aircraft to third parties on the surface (Rome, 1952) signed at Montreal, September 23, 1978. Annals of Air and Space Law 4:29–74
Grotius H (1646) De Jure Belli Ac Pacis, vol 2 (trans: Scott JB). Johan Blaeu, Amsterdam
Hyde C (1928) Concerning damages arising from neglect to prosecute. American Journal of International Law 22:140
Jorgensen NHB (2000) The responsibility of states for international crimes. Oxford University Press, New York
Kelsen H (1951) The law of the United Nations. Praeger, New York
Metz S (2002) State support for terrorism. In: Defeating terrorism, strategic issue analysis. http://www.911investigations.net/IMG/pdf/doc-140.pdf
Osborne PH (2003) The law of torts, 2nd edn. Irvin Law, Toronto
Speiser SM, Krause CF (1978) Aviation tort law, vol 1. LCP BW, New York
Starke JG (1989) Introduction to international law, 10th edn. Butterworths, London
Weishaupt G (ed) (1979) Selected international agreements relating to air law. Butterworths, London, Publications for the Association for the Development of the Academic Institute for Air Transport Education and Research

Chapter 3
Initiatives of the Early Twenty-first Century

A. The Two Liability Conventions

I. *The General Risks Convention*

The world has been debating the issue of damage caused to third parties on the ground by falling aircraft ever since the major catastrophes of 11 September 2001. Many States have found their own solutions, with home grown domestic formulae that offer compensation to victims of such damage, whether the damage is the result of an accident or a violent crime. The international community has also been active in this regard, and the latest initiative of the International Civil Aviation Organization (ICAO)[209] in the nature of two international treaties adopted in May 2009 are good examples of such initiatives.

The Convention on Compensation for Damage Caused by Aircraft to Third Parties[210] (hereafter referred to as the General Risks Convention) and the Convention on Compensation for Damage to Third Parties Resulting From Acts of Unlawful Interference Involving Aircraft[211] (hereafter referred to as the Unlawful Interference Compensation Convention) were adopted at the International Conference on

---

[209]The International Civil Aviation Organization, a specialized agency of the United Nations, was established by Article 44 of the *Convention on International Civil Aviation* (Chicago Convention), signed at Chicago on 7 December 1944 (ICAO Doc 7300/9, Ninth Edition, 2006). The main objectives of ICAO are to develop the principles and techniques of international air navigation and to foster the planning and development of air transport. ICAO has 190 Contracting States. ICAO's Mission and Vision Statement is "to achieve its mission of safe, secure and sustainable development of civil aviation through cooperation amongst its member States." In December 2004, following a decision by the 35th Session of the ICAO Assembly, the Council of ICAO approved six Strategic Objectives for 2005–2010: They are: safety; security; environmental protection; efficiency; continuity; and rule of law. The Strategic Objective applicable to this article is rule of law.

[210]DCCD Doc No. 42, ICAO, Montreal, 1 May 2009.

[211]DCCD Doc No. 43, 1/5/09. This Convention will be discussed in greater detail later in this article.

Air Law (a diplomatic conference held at ICAO in Montreal from 20 April to 2 May 2009). The raison d'être of the former Convention is contained in its Preamble whereby the States Parties to the Convention recognized the need to ensure adequate compensation for third parties who suffer damage resulting from events involving an aircraft in flight. The States Parties also took cognizance of the need to modernize the Convention on Damage Caused by Foreign Aircraft to Third Parties on the Surface, Signed at Rome on 7 October 1952,[212] and the Protocol to Amend the Convention on Damage Caused by Foreign Aircraft to Third Parties on the Surface, Signed at Rome on 7 October 1952, Signed at Montreal on 23 September 1978.

The ICAO initiative to convene the two conferences has its genesis in the effort of the Organization to modernize the Rome Convention of 1952[213] which addressed damage caused to third parties on the surface by foreign aircraft.[214] It must, however be noted that even before ICAO was established in 1944 by the Chicago Convention[215] there were established principles pertaining damage caused by an aircraft in flight to persons or property on the surface which gave rise to a right to compensation

---

[212] The Rome Convention of 1952 entered into force in February 1958 and was ratified by only 46 States Parties, a fact which largely brings to bear its irrelevance to modern day exigencies of liability in air transport. During the 31st Session of the ICAO Legal Committee in September 2000, a formal proposal made by Sweden calling for the modernization of the 1952 Rome Convention under the aegis of ICAO received the endorsement of the Committee. Inspiration for initiating the modernization process was drawn from the adoption of the Montreal Convention of 1999, which the 30th Session of the Legal Committee in 1997 had initiated and which entered into force on 4 November 2003. The Legal Committee, at its 31st Session had recognized that the Montreal Convention enhanced the rights of claimants in respect of death or bodily injury of passengers, and that such rights should also be given formal recognition through treaty with regard to damage to third parties on the surface. Subsequently, in 2002, the Council considered a Secretariat study on the subject and agreed to establish a study group to assist the Secretariat in future work. The Secretariat developed a draft Convention with the assistance of this Study Group.

[213] The Rome Convention of 1952 entered into force in February 1958 and was ratified by only 46 States Parties, a fact which largely brings to bear its irrelevance to modern day exigencies of liability in air transport. During the 31st Session of the ICAO Legal Committee in September 2000, a formal proposal made by Sweden calling for the modernization of the 1952 Rome Convention under the aegis of ICAO received the endorsement of the Committee. Inspiration for initiating the modernization process was drawn from the adoption of the Montreal Convention of 1999, which the 30th Session of the Legal Committee in 1997 had initiated and which entered into force on 4 November 2003. The Legal Committee, at its 31st Session had recognized that the Montreal Convention enhanced the rights of claimants in respect of death or bodily injury of passengers, and that such rights should also be given formal recognition through treaty with regard to damage to third parties on the surface. Subsequently, in 2002, the Council considered a Secretariat study on the subject and agreed to establish a study group to assist the Secretariat in future work. The Secretariat developed a draft Convention with the assistance of this Study Group.

[214] See generally Abeyratne (2006, pp. 185–212).

[215] It is worthy of note that the Chicago Conference of 1944 did not make any reference to the Rome Convention of 1933 although the Conference encouraged States to give consideration to the early calling of an international conference on private air law for the purpose of adopting a convention dealing with transfer of title to aircraft and to ratify or adhere to a Convention for the Unification of Certain Rules Relating to the Precautionary Attachment of Aircraft, also done in Rome in 1933. Following the Chicago Conference, the Interim Assembly of PICAO in 1946 also

A. The Two Liability Conventions

on proof only that damage exists and that it is attributable to the aircraft concerned. These principles were established by the Rome Convention of 1933.[216] This Convention included provisions for damage caused by an object of any kind falling from the aircraft, including in the event of the proper discharge of ballast or of jettison made in case of necessity and instances where damage was caused to any person on board the aircraft.[217] Exceptions were made in the case of an act unconnected with the management of the aircraft which was committed intentionally by a person other than a crew member and where inability of the operator, his servants or agents to prevent such an act was evident. For purposes of the Convention, the aircraft was deemed to be 'in flight' from the beginning of the operations of departure until the end of the operations of arrival.[218] The operator, on whom liability devolved, was considered to be any person who had the aircraft at his disposal and who made use of the aircraft for his own account. The Convention, although not based on principles of fault liability, mitigated damages if the person injured was found to have contributed to the damage by his own negligence.

It must be noted that even before ICAO was established in 1944 by the Chicago Convention[219] there were established principles pertaining damage caused by an aircraft in flight to persons or property on the surface which gave rise to a right to compensation on proof only that damage exists and that it is attributable to the aircraft concerned. These principles were established by the Rome Convention of 1933.[220] This Convention included provisions for damage caused by an object of any kind falling from the aircraft, including in the event of the proper discharge of ballast or of jettison made in case of necessity and instances where damage was

---

made no mention of the Rome Convention of 1933 relating to damage caused by aircraft to third parties on the surface.

[216] Convention for the Unification of Certain Rules Relating to Damage Caused by Aircraft to Third Parties on the Surface, signed at Rome on 29 May 1933, Weishaupt (1979, p. 223).

[217] Convention for the Unification of Certain Rules Relating to Damage Caused by Aircraft to Third Parties on the Surface, signed at Rome on 29 May 1933, Article 2.

[218] The meaning imputed to the words "beginning of the operations of departure until the end of the operations of arrival" is debatable. It is interesting that an earlier treaty, the Convention for the Unification of Certain Rules Relating to International Carriage by Air (Warsaw Convention) signed at Warsaw on 12 October 1929 applies liability for accidents taking place on board the aircraft or any of the operations of embarking or disembarking. The word "on board" has been interpreted judicially in different circumstances. See Abeyratne (2001a, pp. 197–198).

[219] It is worthy of note that the Chicago Conference of 1944 did not make any reference to the Rome Convention of 1933 although the Conference encouraged States to give consideration to the early calling of an international conference on private air law for the purpose of adopting a convention dealing with transfer of title to aircraft and to ratify or adhere to a Convention for the Unification of Certain Rules Relating to the Precautionary Attachment of Aircraft, also done in Rome in 1933. Following the Chicago Conference, the Interim Assembly of PICAO in 1946 also made no mention of the Rome Convention of 1933 relating to damage caused by aircraft to third parties on the surface.

[220] Convention for the Unification of Certain Rules Relating to Damage Caused by Aircraft to Third Parties on the Surface, signed at Rome on 29 May 1933, Weishaupt (1979, p. 223).

caused to any person on board the aircraft.[221] Exceptions were made in the case of an act unconnected with the management of the aircraft which was committed intentionally by a person other than a crew member and where inability of the operator, his servants or agents to prevent such an act was evident. For purposes of the Convention, the aircraft was deemed to be 'in flight' from the beginning of the operations of departure until the end of the operations of arrival.[222] The operator, on whom liability devolved, was considered to be any person who had the aircraft at his disposal and who made use of the aircraft for his own account. The Convention, although not based on principles of fault liability, mitigated damages if the person injured was found to have contributed to the damage by his own negligence.

In adopting the *General Risks Convention* there were several considerations that were taken into account by the ICAO member States other than the need to modernize the Rome Convention.[223] The States Parties also recognized the importance of ensuring protection of the interests of third-party victims and the need for equitable compensation, as well as the need to enable the continued stability of the aviation industry. They saw a compelling need for the orderly development of international air transport operations and the smooth flow of passengers, baggage and cargo in accordance with the principles and objectives of the *Convention on International Civil Aviation*, done at Chicago on 7 December 1944.[224] Furthermore, they were convinced that collective State action for further harmonization and codification of certain rules governing the compensation of third parties who suffer damage resulting from events involving aircraft in flight through a new Convention is the most desirable and effective means of achieving an equitable balance of interests.

1. Some General Features of the Convention

The *General Risks Convention*, which imposes liability on the operator of aircraft, extends that liability to property and environmental damage, and allows for damages to be paid for death, bodily injury and mental injury. It goes on to say that damages due to mental injury shall be compensable only if caused by a "recognizable psychiatric illness" resulting either from bodily injury or from

---

[221] Convention for the Unification of Certain Rules Relating to Damage Caused by Aircraft to Third Parties on the Surface, signed at Rome on 29 May 1933, Article 2.

[222] The meaning imputed to the words "beginning of the operations of departure until the end of the operations of arrival" is debatable. It is interesting that an earlier treaty, the Convention for the Unification of Certain Rules Relating to International Carriage by Air (Warsaw Convention) signed at Warsaw on 12 October 1929 applies liability for accidents taking place on board the aircraft or any of the operations of embarking or disembarking. The word "on board" has been interpreted judicially in different circumstances. See Abeyratne (2001a, pp. 197–198).

[223] See generally Abeyratne (2006, pp. 185–212).

[224] Convention on International Civil Aviation, ICAO Doc 7300, 9th edn, 2006, See *supra* note 5.

direct exposure to the likelihood of imminent death or bodily injury.[225] The term "recognizable psychiatric illness" has not been defined in the Convention, which has seemingly left the matter to judicial interpretation if litigation were to arise where this provision is invoked. Although there is a distinct *cursus curiae* admitting of compensation for mental injury which is caused as a direct result of bodily injury,[226] this is first instance of legislative provision in private air law that allows for compensation for mental injury caused from direct exposure to the likelihood of imminent death or bodily injury.

2. Specific Features of the Convention

The Convention applies to damage to third parties which occurs in the territory of a State Party caused by an aircraft in flight on an international flight, other than as a result of an act of unlawful interference.[227] It is not applicable to damage caused by State aircraft, which are deemed to be aircraft used in military, customs and police services. There is also the possibility that a State may declare that the Convention regime applies to its domestic flights.[228] The operator is liable for damage sustained by third parties upon condition only that the damage was caused by an aircraft in flight.[229]

The operator is strictly liable for each event based on the weight of the aircraft involved. The weight categories and amounts are identical to those found in Article 4 of the Unlawful Interference Compensation Convention. In effect, the overall strict liability of the operator is capped. However, these limits only apply if the operator proves that it was not negligent or that the damage was solely due to the negligence of another person. The consequence of this potential unlimited liability of the operator is that the General Risks Convention does not provide for an International Fund and related provisions. The drafters felt that historically, damage to third parties from general risks has always been compensated, and does not pose a threat to the whole air transport industry.

The Exclusive Remedy provision provides that any action for compensation for damage to third parties brought against the operator can only be brought subject to the conditions in the Convention. The provision is intended to prevent the claimant from invoking or relying on other sources of law to try to circumvent the provisions of the Convention such as those relating to liability. The Convention provides that claims against the operator can only be brought according to the conditions and

---

[225] General Risks Convention, DCCD Doc No. 42, ICAO, Montreal, 1 May 2009, Article 2.1.

[226] See Mankiewicz (1979, pp. 187–211). See also generally, Abeyratne (1999, pp. 193–205). Also Abeyratne (2000, pp. 225–261).

[227] General Risks Convention, DCCD Doc No. 42, ICAO, Montreal, 1 May 2009, Article 2, paragraph 1.

[228] General Risks Convention, DCCD Doc No. 42, ICAO, Montreal, 1 May 2009, Article 2.2.

[229] General Risks Convention, DCCD Doc No. 42, ICAO, Montreal, 1 May 2009, Article 3, paragraph 1.

limits of liability in the Convention, but it does not say that claims can be brought against the operator only. The owner, lessor or financier of an aircraft, not being an operator, is not liable under the General Risks Convention or under the domestic law of States Parties, so there would be no interest in bringing claims against these persons, but other potentially liable persons can be subjected to claims.[230] It is also provided that the Convention shall enter into force 60 days after the deposit of the 35th instrument of ratification.

The quantum of liability of the Operator is based on the weight of the aircraft and ranges from 750,000 Special Drawing Rights (SDR) for aircraft having a maximum mass of 500 kg or less to 700,000,000 SDR for aircraft having a maximum mass of more than 500,000 kg.

At the Conference, the International Air Transport Association and its member airlines expressed their firm belief that the proposed General Risks Convention is not necessary. Their contention was that the domestic laws of ICAO Member States have adequately dealt with major aviation incidents involving damage to third party victims on the ground. Aviation insurance for this type of damage has always been available, and the insurance industry has no record of leaving such claims uncompensated. Furthermore, it was claimed that the casualty rate as a result of third party damages has historically been extremely low.[231]

The manufacturers of aircraft had a much more serious complaint against the *General Risks Convention*. The Aviation Working Group (AWG) which is a group of manufacturers of aircraft and components and lessors contended that in the event of an accident, manufacturers shared an equal burden of liability as operators of aircraft and therefore, singling out operators in the Convention and capping their liability would give the operators more protection than the members of AWG. They further contended that the introduction of the operator's cap fundamentally changed the balance of liability exposure between affected parties. Prior to that change, an operator was potentially liable to all third parties, with a broad recourse against other parties. Following that change, absent fault (for example, in a weather-related incident) an operator would be exonerated from liability above the cap. The AWG concluded that this imbalance has the effect of shifting liability and settlement risk to manufacturers in a number of cases, causing significant, if unintended, potential adverse and unjust consequences.[232]

II. The Unlawful Interference Compensation Convention

The International Conference of Air Law also adopted the Convention on Compensation for Damage to Third Parties Resulting From Acts of Unlawful Interference

---

[230]General Risks Convention, DCCD Doc No. 42, ICAO, Montreal, 1 May 2009, Article 13.
[231]DCCD Doc No. 17, 17 April 2009.
[232]DCCD Doc No. 5, 27 February 2009.

A. The Two Liability Conventions

Involving Aircraft.[233] (hereafter referred to as the Unlawful Interference Compensation Convention). Which imposes strict liability on the operator[234] of an aircraft to compensate[235] for damage caused to third parties within the scope of the Convention[236] if that damage was caused by an aircraft in flight.[237] It will mainly

---

[233] DCCD Doc No. 43, 1/5/09.

[234] According to the Convention, "Operator" means the person who makes use of the aircraft, provided that if control of the navigation of the aircraft is retained by the person from whom the right to make use of the aircraft is derived, whether directly or indirectly, that person shall be considered the operator. A person shall be considered to be making use of an aircraft when he or she is using it personally or when his or her servants or agents are using the aircraft in the course of their employment, whether or not within the scope of their authority. The operator shall not lose its status as operator by virtue of the fact that another person commits an act of unlawful interference. General Risks Convention, DCCD Doc No. 42, ICAO, Montreal, 1 May 2009, Article 1(f).

[235] Under Article 4, the operator's liability is limited or capped, based on the weight of the aircraft, ranging from 750,000 Special Drawing Rights (SDRs) for the smallest aircraft to 700,000,000 SDRs for the largest aircraft. The liability cap may be broken in exceptional circumstances. Under Article 23, where the total amount of damages exceeds the limits of liability of the operator under Article 4, plus the amounts payable by the International Fund under Article 18, paragraph 2 (i.e., the amount of damages exceeds the first and second layers), a person who has suffered damage may claim additional compensation from the operator. To succeed, the person must prove that the operator or its employees have contributed to the occurrence of the event by an act or omission done with intent to cause damage or recklessly and with knowledge that damage would probably result. Where an employee has contributed to the damage, the operator shall not be liable for such additional compensation if it proves that an appropriate system for the selection and monitoring of its employees has been established and implemented. Paragraph 4 of Article 23 sets out the circumstances where the operator or its senior management shall be presumed not to have been reckless.

[236] The Convention's scope covers acts of unlawful interference. As defined in the Convention, an "act of unlawful interference" means an act which is defined as an offence in the *Convention for the Suppression of Unlawful Seizure of Aircraft*, Signed at The Hague on 16 December 1970, or the *Convention for the Suppression of Unlawful Acts Against the Safety of Civil Aviation*, Signed at Montréal on 23 September 1971, and any amendment in force at the time of the event. The Hague Convention of 1971 defines an act of unlawful interference in Article 1 as an act committed by "any person who on board an aircraft in flight unlawfully, by force or threat thereof, or by any other form of intimidation, seizes, or exercises control of that aircraft, or attempts to perform any such act, or is an accomplice of a person who performs or attempts to perform any such act." The Montreal Convention of 1971 defines the offence in Article 1 as an act of violence against a person on board an aircraft in flight if that act is likely to endanger the safety of that aircraft; or destruction of an aircraft in service or damage to such an aircraft which renders it incapable of flight or which is likely to endanger its safety in flight; or the placement or cause for placement on an aircraft in service, by any means whatsoever, a device or substance which is likely o destroy that aircraft, or to cause damage to it which renders it incapable of flight, or damage which is likely to endanger its safety in flight; or destruction or damage to air navigation facilities or interference with their operation, if any such act is likely to endanger the safety of aircraft in flight; or communication of information which the perpetrator knows to be false, thereby endangering the safety of such aircraft in flight. These categories of action are extended to persons who attempt to commit such acts or act as accomplices in the performance of such acts.

[237] An aircraft is considered to be "in flight" at any time from the moment when all its external doors are closed following embarkation or loading until the moment when any such door is opened for disembarkation or unloading. General Risks Convention, DCCD Doc No. 42, ICAO, Montreal, 1 May 2009, Article 1(c).

focus on principles of State liability for private acts of unlawful interference with civil aviation which forms a distinct body of international law, the absence of consideration of which in the Convention has been questioned both by the airline industry, practicing lawyers and the academic world.

The Convention, in its *Preamble* gives its rationale as having emerged through an initial recognition of the States Parties that acts of unlawful interference with aircraft which cause damage to third parties and to property, have serious consequences and that that there are currently no harmonized rules relating to such consequences. The States Parties also recognized the importance of ensuring protection of the interests of third-party victims and the need for equitable compensation, as well as the need to protect the aviation industry from the consequences of damage caused by unlawful interference with aircraft. Accordingly, it was concluded there was a compelling need for a coordinated and concerted approach to providing compensation to third-party victims, based on cooperation between all affected parties. It was therefore reaffirmed that there should be an orderly development of international air transport operations and smooth flow of passengers, baggage and cargo in accordance with the principles and objectives of the Chicago Convention.[238] The approach taken in the convention was therefore to ensure collective State action for harmonization and codification of certain rules governing compensation for the consequences of an event of unlawful interference with aircraft in flight through a new Convention which would achieve an equitable balance of interests.

1. Some General Features of the Convention

The *Unlawful Interference Compensation Convention*, which also imposes liability for property and environmental damage, introduces an interesting dimension of operator liability in that it extends the operator's liability to its "senior management" which is defined in the Convention as members of an operator's supervisory board, members of its board of directors, or other senior officers of the operator who have the authority to make and have significant roles in making binding decisions about how the whole of or a substantial part of the operator's activities are to be managed or organized. This is an implicit recognition of the current trend at common law which admits of corporate negligence and negligent entrustment.

The *Corporate Manslaughter and Corporate Homicide Act* of 2007,[239] provides that an organization[240] is guilty of an offence if the way in which its activities are managed or organized causes a person's death, and amounts to a gross breach of a

---

[238]*Supra*, note 5.

[239]http://www.opsi.gov.uk/acts/acts2007/ukpga_20070019_en_1#pb1-l1g1.

[240]An organization that is a servant or agent of the Crown is not immune from prosecution. *The Corporate Manslaughter and Corporate Homicide Act* of 2007, Section 11.

relevant duty of care owed by the organization to the deceased.[241] The Act applies *inter alia* to a corporation. The offence is termed "corporate manslaughter," in so far as it is an offence under the law of England and Wales or Northern Ireland; and "corporate homicide," in so far as it is an offence under the law of Scotland. An organization that is guilty of corporate manslaughter or corporate homicide is liable on conviction to a fine and the offence of corporate homicide is indictable only in the High Court of Judiciary.[242]

The Act provides that the concept of "relevant duty of care," in relation to an organization, means: a duty owed to its employees or to other persons working for the organization or performing services for it; a duty owed as occupier of premises; a duty owed in connection with the supply by the Organization of goods or services (whether for consideration or not); and the carrying on by the Organization of any construction or maintenance operations, the carrying on by the Organization of any other activity on a commercial basis, or the use or keeping by the Organization of any plant, vehicle or other thing.[243] Section 8 of the Act addresses the issue of "gross breach" and provides that where it is established that an Organization owed a relevant duty of care to a person, and it falls to the jury to decide whether there was a gross breach of that duty, the jury must consider whether the evidence shows that the Organization failed to comply with any health and safety legislation that relates to the alleged breach, and if so how serious that failure was: how much of a risk of death it posed. The jury may also consider the extent to which the evidence shows that there were attitudes, policies, systems or accepted practices within the Organization that were likely to have encouraged any such failure or to have produced tolerance of it; taking into consideration any health and safety guidance that relates to the alleged breach. The provision does not prevent a jury from having regard to any other matters they consider relevant. For purposes of this provision, "health and safety guidance" means any code, guidance, manual or similar publication that is concerned with health and safety matters and is made or issued (under a statutory provision or otherwise) by an authority responsible for the enforcement of any health and safety legislation.

The possible application of this piece of legislation to air transport seems now is a reality, given the nature of the air transport product and the operation of aircraft as well as recognition of the link in the convention. The profession of aeronautics, particularly relating to the piloting of aircraft, remains one of the most responsible, particularly in the context of the many lives that are entrusted to the airline pilot at any given time. Commercial airline pilots operate in a highly complex environment, particularly in single pilot operations. The difficulties faced by pilots in the work environment are compounded by the fact that often inadequate information aggravates the problem. Pilots rely heavily on their visual and auditory senses while flying, and it is of paramount importance that accurate information be available to

---

[241]*The Corporate Manslaughter and Corporate Homicide Act* of 2007, Section 1.

[242]*The Corporate Manslaughter and Corporate Homicide Act* of 2007, Section 1.5.

[243]*The Corporate Manslaughter and Corporate Homicide Act* of 2007, Section 2.1(a)–(c).

the pilot at all times. Most importantly, pilots have usually the predilection to complete their given schedule no matter what, such as competing a flight as planned, meeting schedules, impressing their employees and pleasing the people they carry. Therefore negligent issues concerning the professional conduct of a pilot form quintessential elements for a highly esoteric legal debate.

Another innovation of the *Unlawful Interference Compensation Convention* is that it allows for damages to be paid for death, bodily injury and mental injury and goes on to say that damages due to mental injury shall be compensable only if caused by a "recognizable psychiatric illness" resulting either from bodily injury or from direct exposure to the likelihood of imminent death or bodily injury.[244] The term "recognizable psychiatric illness" has not been defined in the Convention, which has seemingly left the matter to judicial interpretation if litigation were to arise where this provision is invoked. Although there is a distinct *cursus curiae* admitting of compensation for mental injury which is caused as a direct result of bodily injury,[245] this is first instance of legislative provision in private air law that allows for compensation for mental injury caused from direct exposure to the likelihood of imminent death or bodily injury.

2. Specific Features of the Convention

The Convention applies to damage to third parties which occurs in the territory of a State Party caused by an aircraft in flight on an international flight, as a result of an act of unlawful interference.[246] This provision makes sure that damage in any State Party will be compensated, whether or not the operator is from a State Party. In certain instances the Convention has retained the flexibility to apply to such damage that occurs in a State non-Party: where an operator from a State Party causes damage in a State non-Party. In such an instance, the an aggrieved party has recourse, pursuant to Article 8 of the Convention, to an organization called the International Civil Aviation Compensation Fund ("the International Fund"), which is primarily established to pay compensation to persons suffering damage in the territory of a State Party, of providing financial support where an operator from a State Party causes damage in a State non-Party. The Fund, which will have independent legal personality will comprise a Conference of Parties (COP) which will be composed of a Secretariat headed by a Director. It will be the COP that will provide financial support to the operator.[247]

---

[244]Unlawful Interference Compensation Convention, DCCD Doc No. 43, 1/5/09, Article 3.3.

[245]See Mankiewicz (1979, pp. 187–211). See also generally, Abeyratne (1999, pp. 193–205). Also Abeyratne (2000, pp. 225–261).

[246]Unlawful Interference Compensation Convention, DCCD Doc No. 43, 1/5/09, Article 2, paragraph 1.

[247]Unlawful Interference Compensation Convention, DCCD Doc No. 43, 1/5/09, Article 28.

A. The Two Liability Conventions

Although the above discussion implies an international element, the Convention extends further, in that it provides for the possibility for application in essentially domestic situations.[248] Therefore, the Convention may also apply to such damage that occurs in the territory of that Party which is caused by an aircraft in flight other than on an international flight, as a result of an act of unlawful interference. This would however be at the option of the State Party concerned.

As stated above, the liability of the operator to compensate is based solely on strict liability, and the need for the plaintiff to prove negligence of the operator does not arise. The operator is liable to compensate for damage upon condition only that the damage was caused by an aircraft in flight.[249] Again, as was stated earlier, damages due to death, bodily injury and mental injury are compensable, as is damage to property; environmental damage is also compensable, if such compensation is provided for under the law of the State where the damage occurred.

With regard to the quantum of liability, the operator's liability is limited or capped, based on the weight of the aircraft, ranging from 750,000 Special Drawing Rights (SDRs) for the smallest aircraft to 700,000,000 SDRs for the largest aircraft. This liability cap may be broken in exceptional circumstances only.[250] The limits of liability of the operator may be broken in exceptional cases. Where the total amount of damages exceeds the limits of liability of the operator under Article 4, plus the amounts payable by the International Fund (i.e., the amount of damages exceeds the first and second layers), a person who has suffered damage may claim additional compensation from the operator.[251] To succeed, the person must prove that the operator or its employees have contributed to the occurrence of the event by an act or omission done with intent to cause damage or recklessly and with knowledge that damage would probably result. Where an employee has contributed to the damage, the operator shall not be liable for such additional compensation if it proves that an appropriate system for the selection and monitoring of its employees has been established and implemented. The circumstances where the operator or its senior management shall be presumed not to have been reckless are set out in the Convention.[252]

The COP can also decide whether to provide supplementary compensation to passengers on board an aircraft involved in an event. Compensation shall be paid by the International Fund to the extent that the total amount of damages exceeds the Article 4 limits[253] In other words, where there is damage for which the operator is liable, it will pay up to the level of its cap, and the International Fund will pay additional compensation above and beyond the level of the cap. It is expected that

---

[248] Unlawful Interference Compensation Convention, DCCD Doc No. 43, 1/5/09, Article 2.2.
[249] Unlawful Interference Compensation Convention, DCCD Doc No. 43, 1/5/09, Article 3.1.
[250] Unlawful Interference Compensation Convention, DCCD Doc No. 43, 1/5/09, Article 4.
[251] Unlawful Interference Compensation Convention, DCCD Doc No. 43, 1/5/09, Article 23. Also, Article 18.2.
[252] Article 23.4.
[253] Article 18.1.

operators will be able to obtain insurance up to the amount of the cap. If insurance is unavailable, or is only available at a cost incompatible with the continued operation of air transport, the International Fund may pay the damages for which the operator is liable.[254] In general, the maximum amount of compensation that would be available from the International Fund is set at 3 billion SDRs for each event.[255]

The COP could, *inter alia*, establish Regulations of the International Fund, Guidelines for Compensation, Guidelines on Investment, fix the contributions to be made to the International Fund and decide the cases where financial support should be given to the operator in cases of events in States non-Party.[256] The COP is required to meet once a year unless it decides otherwise.

The contributions to the International Fund shall be mandatory amounts collected in respect of each passenger and each tonne of cargo departing on an international commercial flight from an airport in a State Party.[257] Where a State Party has made a domestic opt-in declaration, such amounts shall also be collected in respect of each passenger and each tonne of cargo departing on a commercial flight between two airports in that State Party. The Convention also provides that contributions in respect of each passenger and tonne of cargo shall not be collected more than once in respect of each journey, whether or not that journey includes stops or transfers. It is envisaged that the COP may specify amounts of contribution from general aviation. The operator is required to collect the amounts and remit them to the International Fund. In general, the total amount of contributions collected by the International Fund within two consecutive years shall not exceed 9 billion SDRs.[258]

One of the functions of the COP is to decide the period and rate of contributions in respect of passengers and cargo departing from a State Party to be made from the time of entry into force of the Convention for that State Party.[259] There is also provision that initial contributions shall be paid in respect of passengers and cargo departing on flights covered by a domestic opt-in declaration. Contributions shall be fixed so that the funds available amount to 3 billion SDRs within four years. If the funds available are deemed sufficient in relation to likely compensation or financial assistance to be provided in the foreseeable future and reach the 3 billion SDRs level, the COP may decide to stop collecting.

Where an operator fails to collect or remit contributions, the Director is empowered to take appropriate measures for recovery of the amount due.[260] Each State Party is required to ensure that certain statistics and other data are provided to the International Fund; failure to do so could result in the liability of the State Party for

---

[254] Article 18.3.
[255] Article 18.2.
[256] A full list of the powers and duties of the COP is provided in Article 9.
[257] Article 12.
[258] Article 14.3.
[259] Article 14.
[260] Article 15.2.

A. The Two Liability Conventions

any resulting shortfall in contributions.[261] Where the damage was caused, or contributed to, by the claimant, or the victim, the operator or the International Fund is wholly or partially exonerated from the liability to pay compensation.[262]

The operator has a right of recourse against any person who has committed, organized or financed the act of unlawful interference; and also against any other person.[263] Similarly, where the International Fund has made payments to claimants, it has a right of recourse against any person who has committed, organized or financed the act; against the operator under the conditions established in Article 23; and against any other person.[264] The Convention does not grant a plaintiff a right or recourse against an owner, lessor or financier of the aircraft which is not an operator, or against a manufacturer in certain circumstances.[265]

Another special and unique feature of the Convention is in a special remedy it offers. It provides that essentially, any action for compensation for damage to a third party due to an act of unlawful interference can only be brought against the operator or the International Fund subject to the conditions and limits of liability in the Convention. No claims by a third party shall lie against any other person.[266]

Finally, the Convention provides that it will enter into force 180 days after the deposit of the 35th instrument of ratification on condition that the total number of passengers departing in the previous year from airports in the States that have ratified is at least 750,000,000 as appears from declarations made by these States. A State which has made an opt-in declaration for domestic flights shall declare the total number of passengers that departed on international commercial flights from airports in its territory in the previous year and that number shall be counted toward the 750,000,000.[267]

At the Diplomatic Conference the position of the airline industry with regard to this Convention was clear. In a paper submitted to the Conference the industry maintained that it was important to recognize that terrorists' actions which cause damage to persons and property on the ground are aimed at governments, not airlines. It was also contended that in such instances, prompt compensation should be provided to third party victims on the ground in amounts that would likely exceed the assets of the airline involved and that a punitive approach to dealing with

---

[261] Article 16.

[262] Article 20.

[263] Article 24.

[264] Article 25. Article 26 sets out certain restrictions on the rights of recourse.

[265] Article 27.

[266] Article 29. However, the exclusive remedy provision does not apply to an action against a person who has committed, organized or financed the act. There are other procedural provisions found in Chapter VII. Actions for compensation may be brought in a single forum only, namely, before the courts of the State Party where the damage occurred (Article 32, paragraph 1). Also, judgments entered by a court shall, when they are enforceable in the State Party of that court, be enforceable in any other State Party, although recognition and enforcement of a judgment may be refused under certain specified circumstances (Article 34).

[267] Article 40.

the innocent airline victims of criminal or terrorist interference with their aircraft should be avoided; airlines, which are also victims of terrorism, should be accorded the flexibility of strict but limited liability, capped at insurable amounts. Furthermore it was claimed that if a government, with all of its resources, can fail to prevent a terrorist attack, then any industry failure should not result in punitive liability. The industry also maintained in its paper that industry bankruptcies should be avoided and jobs protected in the face of any terrorist atrocity.[268] The paper admonished that if the Conference did not result in a Convention that provided a balance between compensating innocent third party surface victims and protecting innocent airline victims and other industry stakeholders against acts of terrorism, any other alternative would be a treaty that did not have the support of the aviation industry as a whole.[269]

In a paper submitted to a seminar of the Royal Aeronautical Society on 5 March 2009, George N. Tompkins Jr. stated:

> The costs of terrorism today are usually borne by those States in whose territories the acts of terrorism take place and result in death, damage and destruction of persons and property on the ground and in the air. Take for example the terrorist attacks in the United States of 11 September 2001 ("9/11") (hijacked aircraft of commercial air operators used for terrorist attacks), the terrorist attack in Bali, Indonesia of December 2002 (restaurant/nightclub bombed), the terrorist attack in Madrid, Spain of March 2004 (commuter train bombed) and the terrorist attacks in London, England of July 2006 (buses and underground trains bombed).[270]

Tompkins went on to say:

> Now, however, it is being proposed that the operators and users of commercial aircraft – passengers and consignors of property – should fund the compensation for damage caused by terrorists who utilize commercial aircraft as the instruments of their terrorist attacks. Why, one might ask, should a selected segment of society, be called upon or be expected to fund the costs of acts of terrorism directed at society as a whole? Why indeed![271]

This argument is seemingly shared by the European Union which, in a statement issued on 13 April 2009 stated that in accordance with EU Council Directive 2004/80/EC, all 27 Member States are already obligated to provide compensation for victims of violent crime (including terrorism) committed in their territory. Thus there is no requirement within the EU for a Convention designed solely to compensate victims of aerial terrorism. The EU went on to say that in Europe, many States have for some time had systems of State compensation for crime victims (some stimulated by the 1983 Strasbourg Convention). At the level of the European

---

[268]Joint Industry Paper, DCCD Doc No. 10, 26/3/2009. This paper was presented by the International Air Transport Association (IATA), the International Union of Aerospace Insurers (IUAI), the London & International Insurance Brokers' Association (LIIBA), the Civil Air Navigation Services Organisation (CANSO), the Airports Council International (ACI), the Aviation Security Services Association International (ASSA-I) and the Aviation Working Group (AWG).
[269]Joint Industry Paper, DCCD Doc No. 10, 26/3/2009 at 3.
[270]Tompkins Jr (2009, p. 2).
[271]Tompkins Jr (2009, p. 3).

A. The Two Liability Conventions

Union, collective concerns for the victims of violent crime first found expression in April 2004 in Council Directive 2004/80/EC[272] which requires all Member States (at the latest by 1 July 2005) to "provide for the existence of a scheme on compensation to victims of violent international crimes committed in their respective territories."[273]

Another commentator, who has proposed a treaty that would impose State liability for acts of unlawful interference against aircraft, has commented that the distinctive feature of his proposed treaty is that it would make States, as opposed to operators, answerable for damage caused on the surface of the earth by aircraft as a result of hijacking or other unlawful interference. He goes on to say that holding governments fiscally responsible and accountable and therefore liable for aircraft hijackings and terrorism is not in-and-of-itself a novel idea.[274] One such precedent cited by the author is the Libyan Government's agreement to settle lawsuits concerning the destruction of PANAM flight 103 over Lockerbie, Scotland in 1988.[275]

It will be recalled that, on 21 December 1988, PAN AM Flight 103 exploded over Lockerbie, Scotland, as a result of which all 259 passengers perished together with 11 persons (local residents) on the ground. Indictment of two Libyan officials followed, along with a joint declaration by United Kingdom, France and the United States which declared:

> The three States reaffirm their complete condemnation of terrorism in all its forms and denounce any complicity of States in terrorism acts...They consider that the responsibility of States begins whenever they take part directly in terrorist actions, or indirectly through harbouring, training, providing facilities, arming or providing financial support, or any form of protection, and that they are responsible for their actions before individual States and the United Nations.[276]

International response immediately followed when the United Nations Security Council adopted Resolution 731 which urged Libya to cooperate in surrendering the accused for trial. The Resolution also called upon Libya to accept responsibility and pay appropriate compensation to the victims' families.

Given the availability of domestic remedies in many States against the type of liability envisioned in the two treaties, it is interesting to watch the progress of these instruments toward their journey to coming into effect. If these treaties come into effect, the *General Risks Convention* will serve the purpose of exonerating the operator from unlimited strict liability if he can show that the cause of the accident was attributable to an unlawful interference with civil aviation. However, the bottom line is that States have to be convinced of the need for international treaties

---

[272] OJ L251/15 6.8.2004.
[273] Article 12.2.
[274] Petras (2007).
[275] Smith v. Socialist Peoples Libyan Arab Jamahiriya, 866 F. Supp 306 (1995).
[276] Letter dated 20 December 2001 from the Permanent Representative of France, the United Kingdom of Great Britain and Northern Ireland and the United States of America to the United Nations, addressed to the Secretary General.

that require operators of aircraft to obtain insurance against perils that are already covered by the States, such as the European example cited earlier particularly in terms of the *Unlawful Interference Compensation Convention*.

Of the two Conventions, the sore point certainly seems to be the *Unlawful Interference Compensation Convention*. What seems to be lacking is the explicit or implicit recognition of the reality that the world is interconnected and there are multiple actors who should bear responsibility for acts of terrorism and unlawful interference with social intercourse. Any consideration of responsibility and liability in this context should be based on a responsibility regime which is structured on a causal model[277] that clearly identifies those who have to bear responsibility. Causation, which is the cornerstone of liability in this regard, is well brought to bear in the *Corfu Channel* case,[278] decided in 1949 where, although the ICJ was unable to identify the reprehensibility of the Albanian agents who purportedly lay the mines that caused damage to the British ships patrolling the Channel, the Court opined that such damage could not have been caused without the knowledge of the Albanian Government. The Court went on the basis that Albanian authorities either knew or ought to have known of the impending damage and were consequently guilty of an offence since they did nothing to prevent it. In the context of aviation security, this approach should be a valid consideration, at least to the extent that one party – the operator and his senior management – should not be the only target for compensation. The rationale for this thinking should be, of necessity and logic, based on the incontrovertible premise that terrorism is mainly aimed at States and their instrumentalities and that primary responsibility for security in aviation falls squarely on the States themselves.

This having been said, it is submitted that aviation security needs a more proactive approach that transcends the apportionment of blame and concentrates more on risk avoidance and risk management. The ICAO Security Panel, at its twentieth meeting held in Montreal from 30 March to 3 April 2009 identified several strategic focus areas which related to the need for airlines and airports to strive increasingly to achieve greater efficiency and effectiveness in their operations and processes; address new and emerging threats, enhance international aviation security collaboration; improve human factors and security culture; develop innovative and efficient security measures and promote global compliance through auditing and assistance.[279] In an era in which global aviation is increasingly under attack from new threats of unconventional terrorist attacks, improvised

---

[277] A four pronged approach to set in place a causal model has been suggested. These four steps are: a factual test as to whether an act or omission can be regarded as State conduct, by the operation of attribution principles; a legal test as to whether the attributed act or omission constitutes a violation of an international legal obligation of that State; a causal test to determine the scope of responsibility that potentially arises from a wrongful act or omission of that State; and a policy test to determine whether non-causal considerations justify enhancing or diminishing the responsibility of the State. See Becker (2006, p. 332).

[278]*UK* v. *Albania*, [1949] ICJ Rep. 4 (9 April) at 22.

[279]See AVSECP/20 Report Appendix C to the Report on Agenda item 1 at 1–10.

B. Innovative Security Tools

I. Biometric Identification

Aviation has reached the stage where quantum physics not only assists in the aeronautical aspects of air transport but also contributes to the day to day activities involving passenger clearance, immigration and customs. A brand new technique known as quantum cryptography is on the way, calculated to eliminate the terrifying vulnerabilities that arise in the way digitally stored data are exposed to fraudulent use. This new technique uses polarized photons instead of electronic signals to transmit information along cables. Photons are tiny particles of light that are so sensitive that when intercepted, they immediately become corrupted. This renders the message unintelligible and alerts both the sender and recipient to the fraudulent or spying attempt. The public key directory (PKD) – designed and proposed to be used by customs and immigration authorities who check biometric details in an electronic passport, is based on cryptography – and is already a viable tool being actively considered by the aviation community as a fail-safe method for ensuring the accuracy and integrity of passport information. This article examines the technical and legal consequences that might flow from the use of the public key directory.

Starting from the premise that the passport is primarily a document which establishes the identity of the holder,[280] the various approaches[281] taken by

---

[280] See *Naziranbai v. the State*, 1957 *Madhya Bharat Law Reporter*, at 1, where the court recognized the passport as essentially being a document of identity and nationality issued to citizens or subjects of a state who intend to travel overseas. See also Turack (1972, pp. 20–21). Also, Abeyratne, *infra*, note 294, at 10.

[281] ICAO has been working on the development of passports since 1968. The Seventh Session of the ICAO Facilitation Division in 1968 recommended that a small panel of qualified experts including representatives of the passports and/or other border control authorities, be established: to determine the establishment of an appropriate document such as a passport card, a normal passport or an identity document with electronically or mechanically readable inscriptions that meet the requirements of document control; the best type of procedures, systems (electronic or mechanical) and equipment for use with the above documents that are within the resources and ability of Member States; the feasibility of standardizing the requisite control information and methods of providing this information through automated processes, provided that these processes would meet the requirements of security, speed of handling and economy of operation. See Facilitation Division, Report of the Seventh Session, 14–30 May 1968, ICAO Doc 8750-FAL/564, Agenda Item 2.3, at 2.3-4. See also *AT-WP/1079, 1/12/70*, Attachment A, which sets out the Terms of Reference of the Panel.

ICAO[282] in advancing technologies that facilitate this task at borders have evolved into the use of biometric identification of the passport holder as the ultimate frontier in the identification process. The techniques of biometrics employed in a machine readable travel document (MRTD), be it a visa or passport,[283] enable the user to uniquely encode a particular physical characteristic of a person into a biometric identifier or biometric template which can be verified by machine to confirm or deny a claim regarding a person's identity. Accordingly, biometric identification of a person either correctly establishes his identity as being consistent with what is claimed in the passport he is holding or brings to bear the possibility that the person carrying a particular passport is an imposter. A biometric is a measurable, physical characteristic or personal trait used to recognize the identity, or verify[284] the claimed identity of a person. In the modern context, biometrics are usually incorporated in an MRTD with a view to achieving five goals, the first of which is global interoperability[285] enabling the specifications of biometrics deployed in travel documents across the world to be applied and used in a universally operable manner. This is a critical need if the smooth application of biometric technology were to be ensured across borders. The second goal is to ensure uniformity within States in specific standard setting by States authorities who deploy biometrics in travel documents issued by them. The third is technical reliability, where States are required to ensure that technologies used in deploying biometrics are largely failure-proof and of sufficient quality and standard to ensure a State immigration authority reading documents issued by other States them that the

---

[282]The International Civil Aviation Organization, a specialized agency of the United Nations, was established by Article 44 of the *Convention on International Civil Aviation* (Chicago Convention), signed at Chicago on 7 December 1944 (ICAO Doc 7300/8, Eighth Edition, 2000). The main objectives of ICAO are to develop the principles and techniques of international air navigation and to foster the planning and development of air transport. ICAO has 190 Contracting States.

[283]A passport asserts that the person holding the passport is a citizen of the issuing State while a visa confirms that the State issuing the visa has granted the visa holder the non-citizen privilege of entering and remaining in the territory of the issuing State for a specified time and purpose. The machine readable passport (MRP) is a passport that has both a machine readable zone and a visual zone in the page that has descriptive details of the owner. The machine readable zone enables rapid machine clearance, quick verification and instantaneous recording of personal data. Besides these advantages, the MRP also has decided security benefits, such as the possibility of matching very quickly the identity of the MRP owner against the identities of undesirable persons, whilst at the same time offering strong safeguards against alteration, counterfeit or forgery. Abeyratne (1992, pp. 1–31).

[284]To "verify" means to perform a one-to-one match between proffered biometric data obtained from the holder of the travel document at the time of inquiry with the details of a biometric template created when the holder enrolled in the system.

[285]"Global interoperability" means the capability of inspection systems (either manual or automated) in different States throughout the world to exchange data, to process data received from systems in other States, and to utilize that data in inspection operations in their respective states. Global interoperability is a major objective of the standardized specifications for placement of both eye-readable and machine-readable data in all MRTDs.

B. Innovative Security Tools

details in the document do provide accurate verification of facts. Fourthly, the technology used has to be practical and not give rise to the need for applying disparate types of support technology at unnecessary cost and inconvenience to the user. The final goal is to ensure that the technology used will be sufficiently up to date for at least 10 years and also be backwardly compatible with new techniques to be introduced in the future.

Biometrics target the distinguishing physiological or traits of the individual by measuring them and placing them in an automated repository such as machine encoded representations created by computer software algorithms that could make comparisons with the actual features. Physiological biometrics that have been found to successfully accommodate this scientific process are facial recognition, fingerprinting and iris-recognition which have been selected by ICAO as being the most appropriate. The biometric identification process is fourfold: firstly involving the capture or acquisition of the biometric sample; secondly extracting or converting the raw biometric sample obtained into an intermediate form; and thirdly creating templates of the intermediate data is converted into a template for storage; and finally the comparison stage where the information offered by the travel document with that which is stored in the reference template.

Biometric identification gets into gear each time an MRTD holder enters or exists the territory[286] of a State and when the State verifies his identity against the images or templates created at the time his travel document was issued. This measure not only ensures that the holder of the document is the legitimate claimant to that document and to whom it was issued, but also enhances the efficacy of any advance passenger information (API)[287] system used by the State to pre-determine the arrivals to its territory. Furthermore, matching biometric data presented in the form of the data contained in the template accurately ascertains as to whether the travel document has been tampered with or not. A three way check, which matches the biometrics with those stored in the template carried in the document and a central database, is an even more efficacious way of determining the genuineness of a travel document. The final and most efficient biometric check is when a four

---

[286]The *Convention on International Civil Aviation* (Chicago Convention), signed at Chicago on 7 December 1944 (ICAO Doc 7300/9, Ninth Edition, 2006), defines, in Article 2, "territory of a State" as the land areas and territorial waters adjacent to the State under the sovereignty, suzerainty, protection and mandate of such State.

[287]API involves exchange of data information between airlines and customs authorities, where an incoming passenger's essential details are notified electronically by the airline carrying that passenger prior to his arrival. The data for API would be stored in the passenger's machine readable passport, in its machine readable zone. This process enables customs authorities to process passengers quickly, thus ensuring a smoother and faster clearance at the customs barriers at airports. One of the drawbacks of this system, which generally works well and has proven to be effective, is that it is quite demanding in terms of the high level of accuracy required. One of the major advantages, on the other hand, is the potential carried by the API process in enhancing aviation security at airports and during flight. See Abeyratne (2002b, pp. 631–650).

way determine is effected, were the digitized photograph is visually matched (non electronically) with the three way check described above.[288] In this context, it is always recommended that the facial image (conventional photograph) should be incorporated in the travel document along with the biometric templates in order to ensure that his identity could be verified at locations where there is no direct access to a central database or where the biometric identification process has not entered into the legal process of that location.

In May 2003, The New Technologies Working Group (NTWG) of the Technical Advisory group on Machine Readable Travel Documents (TAG/MTRTD) of ICAO, endorsed its *New Orleans Principle* of March 2003, which resolved that member States will continue to use the facial image as the primary identifier for MRTDs and as such the utilization of standardized digitally-stored facial images should be the globally interoperable biometric to support facial recognition technologies for machine assisted identity verification with MRTDs. Furthermore, the NTWG recognized that in addition to digitally stored facial images, member States of ICAO could also use digitally stored iris images or fingerprints as additional globally interoperable biometrics for purposes of identifying persons through MRTDs.

The challenges facing biometric technology are few, but significant. Biometric technology is evolving so rapidly that it is difficult to maintain consistent standards. The standards themselves are not regularly tested. Some technologies are not adequately established so as to lend themselves to easy decoding and interpretation, particularly when confirming identity on a one-to-one basis with a large central database. More importantly, from a legal perspective, biometric technology brings to bear the compelling need to be aware of privacy issues[289] and data protection legislation of various jurisdictions, as well as liability of the database manager that might emerge pursuant to a breakdown of the database or inaccuracy of information produced as a result of data-matching, which in turn might lead to inconsistencies in the identification process.

II. Public Key Directory

In order to assure inspecting authorities (receiving States) that they would know when the authenticity and integrity of the biometric data stored in the MRTD,

---

[288]Issuing States must ensure the accuracy of the biometric matching technology used and functions of the systems employed if the integrity of the conducted checks are to be maintained. They must also have realistic and efficient criteria regarding the number of travel documents checked per minute in a border control situation and follow a regular biometric identification approach such as facial recognition, fingerprint examination or iris identification system.

[289]Abeyratne (2001b, pp. 153–162; 2002a, pp. 83–115). Also Abeyratne (2002b, pp. 631–650).

B. Innovative Security Tools

which they inspect, are compromised and tampered with, the Public Key Infrastructure (PKI) scheme was developed by the TAG/MRTD, which has been pioneering work on the MRTD for over a decade.[290] The scheme is not calculated to prescribe global implementation of public key encryption, but rather acts as a facilitator enabling States to make choices in areas such as active or passive authentication, anti skimming and access control and automated border crossing, among other facilitative methods. The establishment of a public key directory, through means of public key cryptology and in a PKI environment, is consistent with ICAO's ultimate aim and vision for the application of biometric technology on the fundamental postulate that there must be a primary interoperable form of biometric technology for use at border control with facilities for verification, as well as by carriers and the issuers of documents. This initial premise is inevitably followed by the assumption that biometric technologies used by document issuers must have certain specifications, particularly for purposes of identification, verification and the creation of watch lists. It is also ICAO's vision that States, to the extent possible, are protected against changing infrastructure and changing suppliers, and that a technology, once put in place, must be operable or at least retrievable for a period of 10 years.

The Public Key Directory is a central repository for all public keys that are established individually by States. A key is a string of characters which is used to encrypt or decrypt critical information in a document. Therefore the PKI system ensures that digital signatures assigned to data (and not the data itself) in a MRTD are encrypted or decrypted using both a private key – which is used by the passport issuing authority to encrypt the digital signature – and a public key – to be used by the party reading the document to decrypt the signature. Both the private key and the public key play critical roles in the process of encryption and decryption, which is the essence of the public key directory. It is integral to the programme to have an efficient and commonly accepted means of sharing and updating the public keys in effect for all non-expired passports in existence for all participating countries at a given time. Each participating State will therefore install its own secure facilities to generate key pairs. In each case the private key, used to encrypt digital signatures, will be held secret by the State. The public key, on the other hand, can be released for circulation in the public domain. The

---

[290]ICAO's terms of reference in the development of specifications for machine readable passports stem from the Chicago Convention which provides for ICAO's adoption of international Standards and Recommended Practices dealing, *inter alia,* with customs and immigration procedures. *Convention on International Civil Aviation* (Chicago Convention), signed at Chicago on 7 December 1944 (ICAO Doc 7300/9, Ninth Edition, 2006), Article 37(j). It is interesting that, although passports apply to other modes of international travel as well, ICAO has been singly recognized as the appropriate body to adopt specifications for MRTDs. This alone speaks for the uniqueness of ICAO's facilitation programme. See Machine Readable Travel Documents, *ICAO Doc 9303/3* Third Edition 2005, 1–1 to 1–3.

reading authority at the point of entry would use the appropriate public key to decrypt the information in order to verify whether the data in the MRTD has been altered in any way.

Public key encryption is purely a mathematical process designed to scramble and unscramble messages using two keys (the public key and the private key) and numerical data which contain information the process scrambles the contents of a message. The keys are shared between the scrambler and the un-scrambler. When translated to the e-passport the process works in the following way. The State which issues the passport encrypts information that is placed in the passport using its private key. The State which examines the passport (on arrival of the passenger) obtains the issuing State's public key and uses it to decrypt the information in the passport.

Contrary to popular belief, the PKD is neither a database of e-passports nor a repository of passport information. It is also not a look-out list nor is it a list of persons. Above all, it is not a large database as it remains a database only of public keys. Public keys do not carry personal information but are decoders of information that have been encrypted. The encryption process entitles a reading State to decode the encrypted digital signature on the mandatory passport data which cannot readily be deciphered. Other mandatory data in the machine readable zone of the passport, such as the facial image (photograph) of the passport holder, which is readily visible, do not fall within the process of decryption.

Public keys contain information that can and should be released into the public domain in order to provide for a globally interoperable system that authenticates the contents of integrated circuit chips in passports. There is thus no security issue involved in any potential user's access to public keys, and distribution via the Internet is planned. However, access to the web site will effectively be limited to the users of the system, and specialized system protocols will be required in such transactions. The transmission of key certificates from e-passport issuing States to ICAO, however, will require protection to ensure that bogus keys are not inserted into the system. One of the requirements to be placed on the successful contractor is to demonstrate the capability and competence to build a system with the necessary security measures. The rules and regulations will require adherence to procedures necessary to implement these measures.

The operation of the PKD and the transactions between the PKD and the users will be relatively simple. The PKD will function as a sort of message board, containing "messages" (public key lists) posted by ICAO after ICAO has verified them as genuine. Contributing administrations will be required to send their key lists to ICAO for posting well in advance of their effective date. Accessing the PKD to verify individual passports is not contemplated. Entities using the system will periodically download the whole directory to update the lists in their own systems and use these lists to verify individual passports. This arrangement, together with the redundancy built into the system, is expected to mitigate the risks associated with any system failure. However, the expected level of system performance will be stipulated in the contract with the PKD operator.

B. Innovative Security Tools

1. ICAO's Role Regarding the Public Key Directory

In May 2003, the ICAO Council considered work[291] conducted by its Air Transport Committee[292] and the approval by the Committee of a "Blueprint" for incorporating biometric identification in passports and other MRTDs for the purpose of ascertaining and verifying identity. The Committee had taken into consideration a rigorous and sustained six-year study of technology options for introducing the capability to link a document positively to the rightful holder and to verify the authenticity of the document. The study itself had resulted in a four-part recommendation of the TAG/MRTD. The Blueprint specifies that the primary biometric to be used worldwide will be the face and that the compressed image of the face will be stored, along with the data from the machine readable zone of the passport, in a contact-less Integrated circuit chip. The validity of the data in the chip has to be ensured and, in order to give the reader that assurance, the data in the chip, as well as the facial image, will be digitally "signed". The Committee was apprised that a specially tailored public key infrastructure (PKI) scheme had been specified in order to protect the signed data from counterfeiting or unauthorized alteration by ensuring that any overwriting of data on the chip does not go undetected. The basic premise underlying the study and the recommendation of the TAG/MRTD was that, in the absence of any PKI, the trustworthiness of data in a chip, and hence the global interoperability of the e-passport, cannot be assured.

Based on the above, the TAG-MRTD recommended to the Air Transport Committee that ICAO be the designated Organization to oversee the PKD. This recommendation was based on an interpretation provided to the Council, by the TAG/MRTD, that ICAO had a clear mandate under the Chicago Convention[293] to adopt standards dealing with customs and immigration procedures and to provide for compliance with, *inter alia*, passport laws and regulations, taking into consideration the Organization's sustained and long track record as the developer of MRTD standards, and its international stature as a UN agency. Furthermore, it was claimed that an oversight role in the PKD is deemed particularly appropriate for ICAO due

---

[291] See Establishment of A Public Key Directory (PKD), C-WP/12384, 19/11/04 Revised, 2/2/05, presented to the Council by the Secretary General.

[292] Article 54(d) of the Chicago Convention provides that it shall be a mandatory function of the ICAO Council to appoint and define the duties of an Air Transport Committee, which shall be chosen from among the representatives of the members of the Council and which shall be responsible to it. The Committee is therefore a subordinate body of the Council which largely considers work conducted by the Secretariat in the field of air transport prior to forwarding such work to the Council for final consideration.

[293] *Convention on International Civil Aviation* (Chicago Convention), signed at Chicago on 7 December 1944 (ICAO Doc 7300/9, Ninth Edition, 2006), Articles 13, 23 and 37(j). Although Article 37(j) is directly in point, it is somewhat questionable as to whether Articles 13 and 33 bestow upon ICAO any special mandate to address the need to develop machine readable travel documents and technology related thereto. Article 13 merely states that the laws of States with regard to various aspects of entry and departure should be complied with. Article 23 provides that each Contracting State undertakes, *inter alia*, to establish customs and immigration procedures.

to its substantial interest in document security as an essential component of the aviation security and facilitation programmes elaborated in Annexes 9 and 17. It was the view of the TAG/MRTD that a politically neutral site overseen by ICAO and funded by the e-passport issuing States would provide a trusted resource from which government inspection agencies, airlines, and other entities in all member States might download all public keys in circulation for the purpose of verifying the authenticity of passports as documents of identity, with full confidence that the keys were genuine. It was further contended that, in this regard, an important function of ICAO would be to receive the public keys sent in by issuing States by diplomatic means and perform a technical "due diligence" procedure to verify their authenticity before uploading them to the data base.

The Council was also advised that, in playing an oversight role, ICAO would not be authenticating individual passports or their content. Authentication of a passport remains the function and responsibility of the government agency or aircraft operator examining it.

The envisioned scheme involved the oversight of a central public key directory by ICAO, which was deemed essential for a cooperative, interoperable regime for passport security that will be accessible by all member States. Furthermore, it was contended that a central PKD would be accessible by aircraft operators, who are on the "front lines" as the first to examine the passports of travellers. As a deterrent to the fraudulent alteration or counterfeiting of passports, or the use of stolen passports by imposters to gain access to aircraft, PKI is potentially a most effective anti-terrorism and aviation security measure.

In terms of organizational matters, the proposal for ICAO's oversight role involves two components, i.e., maintaining and administering the PKD, both of which would be funded by the fees collected from States issuing e-passports and uploading their public keys. As the supervisory authority, ICAO would act on behalf of e-passport issuing States; be responsible for establishment of the PKD system, appointment of the PKD operator; and providing oversight of the system operation, financial matters and policies as decided or approved by the Council. In this regard ICAO's functions would include: receipt of new key certificates from e-passport issuing States, verification of their authenticity, and formal acceptance and uploading to the PKD; liaison with all country contributors and users, and with contractor operational staff, in administrative and operational matters such as new country sign-up and collection of fees; calculation of proposed fee schedules; distribution of revenue to the PKD operator and relevant ICAO units, and development of the regulations and procedures manuals; and periodic reporting to the Council on all of the above matters.

Separately, the contractor chosen or the PKD operator would have the responsibility to design, install and operate the PKD system in accordance with the contractual agreement made with ICAO. The PKD operator would provide data base services not only to contributing States but also to States and other entities using the keys to verify e-passports presented to them.

As for financial management and outlay, the proposal for ICAO involvement in the PKD as outlined above will be based and carried out on the principles of

B. Innovative Security Tools 117

cost-recovery, whereby fees from the States that produce e-passports and send their public keys to ICAO for uploading to the Directory will support administrative and other expenses incurred. At the time of writing, ICAO had already received an advance contribution from one member State and had received letters of intent and requests for invoices from several others. The cost-income formula will be calculated on a schedule of country sign-up fees and annual user fees based on the total estimated cost of a five-year operation and the number of countries expected to sign up in each year. A special account would be set up in ICAO for the receipt and distribution of contributions and assessments.

Essentially, there will be three main protagonists in the PKD process. Firstly there would be the "group" of e-passport issuing States, comprising a group that would be constituted as a legal body with its own governance structure. This body would be the owner of the PKD and determine independently its mode of operations – including membership, and financial operations. Secondly, there would be ICAO, duly authorized by the Council to act as an agent of the Group, with defined responsibility for providing advice to the Group and executing the work of the group based on agreed terms and conditions. It is envisioned that such an arrangement would cover ICAO against any financial liability arising either from contracting with a third party or a shortfall in the finances of the group. The Group, as a whole, should underwrite the financing of the activities undertaken by ICAO. The last person in the triangle is the contractor, who is appointed either by the group or by ICAO on authority granted by the Group. ICAO's responsibility for the management of the contractor's activity would be defined by the Group.

2. Legal Liability of ICAO

As stated above, ICAO's responsibility with regard to oversight of the PKD process would involve two areas, i.e., maintenance and administration. A host of functions are attached to these two supervisory functions, such as acting on behalf of e-passport issuing States, and being responsible for establishment of the PKD system, appointment of the PKD operator and providing oversight of the system operation, financial matters and policies as decided or approved by the Council. The first question that arises in regard to ICAO's legal status is whether the Organization has the legal capacity to perform the abovementioned functions and be responsible for them. In other words, if ICAO's legal liability were to be questioned in a court of law in any jurisdiction of an ICAO member State, would the courts recognize ICAO as having the legal capacity to assume these functions and be legally accountable for them?

3. ICAO's Capacity to Conduct Business

The basic issue regarding ICAO's legal status lies in Article 44 of the Chicago Convention, which recognizes that ICAO's aims and objectives are to develop the principles and techniques of international air navigation and to foster the planning and development of international air transport so as to, *inter alia*, meet the needs of the peoples of the world for safe, regular, efficient and economical air transport.[294] This general proviso is qualified by Article 37 (j) of the Convention which provides that ICAO shall adopt and amend from time to time, as may be necessary, international standards and recommended practices and procedures dealing with, *inter alia*, customs and immigration procedures. It is arguable that, if the PKD process were to be classified as a "procedure" dealing with customs and immigration procedures, ICAO can have oversight, maintain and administer the PKD.

The second issue is, can ICAO be recognized as having legal capacity, firstly in Canada, which is home to ICAO and secondly in any of ICAO's member States. The Headquarters Agreement between ICAO and Canada,[295] in Article 2, explicitly provides that ICAO shall possess juridical personality and shall have the legal capacities of a body corporate including the capacity to contract; to acquire and dispose of movable and immovable property; and to institute legal proceedings. With regard to the question as to whether ICAO can be sued in Canada, Article 3 of the Agreement provides that the Organization, its property and its assets,[296] wherever located and by whomsoever held, shall enjoy the same immunity from suit and every form of judicial process as is enjoyed by foreign states. Canada's recognition of ICAO having legal capacities of a body corporate is consistent with Article 104 of the United Nations Charter which provides that the United Nations shall enjoy in the territory of each of its member States such legal capacity as may be necessary for the exercise of its functions and the fulfilment of its purposes.[297] The question which naturally arises from these provisions is "what effect does the Headquarters Agreement between ICAO and the Government of Canada have as a legally enforceable document before the local courts"? In the 1988 *Applicability of the Obligation to Arbitrate Case*,[298] where the International Court of Justice had to consider whether United States anti-terrorism legislation necessitated the closure of the Palestine Liberation Organization's observer mission to the UN in New York, the Court held that the United States was obligated to respect its obligation, contained in Article 21 of the UN Headquarters Agreement with the United States, that the United States had to enter into arbitration in case of a dispute on the interpretation of the Agreement. The court laid particular emphasis on the fact that provisions of a treaty must prevail over the domestic law of a State Party to that treaty.[299] Therefore, there is no room for doubt that ICAO is able to conduct

---

[294] Article 44(d).

[295] Headquarters Agreement Between the International Civil Aviation Organization and the Government of Canada, ICAO Doc 9591.

[296] "Assets" include funds administered by ICAO in furtherance of its constitutional functions.

[297] By virtue of Article 57 of the United Nations Charter, which provides that the various specialized agencies shall be brought into relationship with the United Nations, the acknowledged status of the United Nations as per Article 104 can be applied to ICAO.

[298] *ICJ Reports* 1988, 12; 82 *ILR* 225.

[299] *Id*. 33–34.

business both in Canada and in the territories of any of its member States as a juridical person.

4. ICAO's Immunities and Liabilities

At customary international law, the position of an international organization regarding immunity from suit and other judicial process is unclear[300] and falls within applicable treaty provision, such as the United Nations Charter, Article 105 of which clearly stipulates that the United Nations Organization shall enjoy in the territory of each of its members such privileges and immunities as are necessary for the fulfilment of its purposes. Immunities of the United Nations system are also addressed in the *General Convention on the Privileges and Immunities of the United Nations* of 1946, which sheds some light as to the rights and liabilities of the United Nations and its various entities.[301] ICAO's legal liability within Canada may well hinge on the recognition by the Canadian government that ICAO shall enjoy the same immunity from suit and every form of judicial process as is enjoyed by foreign States. Should the matter of ICAO's immunity be brought before a court within Canada, such court might well look into the true worth of the statement.

Immunity of foreign States in a local jurisdiction has undergone an interesting metamorphosis, from the recognition of personal sovereignty to acceptance of more abstract concepts of State sovereignty. The immunity accorded to ICAO by Canada would impute to the Organization the independence and equality of a State, which municipal courts would be reluctant to impugn or question unless with the consent of ICAO.[302] The United States courts have held that some acts deserve exclusive and absolute immunity, such as internal administrative acts, diplomatic activity and the grant of public loans.[303] In the 1988 case *International Tin Council* v. *Amalgamet Inc.*,[304] The plaintiff ITC averred that it was not obliged to go in for arbitration on the ground that it was an international organization and action under the litigation was performed by the plaintiff as an act of State. The court found this argument untenable as it could not find a "sovereign" character in the contract in

---

[300] Shaw (2003, p. 692).

[301] For the military analogy, see Lazareff (1971). Also Brownlie (1990, p. 372). These authors refer to the NATO *Status of Forces Agreement* of 1951, the provisions of which exclusively governed the relations between the State sending troops and the state receiving them. The courts held that the state sending troops to another State has overall jurisdiction of the troops in terms of offences committed in the receiving State, although the latter may prosecute foreign troops in its own soil if an offence were to be committed which was illegal in that State's jurisdiction. However, the overall principle recognized by the courts was that the sending State has primary jurisdiction over its subjects (or troops) sent on mission if the offence committed related to the performance of duty. See also Woodliffe (1992, p. 298).

[302] See *Ex parte Pinochet* (No. 3) [2000] 1 A C 147 at 201 (*per* Lord Browne-Wilkinson) and 268-9 (*per* Lord Millett).

[303] The Victory Transport Case, ILR 35 at 110.

[304] New York County, 25 January 1988, 524 NYS 2d. 971 (1988); (1989) 80 ILR 31–38.

question. This decision can be distinguished from the ICAO situation as the ITC had not been given the status of a foreign States as has ICAO under its Headquarters agreement with Canada.

5. Waiver of Immunity

There are instances where the courts might deem immunity granted by treat or other agreement to be waived. Waiver of immunity might result either from express agreement between the parties to a contract or by implied acquiescence of the party purporting to enjoy immunity through overt or covert acts. The leading case in this area concerns the 1967 decision[305] of the District of Columbia Circuit Court ruled that the Inter-American Development Bank did not enjoy immunity as any immunity given to the bank had been waived by the Bank by virtue of Article XI (3) of its Articles of Agreement with a Brazilian Corporation who was the other party to the action. An advance waiver, incorporated in a commercial agreement, even though it is calculated to apply only to a particular situation, cannot be deemed invalid and will be generally applicable according to the merits of the case. In *Standard Chartered Bank v. International Tin Council and others*[306] The Queen's Bench in England rejected the claim that an advance waiver is inapplicable to a dispute if it were meant specifically in the contract to apply to "a particular case," which was interpreted by the court as a particular transaction and not a whole dispute. A choice of forum clause in a specific agreement could also be interpreted as a waiver of immunity from suit that could be effectively performed in advance.[307]

In the particular case of the public key directory, ICAO is not merely an overseer of the maintenance and administration of the PKD but has other functions such as being the agent of the group of States who own the directory as well as being a party to possible contracts with a provider of services and technology aimed at running the directory. The status of ICAO would clearly be bifurcated into that of an international organization bestowed with immunity similar to that enjoyed by a sovereign State in its overall role in being responsible for the maintenance of the directory and, on the other hand to being an organization which is a legal person having the capacity to enter into legally enforceable contracts. From the above discussion it could well be subsumed that ICAO would enjoy jurisdictional immunity and immunity for any act perceived as a sovereign act performed by a foreign State. With regard to any local contract that ICAO may enter into, courts may consider restrictive immunity depending on the merits of the case. With regard to liability, it is clear that courts would view with serious apprehension any claim to

---

[305] Lutcher SA Cellulose e Papel v. Inter-American Development Bank, 382 F.2d. 454 (DC Cir. 1967).
[306] [1986] 2 All ER 257; [1987] 1 WLR 641(1988) 77 ILR 16.
[307] See Arab Banking Corporation v. International Tin council and Algemene Bank Nederland and Others (Interveners) and Holo Trading Company Ltd. (Interveners) (1988) 77 ILR 1–8.

C. Advance Passenger Information

One of the most dramatic events pertaining to aviation security occurred in July 2005 when United States air traffic controllers turned back a KLM flight en route to Mexico City from Amsterdam, which was flying over US airspace. The action was grounded on the basis that two of the passengers in the passenger list earlier provided to the US authorities were on a "no fly" list. The importance of this drama to modern day aviation is that the aircraft was merely over-flying the territory of a State. Even more important is the fact that at the time of the incident, there was no US legislation covering the act of refusal to grant over-flying permission to an aircraft in that situation.[308] However, within days, The US Transportation Security Administration (TSA) announced that rules will be adopted to require that passengers on all flights landing in and overflying US territory will be screened against a "no fly" list.[309]

The Passenger Name Record (PNR) is a subject that has been under intense scrutiny by the Council of ICAO,[310] which has developed PNR Data Guidelines that

---

[308] Consequent upon the events of 2001, President George Bush signed a new *American Transportation & Security Act* on 25 November 2002 making mandatory API transmission and the provision of PNR data pertaining to all passengers arriving in the United States. Such information, required prior to departure and arrival in the United States should include in the passenger and crew manifest for each flight, in accordance with Section 115 of the *Transportation & Security Act,* is:

The full name of each passenger and crew member.
The date of birth and citizenship of each passenger and crew member.
The sex of each passenger and crew member.
The passport number and country of issuance of each passenger and crew member if required for travel.
The United States visa number or resident alien card number of each passenger and crew member, as applicable.
Such other information as the under Secretary, in consultation with the Commissioner of Customs, determines is reasonably necessary to ensure aviation safety.

[309] Crossing the Line, *Airline Business*, August 2005, at 9.

[310] The International Civil Aviation Organization (ICAO) is the specialized agency of the United Nations on the subject of international Civil Aviation. ICAO derives its existence through Articles 43 and 44 of the Convention on International Civil Aviation (Chicago Convention), signed at Chicago on 7 December 1944. See ICAO Doc. 7300/8 (Eighth Edition: 2000). Article 44 lists, *inter alia* as ICAO's objectives, insuring the safe and orderly growth of international civil aviation throughout the world and meeting the needs of the peoples of the world for safe, regular, efficient and economical air transport. Chicago Convention, Article 44(a) and Article 44(d). ICAO has 189 Contracting States, all of whom have ratified the Chicago Convention and gained ICAO membership *ipso facto*.

have been transmitted to Contracting States for their comments[311] This exercise was carried out on the understanding that, in the present context of the compelling need for the enhancement of aviation security, the global aviation community has shown an increased interest[312] in adding the PNR data as a security measure in addition to the already existing Advanced Passenger Information (API)[313] and the Machine Readable Travel Document (MRTD), which, although primarily are facilitation tools, greatly assist States authorities in ensuring border security.

One of the issues that emerge from PNR data collection is extraterritoriality and the question as to whether at law a State can require information held by other States relating to flights that originate and end in the latter States. An example is Canada, which may be required by the US to divulge information pertaining to passengers on domestic flights operating within the territorial limits of Canada but over-fly United States' territory for reasons of expediency and fuel efficiency. While there is no room for doubt that usually, requirements for safety and security of a State are based on sound legal justification with a view to protecting A State's integrity and internal security, a requirement for information by a particular State of those that do not enter the territory of that State might open itself to question, as to whether such would impinge upon another sovereign State's right to privacy[314] and dignity.

D. The Passenger Name Record

A new Recommended Practice for inclusion in Annex 9 to the Chicago Convention (Facilitation) was adopted by the ICAO Council in March 2005. This Recommended Practice, which supplements an already existing Recommended Practice,

---

[311] Attachment to State Letter EC 6/2-05/70, Passenger Name Record (PNR) data, 9 June 2005.

[312] The advantage of collection by States of PNR Data was first discussed by the global aviation community at the 12th Session of the ICAO Facilitation Division that was held in Cairo, Egypt from 22 March to 1 April 2004. Consequently, the Division adopted Recommendation B/5, that reads as follows:

> It is recommended that ICAO develop guidance material for those States that may require access to Passenger Name Record (PNR) data to supplement identification data received through an API system, including guidelines for distribution, use and storage of data and a composite list of data elements [that] may be transferred between the operator and the receiving State.

Pursuant to this recommendation, in June 2004, the Air Transport Committee of the ICAO Council requested the Secretary General to establish a Secretariat Study Group to develop Guidelines on PNR data transfer. The Council, in endorsing Recommendation B/5, directed that these Guidelines were to be submitted early in 2005.

[313] See Abeyratne (2002b, pp. 631–650). Also by Abeyratne (2001b, pp. 153–162; 2003, pp. 297–311).

[314] See Abeyratne (2001b, pp. 153–162).

provides that Contracting States requiring Passenger Name Record (PNR) access should conform their data requirements and their handling of such data to guidelines developed by ICAO. It is worthy of note that Article 13 of the Chicago Convention provides that the laws and regulations of a Contracting State as to the admission to or departure from its territory of passengers, crew or cargo of aircraft, such as regulations relating to entry, clearance, immigration, passports, customs, and quarantine shall be complied with, by or on behalf of such passengers, crew or cargo upon entrance into or departure from, or while within the territory of that State. This provision gives a State the discretion to specify the information it requires relating to persons wishing to gain entry into its territory. Accordingly, a State may require aircraft operators operating flights to, from or in transit through airports within its territory to provide its public authorities, upon request, with information on passengers such as PNR data.

The philosophy underlying the importance of PNR data and their efficient use by States for enhanced expediency in border crossing by persons is embodied in the General Principles set out in Chapter 1 of Annex 9 which require Contracting States to take necessary measures to ensure that: the time required for the accomplishment of border controls in respect of persons is kept to the minimum;[315] the application of administrative and control requirements causes minimum inconvenience; exchange of relevant information between Contracting States, operators and airports is fostered and promoted to the greatest extent possible; and, optimal levels of security, and compliance with the law, are attained.

Contracting States are also required to develop effective information technology to increase the efficiency and effectiveness of their procedures at airports.[316]

I. Definition and Application of PNR

The air transport industry regards a *Passenger Name Record* (PNR), as a generic term applicable to records created by aircraft operators or their authorized agents for each journey booked by or on behalf of any passenger. The data is used by operators for their own commercial and operational purposes in providing air

---

[315]There is an abiding symbiosis between security and facilitation in the field of air transport. While security is of paramount interest to the global aviation community, it must not unduly disrupt or in any adversely affect the expediency of air transport. To this end, Recommended Practice 2.2 of Annex 9 – Facilitation – to the Chicago Convention suggests that Each Contracting State should whenever possible arrange for security controls and procedures to cause a minimum of interference with, or delay to the activities of civil aviation provided the effectiveness of these controls and procedures is not compromised. See McMunn (1996, p. 7).

[316]It must be noted that Annex 9 specifies that the provisions of the Annex shall not preclude the application of national legislation with regard to aviation security measures or other necessary controls.

transportation services.[317] The definition applicable in the United States identifies a PNR as a repository of information that air carriers would need to make available upon request under existing regulations and refers to reservation information contained in a carrier's electronic computer reservation system.[318]

The above definitions and identifiers go to show that a PNR is developed and constructed from data that has been provided by or on behalf of the passenger concerning all the flight segments of a journey.[319] This data may be added to by the operator or his authorized agent, for example, changes to requested seating, special meals, additional services requested, *etc*. PNR data could be obtained in many ways. For example, information captured through reservations created by international sales organizations (global distribution systems "GDS" or computer reservation systems "CRS") with pertinent details of the PNR could be transmitted to the operating carrier(s). When reservations are made directly by the aircraft operator and the complete PNR is stored within the operator's automated reservations systems, the information therein could be a useful repository of PNR data. Information contained in records of some operators who may hold sub-sets of the PNR data within their own automated departure control systems (DCS), for their information or for onward transmittal to contracted ground handling service providers, calculated to support airport check-in functions would be another way in which PNR data could be provided. However, it must be noted that in each case, operators (or their authorized agents) will have access to, and be able to amend only that data that has been provided to their system(s). An important consideration in this regard is that some DCS systems are programmed such that details emerging from check-in (i.e., seat and/or baggage information) can be overlaid into the existing PNR for each passenger. However, that capability is limited – covering less than 50% of operating systems today.

The time element, with regard to the capture and relevance of PNR data, is relevant to the use of such data. For instance, Data could be entered into a reservation system many days or weeks in advance of a flight. This could extend to as long as 345 days in advance of departure. Under such circumstances, both the provider and the receiver of PNR data must bear in mind that Information in reservation systems is dynamic and may change continuously from the time when the flight is open for booking. On the other hand, passenger and flight information in the DCS, becomes available only from the time the flight is "open" for check-in (up

---

[317]The Industry Standards related to PNR creation are detailed in IATA's Passenger Services Conference Resolutions and in the ATA/IATA Reservations Interline Message Procedures (AIRIMP) Manual.

[318]Passenger Name Record Information Required for Passengers on Flight in Foreign Air Transportation to or from the United States of 2001, 66 *Fed. Reg.* 67482 (2002).

[319]There are two possible methods of PNR data transfer currently available: (a) the "pull" method, under which the public authorities from the State requiring the data can reach into the aircraft operator's system and extract ("pull") a copy of the required data into their database; and (b) the "push" method, under which aircraft operators transmit ("push") the required PNR data elements into the database of the authority requesting them.

to 48 hours prior to departure). In such an instance, departure control information for a flight will be finalized only upon flight closure, and may remain available 12–24 hours after arrival of a flight at its final destination.

In the case of aircraft operators specializing in charter air services, who often do not hold PNR data in an electronic form, but still use a DCS which will only enable them to have a limited PNR record after the flight has closed, they would still be required to provide any captured data to States requesting it regardless of the process by which they receive PNR data. States could also require supplemental or "requested service" information which may be contained in the PNR, such as information relating to special dietary and medical requirements, "unaccompanied minor" information, requests for assistance etc.

Operators should take particular care in refraining from incorporating in PNR data any information that is not essential to facilitate the passenger's travel. Such information would include, but not be necessarily restricted to details of the passenger's racial or ethnic origin, political opinions, religious or political beliefs, trade-union membership, marital status or data relating to a person's sexual orientation. The ICAO guidelines make specific mention of the fact that Contracting States should not require aircraft operators to collect such data in their PNRs.

The above notwithstanding, any information which would legitimately facilitate the carriage of the passenger, such as details of meal preferences and health issues as well as free text and general remarks, could comprise the PNR. Sensitive data contained in the PNR and is submitted in compliance with a regulation of a State should not be used as the primary source for assessment of risk that the passenger might present to the State concerned.

II. *The Importance of PNR Data to States*

From a regulatory perspective, the two main areas to which PNR data make a contribution are expedition of customs and immigration processing at airports; and facilitation of passenger traffic and the safeguard of the legitimate rights of the passenger. The Chicago Convention provides a sound basis for States to require PNR data in the current context. The Convention, in Article 22, recognizes the importance of facilitating the passage of a person through borders by requiring each contracting State to adopt all practicable measures, through the issuance of special regulations or otherwise, to facilitate and expedite navigation by aircraft between the territories of contracting States, and to prevent unnecessary delays to aircraft, crews, passengers and cargo, especially the administration of the laws relating to immigration, quarantine, customs and clearance.

The main reason for States to require the advance submission of PNR data is that such data could prove to be a valuable tool in ensuring aviation security. PNR data are critically important for the threat assessment value that can be derived from the analysis of such data, not only in possible instances of unlawful interference with civil aviation but also in relation to the fight against terrorism. This critical value of

PNR data has prompted some States to enact legislation or develop draft legislation for approval by their Legislatures requiring that aircraft operators provide their public authorities with PNR data.

PNR data primarily enable States, through the identification of potentially high-risk passengers through PNR data analysis, to improve aviation security; enhance national and border security; prevent and combat terrorist acts and related crimes and other serious crimes that are transnational in nature, including organized crime; and to enforce warrants and prevent flight from custody for such crimes. Such data could also protect the vital interests of passengers and the general public, including their health.

States are aware that, if the guidelines are implemented in a uniform manner, would provide a global framework enabling all States to benefit from the value-added analysis of PNR data for shared security/safety purposes. Air carriers would also benefit from having to comply with only one set of common requirements for PNR data transfer. As for the consumer of air transport, all passengers would benefit from basic protection afforded to them by the exchange of PNR data between air carriers and State authorities.

The above notwithstanding, there are certain fundamental obligations that the State receiving the data has to meet. Firstly, States should require PNR data only of those passengers on flights that are scheduled to enter, depart or transit through airports situated in their territories. Secondly, a State obtaining PNR information should, as a minimum limit the use of data to the purpose for which it collects it. States must restrict access to such data, ensure that the data is adequately protected, and limit the period of data storage, consistent with the purposes for which data is transferred. States must also ensure that individuals are able to request disclosure of the data that is held on them, consistent with the guidelines, in order to request corrections or notations, if necessary. More importantly, they must ensure that individuals aggrieved by the PNR data collection and usage process have an opportunity for redress.

The responsibility of ensuring that their public authorities have the appropriate legal authority to process PNR data requested from aircraft operators, in a manner that observes the guidelines, devolves entirely upon the States. They have been requested by ICAO to forward the full texts of legislation pertaining to PNR data dissemination and use to ICAO for online dissemination to other States, for information. The State concerned will be responsible for responding to any queries arising from such legislation.

III. Advantages of Unified Guidelines

Through the PNR Data Guidelines ICAO has introduced uniform measures for PNR data transfer and the subsequent handling of that data by the States concerned. The guidelines are both durable and easy to follow, making them cost effective for the parties concerned. They would ensure accuracy of information, while at the same time protecting the data subject against encroachment of his privacy. The Guidelines call for completeness of data and the need for timely submissions and effective collection of data. They also ensure that data management will be efficient and efficacious. From a practical perspective, the guidelines also provide useful directions assisting States in designing data requirements and procedures, in order to minimize technical difficulties that might prove too onerous and may impair the implementation of the uniform measures suggested. The Guidelines also contain detailed instructions with a view to assisting both air carriers and States on PNR data transfer from an operator's system to a State and the management of the data including arrangements for storage and protection.

States are enabled, by the guidelines, to design systems and establish arrangements that are compatible with the guidelines while not impairing their ability to implement their laws and enforce them. The guidelines do not interfere with the preservation of national security and public safety of a State. Arguably, one of the most important features of the unified PNR data guidelines is that, by their very nature, they would effectively obviate the complexities that aircraft operators could face with regard to legal, technical and financial issues if they were to be required to respond to multiple, unilaterally imposed or bilaterally agreed PNR data transfer requirements that differ substantially from one another.

1. Extra Territoriality

States also have the responsibility of enacting explicit legal provisions concerning data transfer. Such legislation should clearly elaborate on the reasons for requiring PNR data, or provide explanatory material accompanying such laws or regulations, as appropriate. Since an aircraft operator is obliged to comply with the laws of both the State from which it transports passengers (State of departure) and the State to which these passengers are transported (destination State), when a destination State legislates with regard to its PNR data transfer requirements, it should do so cognizant of the fact that *existing* laws of other States may affect operators' ability to comply with these requirements. Therefore, where there could be an inconsistency between two legal regimes of the departure State and the destination State, or where a conflict arises between any two States, or where an operator advises of a conflict, The ICAO guidelines suggest that the States involved should consult each other to determine what might be done to enable affected operators to continue to operate within the bounds of the laws in both States.

Strictly interpreted, extra-territoriality at international law means the attempt of one State to apply its laws outside its territory[320] and there is a general presumption against the application of extra-territoriality.[321] In the 1979 case of *Mannington*

---

[320]Shaw (2003, pp. 611–612).

[321]*Holmes* v. *Bangladesh Biman Corporation*, [1989] 1 AC 1112 at 1126. Also, *Air India* v. *Wiggins* [1980] 1 WLR 815 at 819. In the 1991 case of *EEOC* v. *Arabian American Oil Company and ARAMCO Services* 113 L E 2d 274, the US Supreme Court held that the practice of extra

Mills v. *Congoleum Corporation*[322] the United States Supreme Court extended the concept of extra territoriality by introducing a test of balance that ensured consideration by one State for the interests of another State.

The above principle of extra-territoriality might not sit comfortably in the instance of a State requiring PNR data from a flight over-flying its territory as there is no *stricto sensu* application of a requirement in a foreign territory. The most fundamental principle of public international law, that of State sovereignty, is embodied in Article 1 of the Chicago, thus importing the principle into the tenets of air law. This Article provides that Contracting States recognize that every State has complete and exclusive sovereignty over the air space above its territory. The territory of a State, for the purposes of the Convention, covers the land areas and territorial waters adjacent to and under the sovereign, suzerainty, protection and mandate of the State concerned.[323] Arguably, these provisions would give the United States the right *in limine* to prescribe requirements on aircraft flying over its territory. Article 12 of the Chicago Convention provides, *inter alia*, that each contracting State undertakes to adopt measures to insure that every aircraft flying over or maneuvering within its territory and that every aircraft carrying its nationality mark, wherever that aircraft may be, shall comply with the rules and regulations relating to the flight and maneuver of air raft there in force. This rule can apply to a foreign carrier who is over-flying the territory of any State having a regulation that certain data pertaining to a flight that over-flies its territory has to be submitted to that State. Also important is Article 9 of the Convention, which allows a Contracting State to restrict or prohibit an aircraft from flying over its territory for reasons of military necessity or public safety. The provision goes on to say that each contracting State could also reserve the right, in exceptional circumstances or during a period of emergency, or in the interest of public safety and with immediate effect, temporarily to restrict or prohibit flying over the whole or part of its territory, provided such action would apply without distinction of nationality to aircraft of all States.[324]

At the 28th Session of the International Law Association held in Madrid in 1913, the meeting drew up text which stated that it was the right of every State to enact prohibitions, restrictions and regulations as it may think proper in regard to passage of aircraft through the airspace above its territory and territorial waters.[325]

---

territoriality by one State against the other cannot in any way be justified under the principles of public international law.

[322] 595 F.2d 1287; 66 ILR at 487. See also *Timberlane Lumber Company* v. *Bank of America*, 549 F. 2d 597 (1976); 66 ILR at 270.

[323] *Convention on International Civil Aviation* (Chicago Convention), signed at Chicago on 7 December 1944 (ICAO Doc 7300/9, Ninth Edition, 2006), Article 2.

[324] *Convention on International Civil Aviation* (Chicago Convention), signed at Chicago on 7 December 1944 (ICAO Doc 7300/9, Ninth Edition, 2006), Article 9(b).

[325] International Law Association, 28th Report, Madrid, 1913, 533–545 at 540.

D. The Passenger Name Record

However, the text contained a caveat that such restrictions should be subject to the rights of subjacent States and the liberty of passage of aircraft of every nation.[326] The balance advocated at the Madrid meeting of the ILA goes to show that even as early as the beginning of the last century, the thinking was that a State ought to allow other States free passage for their aircraft through the airspace above its territory. There is no doubt that the same position prevails even now, particularly through the currently applicable International Air Services Transit Agreement (IASTA) which was concluded at the same time as the Chicago Convention in December 1944 and has been ratified by many ICAO Contracting States. IASTA allows aircraft of foreign States freedom of peaceful transit (over the airspace of a State) and freedom of making non-traffic (non-revenue) stops for such purposes as refuelling and repair. It has been acknowledged that without these two freedoms, the air transport industry could not survive.[327]

The above discussion brings one to the inexorable conclusion that there are two major issues at stake. The first is whether the OPNR is an acceptable tool which helps in enhancing facilitation and security measures in air transport. The answer to this question, as provided by the 12th ICAO Facilitation Division in March/April 2004[328] and subsequently by the ICAO Council[329] is a resounding "yes". This affirmation brings to bear the need to consider whether the PNR should be used strictly as intended, firstly to facilitate customs and immigration procedures regarding persons and secondly to advise States in advance of persons on board an aircraft approaching their territory for purposes of landing there, enabling States to determine appropriate security clearance measures. The security angle of the PNR brings one to the second issue, as to whether a State can use information contained in the PNR to disallow the right of passage to an aircraft flying over its territory, thereby denying that aircraft a fundamental right acknowledged by States through IASTA.

The second issue raises the question of extra territoriality, which can be answered by invoking Articles 9 and 12 of the Chicago Convention, as earlier discussed. These provisions clearly give a State the right to prohibit an aircraft from over-flying its territory if it believes that such over-flying could be a security hazard. The final issue would be to determine the extent to which a State could exercise its right without touching the sensitivities and dignity of a State in an instance where an aircraft plying domestic services within two points in its territory but passes through the airspace of the prohibiting State is disallowed from using the rite of passage.

The entire issue of diversion of an aircraft which is exercising its fundamental rite of passage and the justification of a State for disallowing that aircraft from using that fundamental right hinges on the circumstances prevailing at the time. As was

---

[326] International Law Association, 28th Report, Madrid, 1913, 533–545 at 538.
[327] Honig (1956, p. 29).
[328] *Supra*, note 5.
[329] *Ibid*.

mentioned earlier, this is no legal issue as the question of extra-territoriality does not arise with regard to action taken by a State within its territory. The fundamental postulate in the debate is that sovereignty should no longer mean the mere exercise by one State of rights over its territory but should also mean the right of that State to ensure the safety and security of its citizens as well as the integrity of the State.

Public international law is increasingly becoming different from what it was a few decades ago. It can be said with some justification that international law is the thread which runs through the fabric of international politics and provides the latter with its abiding moral and ethical flavour. Without principles and practices of international law, foreign policy would be rendered destitute of its sense of cooperation and become dependent on a nation's self interest. As President Woodrow Wilson once claimed:

> It is a very perilous thing to determine the foreign policy of a nation in the terms of material interests ... we dare not turn from the principle that morality and not expediency is the thing that must guide us, and that we will never condone equity because it is convenient to do so.[330]

This statement, made in 1950, has great relevance today, when continued progress is being made in technological and economic development and policy decisions of States have far reaching consequences on a trans-boundary basis. Nation States are becoming more interdependent, making decisions made by a particular State in its own interest have a significant negative impact on the interests of other States. Therefore ethics in foreign policy has largely become a construct which combines cultural, psychological and ideological value structures. Within this somewhat complex web of interests, decisions have to be made, which, as recent events in history have shown, require a certain spontaneity from the international community. For example, when Iraq invaded Kuwait in 1990, the members of the United Nations chose economic sanctions against Iraq, claiming that war was the last resort to be embarked upon against Iraq if economic sanctions did not prove to have any effect. In hindsight, one could argue one way or another, firstly, as did the United States, that the use of force bore quick results and, on the hand, as did many officials in Paris, Moscow, Ottawa and Washington, that the decision to wage war against Iraq was too precipitous as not enough time had been given to economic sanctions to compel Iraq to retreat from Kuwait. The precipitous but quick action taken in going to war with Iraq might be justified by some with the analogy of Britain appeasing Hitler in the 1930s without adopting a more aggressive and perhaps belligerent attitude toward German atrocities. This action, which was later airobi as folly by most political scientists, was applauded and endorsed at that time in the British Parliament.

In the absence of extra territoriality the only balancing factor in of State which orders the diversion of an aircraft over-flying its territory, on the basis that persons therein are unacceptable is to have sound justification for doing so in the interests of security and safety.

---

[330]Quoted in Morganthau and Thompson (1950, p. 24).

D. The Passenger Name Record

Advance passenger information and other methods of data processing of air travel find their fundamental legal roots within the *Convention on International Civil Aviation* of 1944[331] by promoting safety of flight in international air navigation and by the promotion of the developments of all aspects of civil aeronautics. This Convention was signed in Chicago and created the International Civil Aviation Organization. The objectives of this organization are set forth in article 44 of the Convention which includes the more specifics aspects of facilitation and aviation security:

The aim and objectives of the Organization are to develop the principles and techniques of international air navigation and to foster the planning and development of international air transport so as to:

(a) Insure the safe and orderly growth of international civil aviation throughout the world
(b) Encourage the arts of aircraft design and operation for peaceful purposes
(c) Encourage the developments of airways, airports, and air navigation facilities for international civil aviation
(d) Meet the needs of the peoples of the world for safe, regular, efficient and economical air transport
(e) Prevent economic waste caused by unreasonable competition
(f) Insure that the rights of contracting States are fully respected and that every contracting State has a fair opportunity to operate international airlines
(g) Avoid discrimination between contracting States
(h) Promote safety of flight in international air navigation
(i) Promote generally the development of all aspects of international aeronautics[332]

In addition to ICAO's objectives, the modern traveler requires rapid processing through the different stages of air transport, whether it implicates the air carrier's processes or those set forth by the border control agencies. In article 22, the Chicago Convention recognizes the importance of facilitating its formalities with respect to each passenger:

> Each contracting State agrees to adopt all practicable measures, through the issuance of special regulations or otherwise, to facilitate and expedite navigation by aircraft between the territories of contracting States, and to prevent unnecessary delays to aircraft, crews, passengers and cargo, especially the administration of the laws relating to immigration, quarantine, customs and clearance.[333]

To achieve such conditions of travel, States have adopted ICAO's recommendations into national laws but have as well implemented some extraterritorial applications to the existing legislations of legality. For example, the United States by

---

[331] *Convention on international Civil Aviation*, 7 December 1944, ICAO Doc. 7300/8 (entered into force 4 April 1947) (hereinafter: Chicago Convention).
[332] Chicago Convention, Art. 44.
[333] Chicago Convention, Art. 22.

Enhanced Border Security and Visa Reform Act,[334] which objectives are to reform the *US Patriot Act*[335] by different new technologies of data capturing such as biometrics. In fact, the government acknowledges the need for expedited clearance of passengers at airports using machine readable travel documents (MRTDs) technology. This should be used in a secure environment in order to ensure that the proper border control authorities remain the only agency in possession of such data. Furthermore, these procedures have been standardized on a worldwide basis and are currently being developed with the initiative of ICAO and ISO with new biometric procedures:[336]

With modern methods comes the inextricable discussion on privacy rights and their possible violations. Most prominent authors recognize four rights of privacy relating to the storage and use of personal data, which can be classified into four sections:

(a) The right to determine what personal information to share with others, and to control the disclosure of such data
(b) The right to know what personal data is disclosed, collected and stored
(c) The right to dispute incomplete or inaccurate data
(d) The right of those who possess legitimate reasons for information on data pertaining to health and safety of society[337]

As a counterpart, air carriers are stricken with the possible burden of financially assuming new technologies of data processing. According to the *Simplifying Travel Organization*,[338] the technology implemented will entail capturing of the passengers information by additional manpower at check-in wit purchase of hardware in order to comply with Annex 9's provision of machine-readable documents.[339]

---

[334]*Enhanced Border And Visa Entry Reform Act of 2002*; 107 established by the Congress of the United States of America at the second session, 22 January 2002.

[335]US Patriot Act. . ..

[336]Heitmeyer (2000, pp. 18–20).

[337]Abeyratne (2002a, p. 86).

[338]Refer to the SPT Brochure 2002. The Simplifying Travel Group is a joint venture with IATA in order to develop new technologies in biometrics for the screening of passengers: "The SPT Program is a joint initiative amongst a number of organizations, representing passengers, airlines, airports, control authorities, travel agents and broad government interests, *to measurably improve the passenger experience and enable security enhancement by*: – Implementing biometrics and other new technologies; – Sharing information amongst service providers; – Enabling controls and services to be effected more efficiently."

[339]"6.5.1 The principal costs for carriers are associated with system development/integration and capture of passenger details for transmission to the destination country of a flight. Costs will likely be incurred in other areas as well; e.g., additional check-in staff to cope with the extended period of time required to complete check-in formalities, additional check-in desks, hardware acquisitions, etc. Various techniques can be used to offset these costs to some degree; e.g., agreements with governments, as is the case in Australia, machine-readable passports, 'up-stream' capture of passenger data at the time of booking, etc. [. . .]" World Customs Organization, "Advance Passenger Information: Guidelines for Customs and Air Carriers" (2003) WCO Annex I to Doc PW0072E1 11.

D. The Passenger Name Record

Furthermore, the usage of advance passenger information not only can be considered as a facilitation aspect, but it is also one of aviation security. The Chicago Convention stipulates at its article 44(d) the necessity of safe and efficient air transport. ICAO has recognized the fact that security and facilitation must act at a joint venture.

A recent organizational change at ICAO, in which the administration of the security and facilitation programmes was merged, recognizes formally the importance of establishing a good balance between the need for effective aviation security and the need to facilitate air travel.[340]

In fact, by transmitting data in advance to a border control agency, it becomes more and more probable to control inadmissible passengers, such as potentially high-risk passengers who have been banned into entering the State.[341] The information shared consists of identifying these individuals that could cause a potential threat to national security. As the Fourth Panel Meeting Facilitation Panel stated:

> Moreover, the events of 11 September 2001 and afterwards have demonstrated that national programmes of travel document issuance and security, and the efficacy of inspection systems in controlling smuggling and illegal migration, can have a significant effect on the security of civil aviation.[342]

In addition, due to the fact that security emphasized at article 2.2 of Annex 9 on Facilitation, Annex 17 on Security also stipulates the importance of its collateral concept at the recommended practice 2.2: "Each Contracting State should whenever possible arrange for security controls and procedures to cause a minimum of interference with, or delay to the activities of civil aviation provided the effectiveness of these controls and procedures is not compromised,"[343] which corresponds to the obligation by States for proper control set forth within the Chicago Convention at its article 13.[344] Each State can therefore exercise an effective control on the individuals crossing the border. However, the fundamental right of privacy of mankind is governed on principles of the right to be informed as to which public

---

[340] McMunn (1996, p. 7).

[341] "The Facilitation programme has taken a proactive stance against law enforcement problems, particularly narcotics trafficking and travel by inadmissible passengers. [...] At its first meeting in 1997, the ICAO Facilitation Panel will review all of the Annex 9 provisions related to inadmissible passengers and will attempt to devise some means to implement them more effectively". McMunn (1996, p. 9).

[342] ICAO Secretariat, "Facilitation Panel Fourth Meeting Information Paper" (Montreal, 2–5 April 2002), ICAO Doc FAL/4-IP/3. This paper was first introduced to the High-Level Ministerial Conference of February 2002 (1P/1).

[343] Abeyratne (1998, p. 78).

[344] Chicago Convention, Article 13: "The laws and regulations of a contracting State as to the admission to or departure of its territory of passengers, crew or cargo of aircraft, such as regulations relating to entry, clearance, immigration, passports, customs, and quarantine shall be complied with by or on behalf of such passengers, crew or cargo upon entrance into or departure from, or while within the territory of that State".

agency should be entitled to dispose of such information as well as the content of such data tracing versus the public's recognized right to justify under national security such a process.[345] It is therefore tantamount to conduct proper automated procedures rather than collect manually data by ground staff.

This API summary will serve as an introduction to the concept of advance passenger information and biometrics procedures, a comparative study of its applications both by public entities as well as the financial and legal implications of such transmission of data on air carriers and border control agencies as well as these new methods of identification.

The conclusion will be to demonstrate that security and facilitation are both complimentary concepts. In fact, as Annex 9 and Annex 17 of the Chicago Convention explicitly state, the concept of aviation security and facilitation are inter-related and form an intertwined relationship: "[...] the relationship between facilitation and security at airports should not be seen as a 'trade-off' or a 'balance' between adversarial programmes. Rather, the enhancement of one results in enlargement of the success of the other."[346] In addition, the composite opinion objective by ICAO's Facilitation Section in regards to security and facilitation annexes is: "[...] that legitimate – and only legitimate – traffic be boarded on aircraft, carried by air and cleared to cross international boundaries, safely and in good time."[347]

2. Definition

The concept of advance passenger information involves the capture of passport details by the carrier prior to departure and the transmission of the details by electronic means to the authorities at destination. With this capture, the authorities can screen the passengers by their databases in order to identify potentially high-risk individuals. The positive aspect is to reduce congestion at airports and consequently decrease delays in border control processing.[348]

---

[345]"One of the issues as important in the API process is that the data required must be collectable by machine or already contained in the airline's system. Manual collection and data entry at the check-in desk for a scheduled flight is time-consuming and prone to errors, and or life. The foundations of 'information privacy,' whereby the individuals would determine when, how, and to what extent information about themselves would be communicated to others, inextricably drawing the right of control of information about oneself, is a cornerstone of privacy." Abeyratne (2001b, p. 153).

[346]McMunn M.K. for ICAO Secretariat, "Facilitation and Security – Not a Zero-Sum Game" (March 1999) ICAO Doc AFCAC/ATC/4-IP at point 9.

[347]Unofficial statement given by Mary K. McMunn, Chief of the Facilitation Section at ICAO.

[348]"[...] This technique is beginning to be used by Border Control Agencies and it has the potential to reduce considerably the inconvenience and delay experienced by some travellers due to border controls". Facilitation Division-11th Session, (1995) ICAO Doc FAL/11-IP/2.

D. The Passenger Name Record   135

API has begun to be for certain States a compulsory method for public authorities to manage risk prior to arrival in order to expedite clearance.[349] The implementation of such a system requires a great deal of regulation as it involves data capturing and processing.

3. History

Since 1948, ICAO's Facilitation Division invoked the presence of reducing exit visas and granting a time constraint to any visa that has been given to a traveler[350] with criteria in order for public authorities to manage the number of entries of a passenger, for example, and to standardize the required information on each visas. In 1959, during the Fifth Session,[351] Rome determined additional criteria, which is not to apply different procedures that would be less favorable to the airline industry in comparison to any other means of transport.

API was first brought to life by the recommendations of adopting Annex 9 in 1963 in Mexico.[352] Following the United Nations Conference on International Travel and

---

[349] Refer to Recommendation Practice 3.34 of Annex 9: "Where appropriate Contracting States should introduce a system of advanced passenger information which involves the capture of certain passport or visa details prior to departure, the transmission of the details by electronic means to public authorities, and the analysis of such data for risk management purposes prior to arrival in order to expedite clearance. To minimize handling time during check-in, document reading devices should be used to capture the information in machine readable travel documents. When specifying the identifying information on passengers to be transmitted, Contracting States should only require information that is found in the machine readable zones of passports and visas that comply with the specifications contained in Doc 9303 (series), Machine Readable Travel Documents. All information required should conform to specifications for UN/EDIFACT PAXLST message formats".

[350] "8.1 (RP) In order to facilitate the unilateral and bilateral elimination of entrance visas for non-immigrants, but at the same time to provide a simplified form of control with respect to the movement of non-immigrants where such control is deemed necessary, the following uniform system should be adopted [...] 8.4 (RP) Each State should abolish exit visas, and reduce any other emergency exit formalities to an absolute minimum". Facilitation Division, "Final Report Of The Second Session" (Geneva, June 1948) ICAO Doc 5464-FAL/535.

[351] "3.1(ST) Governmental regulations and procedures applied to persons travelling by air shall be no less favorable than those applied to persons travelling by other means of transport. 3.2(ST) Contracting States shall make provisions whereby the procedures for clearance of persons travelling by air will be applied and carried out in such a manner to retain advantage of speed inherent in air transport. 3.3(ST) No documents other than those provided for in this Chapter shall be required by Contracting States for the entry into and departure from their territories of tourists. And other temporary visitors". Facilitation Division, "Report of The Fifth Session" (Rome, December 1959) ICAO Doc 8043-FAL/562 Recommendation A-17.

[352] Facilitation Division, "Report Of The Sixth Session" (Mexico, March–April 1963), ICAO Doc 8324-FAL/563.

Tourism, a recommendation[353] was set forth in order for the UN to introduce the implementation of the different member States to Annex 9. This conference took place in Rome and began to define concepts such as visitors, tourist and other facilitation aspects that eventually lead to the implementation of Annex 9.[354]

Furthermore, we can notice the willingness to incorporate security awareness within the concept of facilitation where a recommendation[355] stated the obligation for the air carrier to return an individual, which has been refused by the said State. With this new proposition, it is foreseeable to notice that airlines will need to be much more vigilant when verifying if a passenger can travel.[356]

The concept of API was discussed during the Tenth Session of the Facilitation Division in Montreal in 1988. It suggested a recommendation[357] that was only

---

[353]"Recommendation B-6: WHEREAS the UN Conference on International Travel and Tourism, to be held later in the year 1963, will consider the question of formalities to be complied with by tourists on entry and departure; WHEREAS the provisions of Annex 9 relating to the movement of persons have been carefully developed throughout the years and have been thoroughly reviewed at the Sixth Session of the Facilitation Division, the conclusions of which will be communicated to the Secretary General of the UN Conference; and WHEREAS it is essential that any action taken by the UN Conference should not be inconsistent with the pertinent International Standards and Recommended Practices contained in Annex 9 to the Convention on International Civil Aviation and should actually encourage States to implement that Annex; THE DIVISION RECOMMENDS that the Council request the UN Conference to preface any recommendations it will ultimately adopt by a reference, in a preamble or otherwise, to the continuing obligations of the Contracting States of the International Civil Aviation Organization to implement the provisions of Annex 9." Facilitation Division, "Report Of The Sixth Session" (Mexico, March–April 1963), ICAO Doc 8324-FAL/563 at 32.

[354]UN Conference on International Travel and Tourism, online: http://www.oas.org/TOURISM/docnet/Iatc2en.htm (date accessed: 15 January 2003).

[355]"3.25 (ST) Upon refusal of admission and transfer back of any person, the operator shall be responsible for promptly returning him to the point where he commenced the use of the operator's aircraft or to any other place where the person is admissible." Facilitation Division, "Report Of The Sixth Session" (Mexico, March–April 1963), ICAO Doc 8324-FAL/563 at 40.

[356]Refer to Article 3.58 of Annex 9: "The public authorities shall without delay inform the operator when a person is found inadmissible and consult the operator regarding the possibilities for removal. Note 1. – A person found inadmissible shall be transferred back into the custody of the operator who transported that person directly to the final destination or, where appropriate, into the custody of one of the operators who carried the person to one of the transit destinations. [. . .]."

[357]Facilitation Division, "Report Of The Tenth Session" (Montreal, September 1988) ICAO Doc 9527, FAL/10 at 54: Recommendation B-11: "IT IS RECOMMENDED THAT:

(a) Contracting States, where possible, undertake projects to examine the effects of various advance passenger information programmes (including as appropriate various manual and electronic collection and transmission methods) in facilitating the clearance of arriving passengers through the inspection processes at major international airports;
(b) Where data are transmitted by Electronic Data Interchange, procedures should conform to international message standards and formats;
(c) ICAO would undertake a study of Contracting States' experiences from the projects undertaken under (a) above in the advance passenger information privacy issues and the facilitation and other benefits and costs, by types of programmes, for passengers, air carriers and Contracting States; ICAO should liaise with the Customs Co-operation Council and other

D. The Passenger Name Record

introduced during the 11th Session following a conference in Djerba in 1997, which gave background information on API and comments from Member States. The recommendation on the format of API was implemented within the 10th edition of Annex 9.[358]

In its report, it was stated that the Members of ICAO were concerned about privacy issues that could arise from the usage of electronic information provided by the API system.[359] It is also noted in this report that any electronic messaging should be processed under the Electronic Data Interchange [E.D.I.] format, and become international practice, therefore being common between Contracting States.

One of the WCO mission, through the Permanent Technical Committee was to develop a convention in order to adapt the changing structure of international trade and the evolution of Customs techniques and therefore facilitate States adopting national legislation. In 1973, the Council of the WCO adopted in Tokyo the Convention on the Simplification and Harmonization of Customs Procedures.[360]

---

appropriate international bodies to ensure proper co-ordination in this area, and to safeguard the interests of immigration authorities;
(d) ICAO would keep Contracting States fully informed of developments; and
(e) ICAO would, no later than 1992, report on the study to the Council, which would decide whether the findings and recommendations should be recommended to Contracting States."

[358]ICAO Secretariat, "Informal Facilitation Area Meeting in Consultation with ACI on Advance Passenger Information" ICAO Doc INF/FAL/DJE WP/11 (2 July 1997):

2.1 Article 29 of the Chicago Convention requires every aircraft engaged in international navigation to carry certain documents, including, for passengers, "a list of their names and places of embarkation and destination." Annex 9 specifies, in Standard 2.7, the presentation of a passenger manifest document shall not normally be required, and notes that if the information is required it should be limited to the data elements included in the prescribed format, i.e., names, places of embarkation and destination, and flight details.

2.2 It should be noted that the opinion of this Standard contemplated the passenger manifest as a paper document which would have to be typed or written and delivered by hand. [...] It is widely recognized that in any system involving the exchange of information (automated or not), it is the collection of data which is the major expense. Increases in data collection requirements should result in benefits which exceed the additional costs. This principle was a central issue during the debate over API in the Tenth Session of the Facilitation Division (FAL/10) and the eventual adoption by FAL/11 of API systems as a Recommended Practice.

(Refer to Article 3.14.2 of the 10th Edition of Annex 9.)

[359]"There was, however, considerable support for both B-type Recommendations although several delegates pointed out that there would be a need for the programmes concerned to take into account the importance of the privacy of the individuals reflected in the data protection laws already adopted in many States". ICAO Secretariat, "Informal Facilitation Area Meeting in Consultation with ACI on Advance Passenger Information" ICAO Doc INF/FAL/DJE WP/11 (2 July 1997) at 53.

[360]*Convention On the Simplification And Harmonization Of Customs Procedures* (here in after referred to as Kyoto Convention), online: http://www.unece.org/trade/kyoto/ky-01-e1.htm#Historica (date accessed: 3 January 2003).

The WCO's main objective is to simplify travel and create effective border control for the rapid clearance of passengers. It is stipulated in its recommended practice in the Kyoto Convention as well as in the associated benefit:[361]

> The benefit to Customs is the receipt, in advance of the arrivals of travelers, of information that will aid risk management with the objective of more precise targeting of Customs control. A benefit to travelers is that, on the basis of Customs analysis and evaluation of API, their risk status can be determined prior to arrival in the country concerned. Greater precisions in Customs targeting should result in the vast majority of travelers being assessed as presenting negligible or no risk and thus subject to minimal or no Customs control on their arrival.[362]

It is also noted in the general field of applications that the Convention is aimed at developing a system of pre-clearance to utilize wanting time prior to the departure of an aircraft in order to carry out formalities, which might otherwise delay passengers upon arrival of that aircraft at destination.

IV. *Advance Passenger Information Guidelines*

During the 11th Session of Facilitation Division held in Montreal in 1995, the position of the WCO, formerly CCC in 1992, was stated into guidelines for API mainly due to:

- Information Technology
- Greater co-operation between Border Control Agencies domestically
- Greater international co-operation between Customs administrations and with other Border Control Agencies
- Greater co-operation between Border Control Agencies and carriers[363]

In order to fulfil the roles of the CCC, the system of API can facilitate such an information system by:

> *4.1.4[...]* (a) Providing its Members with information on the technique of API benefits it can bring;
> (b) Providing a forum in which the constraints on API can be discussed and hopefully resolved; and

---

[361] Refer to the Kyoto Convention, at Annex J at Article 5.5: "Recommended Practice 8: The Customs, in co-operation with other agencies and the trade, should seek to use internationally standardized advance passenger information, where available, in order to facilitate the Customs control of travellers and the clearance of goods carried by them".

[362] Kyoto Convention, at Annex J at Article 5.5.

[363] Facilitation Division, "Eleventh Session Information Paper on Advance Passenger Information (API) Guidelines adopted by the WCO" (Montreal, April 1995) ICAO Doc FAL/11-IP/2 at point 3.

D. The Passenger Name Record

(c) Seeking to jointly agreed standards with the airline industry so that API does not develop and proliferate in an inconsistent or unstructured way.[364]

In April of 2002, during a Facilitation Panel in Montreal on API, it recommended:

The usage of API for immigration, quarantine and aviation security (AVSEC) applications to customs.

The internet or other PC-based systems and wireless technologies should be considered for the exchange of data rather than specify UN/EDIFACT syntax for data interchange.

API should be part of a border system management, machine readable passports with electronic visas, automated entry/exit records instead of embarkation or disembarkation cards and as well as interoperability of API systems with other States.

Applicable Standards and Recommended Practices (SARPs) should leave the possibility of including biometrics into article 3.34 of Annex 9 (11th Edition).

ICAO should measure the programme's success in operational efficiency and reduction of airport congestion.[365]

The API topic became more and more a priority during the 1993/1994/1995 triennium and IATA comprehended in a greater capacity the necessity for API implementation.[366] IATA and the WCO formally introduced the formal WCO/IATA guidelines in 1993 following the Working Paper presented by the ICAO Secretariat during the 11th Session.[367] In the preamble of the guideline,[368] it

---

[364]Facilitation Division, "Eleventh Session Information Paper on Advanced Passenger Information (API) Guidelines adopted by the WCO" (Montreal, April 1995) ICAO Doc FAL/11-IP/2 at point 1.3.

[365]Facilitation Division, "Eleventh Session Information Paper on Advanced Passenger Information (API) Guidelines adopted by the WCO" (Montreal, April 1995) ICAO Doc FAL/11-IP/2 at point 4.1.4.

[366]Facilitation panel presented by the Secretariat, "Advance Passenger Information Further Development of ICAO Doctrine" ICAO Working Paper FALP/4-WP/2 (Montreal, 2–5 April 2002).

[367]"4.2.4 Furthermore, given the practical and cost constraints of data capture and transmission, limiting the required information to that which can be captured by machine reading passports and visas, augmented by basic flight details, is a prerequisite. To this end, IATA sees particular benefit in co-operating with the CCC to define the data and message sets for API within the UN/EDIFACT PAXLST development, and in establishing jointly agreed principles which can expand the benefits of automating and integrating all elements of the passenger process from origin to destination". See *infra*, note 412 at point 4.2.4.

[368]*Ibid* at clause no. 4: The Customs Co-operation Council recommended a standardization for API interoperability and an objective to control costs to airlines. It also: "[...] requests Members of the United Nations Organization or its specialized agencies, and Customs or Economic Union which accept this Recommendation to notify the Secretary General of the Council of the date from which they will apply the Recommendation and of the conditions of its application. The Secretary General will transmit this information to the Customs administrations of all Members of the United Nations Organization or its specialized agencies and to Customs or Economic Unions which have accepted this Recommendation".

stipulates that because of the increase of passenger traffic, Customs are strained to process much more additional data when it clears border control. Furthermore, in order to prevent increase in delays, the need for efficient automated processing has become a necessity. This position has also been supported by IATA.[369] Where API should be considered uniform electronic text capturing by the UN/Edifact PAXLST Messaging system. In fact: "API permits a very thorough and rigorous screening of inbound passengers to be carried out, targeting those that present the highest risk and allowing for the faster throughput of low risk."[370] IATA also notes the necessity to create a limitation of standardization to identify data would prevent abuse in the transfer of data. As a suggestion, the data pertaining to the flight should consist of:

- Flight Identification
- Scheduled departure date
- Last place/port of call of aircraft
- Place/port of aircraft initial arrival

V. *Contracting States' Positions*

1. The United States Legislation Pertaining to API and PNR

Due to the most recent events of 2001, President George Bush signed a new *American Transportation & Security Act* on 25 November 2002 making mandatory API transmission and PNR access to all passengers arriving in the United States. The Department of Homeland Security will therefore ensure that the air carriers, airobia and other governmental agencies comply with this new bill.

---

[369]Ibid. at attachment clause no. 5: "IATA has constantly sought to eliminate unnecessary forms and procedures min international air transport and the abolition of the passenger manifest has been an important policy objective for the Association. Recent opportunities to automate government control processes have, however, let to a close look at the concept of API and its potential for facilitation improvements.Collection of passenger details at departure presents a problem of additional workload for airlines at point in the system where staff and facilities are frequently already stretched to maximum capacity and beyond. Consequently, carrier support of API depends heavily on there being truly realizable benefits for airline passengers on arrival at destination. Furthermore, given the practical cost constraints of data capture and transmission, limiting the required information to that which can be captured by machine reading passports and visas, augmented by basic flight details, is a prerequisite. To this end, IATA sees particular benefit in co-operating with the CCC to define the data and message sets for API within UN/EDIFACT PAXLST development, and in establishing jointly agreed principles which can expand the benefits of automating and integrating all elements of the passenger process from origin to destination".

[370]*Ibid.* at attachment clause 9.

D. The Passenger Name Record

According to the WCO/IATA guideline,[371] there is a stipulation that API transmissions should originate from the last port before entering into the port of arrival. However, the US concluded agreements between different States that seem to violate the general guidelines of IATA and the WCO[372] in the sense that under these guidelines no data from an API transmission would only be provided to the port of entry. Under this Act, API data submissions to the US now have been made mandatory on flights bound for another State. These agreements find extra-territorial applications of American legislation where it imposes to another State submitting API as well as PNR passenger information. According to the US Customs Service, it implemented a Canada Smart Border/30 Point Action Plan, better known as the Manley Ridge Agreement. According to this plan adopted in December 2001, the United States and Canada agreed to share API and passenger name records as of spring 2003.[373]

The new American Transportation & Security Act stipulates as section 115 the required information from each flight prior to departure and arrival in the United States:

> A passenger and crew manifest for a flight required under paragraph (1) shall contain the following information:
>
> > The full name of each passenger and crew member;
> > The date of birth and citizenship of each passenger and crew member;
> > The sex of each passenger and crew member;
> > The passport number and country of issuance of each passenger and crew member if required for travel;
> > The United States visa number or resident alien card number of each passenger and crew member, as applicable;
> > Such other information as the under Secretary, in consultation with the Commissioner of Customs, determines is reasonably necessary to ensure aviation safety.[374]

Furthermore, according to sub-section three and four of the same section on Passenger Manifests, the Customs service also can prescribe the time frame it can

---

[371] *Ibid.* at attachment clause 8.1.5: "It should be noted that API transmissions will contain data for passengers carried into a country (initial place/port of arrival) from the last place/port of call of that aircraft abroad. API transmissions will not provide information of passengers' previous flights or ports of call before joining the flight at the last foreign port of call. Neither will API transmissions provide information on onward flights to other countries. Put simply, the API transmission contains only details of passengers carried from last port of call to the first port of call in the country of arrival without regards for the passengers' initial point of departure or their ultimate destination".

[372] Refer to the "US–Mexico Border Partnership Action Plan," online: http://www.whitehouse.gov/infocus/usmxborder/22points.html (date accessed: 17 December 2002).

[373] Refer to the "US–Canada Smart Border/30 Point Action Plan," online: http://wwww.whitehouse.gov/news/2002/12/20021206-1.html (date accessed: 17 December 2002): The United States and Canada have agreed to share Advanced Passenger Information.

[374] One Hundred Seventh Congress of the United States, Aviation and Transportation Security Act, HR 5005 EAS, Chapter 1 of title 49 S. 1447 at section 115 sub-section 2.

expect to receive electronic messaging from air carriers as well as passengers name records and all pertinent identification necessary for screening.

As a response to this new API/PNR data transmission, American Airlines and Continental Airlines have agreed to comply with the new legislations but have requested the US Customs Service to review its penalty procedures if issued erroneously.[375]

2. The Canadian Position

The *Canadian Immigration and Refugee Act*[376] came into force as of June 2002 dealing with required documentations and obligations on air carriers in conjunction with Part 17 of the Regulations issued by Citizenship and Immigration Canada.[377]

As first point of interest, in section 148 of the IRPA, air carriers are required not to carry any person that is not in possession of required documents of travel. In the event that such obligations are not fulfilled, section 278 describes the different penalties, which will be imposed to the transportation companies.

Within Part 17 of the implemented regulations by the CIC, section 269 contains relevant advance passenger information legislation including:

> 269: Details data elements that will be required under the Canadian Advance Passenger Information programs, including;
>
> Surname, first name and initial(s) of any middle names;
> Date of birth;
> Country that issued a passport or travel document, the citizenship or nationality of the airobia;
> Gender;
> Passport number or, if a passport is not required, the number on the travel document that identifies them; and,
> Reservation record locator or file number.

This part also provides for government access to airline reservation systems at 269(2), and seemingly indicates that the government shall have access to any record at any following its creation.[378]

Furthermore, it is important to note that paragraph 2 of the same legislation includes a disposition where any electronic messaging follows the existing UN EDIFACT PAXLST format.

According to the Privacy Commissioner of Canada, complying with the United States new interim rule would infringe upon fundamental privacy rights and could possibly be used for other purposes, such as verifying income tax and other criminal

---

[375] Unofficial letter by American Airlines dated 28 February 2002 and unofficial letter by Continental Airlines dated 28 February 2002.

[376] *Immigration And Refugee Protection Act*, L.c. 2001, c.27 (hereinafter referred to as IRPA).

[377] *Citizenship and Immigration Canada* (hereinafter referred to as CIC).

[378] *IRPA, Part 17 Transportation*, supra note 51 at section 269.

D. The Passenger Name Record

information. It views this as a comparison as a "Big Brother" database.[379] However, it should be mentioned that when a passenger travels, he or she implicitly gives up a certain amount of privacy in order to receive clearance at different border control authorities.

As for the air carrier, a leading charter Montreal based airline, Air Transat A.T. requested from the US Customs a delay until 15 December 2003 in order to fully comply with the new Interim Rule.[380] The airline's representative in government affairs indicated that the airline does not possess at this time any central reservation system as most of its bookings are done through tour operators and other travel agencies. The costs relating to changing to a fully electronic method would represent an investment of 1.3 million dollars. It further criticized the deadlines imposed by the United States:

> We trust that such best efforts to date will be properly considered and that the Final Rule will not unduly penalize or burden smaller or less sophisticated air carriers such as Air Transat, in terms of passenger reservation and seat inventory management, with an unreasonably expeditious effective date.[381]

As a response to the United States Customs Directive on API, the legal director of ICAO gave an opinion and confirmed that although punitive recourses are at this time being imposed to different carriers, it appears to follow the guidelines set forth by the Chicago Convention:

Another essential feature of API as an effective facilitation measure is the accuracy of the information provided. The accuracy of the data contained in a Passenger Manifest is an essential requirement of this document, whether it is transmitted in advance (API) or not, and such requirement should be equally enforceable. Based on this principle, the requirement that the data provided by airlines must comply with an increasing percentage of accuracy only means that the US authorities intend to reduce their degree of tolerance of errors, possibly aiming at tolerance zero which is consistent with such principle. Punitive measures against airlines failing to comply with the required accuracy, the level of severity of such measures and the empowerment of the authorities to apply them are matters of national policy and law, provided that the applicable measures are enforceable within the territory of the State concerned, which appears to be the case [...].

In view of the foregoing, it is concluded that the US Customs Directive on API, in spite of worsening the airline's burden, is consistent with the relevant provisions of the Chicago Convention and its Annex 9, and therefore does not raise extraterritoriality issues.[382]

---

[379]"Privacy Commissioner of Canada: News Release," online: http://www.privcom.gc.ca/media/nr-c/02_05_b_020926_2_e.as (date accessed: 8 November 2002).

[380]Unofficial letter dated 24 January 2003 (not published).

[381]*Ibid.* at page 4.

[382]Weber L, "Inter-office memorandum on United States customs directive on advanced passenger information," 7 June 2002 (not published).

3. The United Kingdom Position

A standing committee on the *Draft Immigration (Leave to Enter Remain) Order 2000*[383] was first introduced by Mrs. Barbara Roche, Minister of State, Home Office. During this parliamentary discussion, it was made clear that API systems were to have a dual positive impact: not only would it permit a rapid clearance process within an airport for the possibility for officials to detect the presence of potentially high-risk individuals. Furthermore, according to Roche, this will in no way diminish the role of customs officers who will have the possibility of examining each passenger as well as their baggage and other belongings.[384] She also stated that the immigration officer could still at any point of border control monitor the passenger traffic and inspect the traveller.[385]

The UK, under this new legislation has enabled many different enforcement agencies in order to collect an intelligence map of potential high-risk individuals and prevent entry.[386] Furthermore, under the assurance of such agencies, both carriers and government agencies determined that such API legislation would be applied through a very rigorous and fair process[387] and compliance would have to be effective within the next 6 months period of time.

We can as well study the position of the major UK carrier, British Airways, towards the new American legislation pertaining to API and PNR access. In a letter

---

[383] House of Commons Standing Committee on Delegated Legislation, *Draft Immigration (Leave to Enter and Remain) Order 2000*, online: http://www.hmso.gov.uk (date accessed: 4 March 2003).

[384] House of Commons Standing Committee on Delegated Legislation, *Draft Immigration (Leave to Enter and Remain) Order 2000*, p. 3: "The power to grant or refuse leave to enter before a person arrives in the UK has two benefits. Advance passenger information could pre-clear certain low-risk school groups and recognized reputable tour groups, thereby speeding their progress through immigration control and removing the need for detailed, individual examination on arrivals. Alternatively, we might send immigration officers overseas, with the agreement of the Government concerned, to address particular pressure points. It also allows us to take advantage of future technological developments such as biometrics. Such measures will benefit the travelling public, carriers and the immigration service".

[385] House of Commons Standing Committee on Delegated Legislation, *Draft Immigration (Leave to Enter and Remain) Order 2000*, p. 3: "As I said, the role of the immigration officer is not diminished, as he or she can still examine a person with continuing leave".

[386] Regulatory Impact Assessment: Introduction to Extended Powers of Information Collection On Passenger and Goods, Schedule 7 to the Terrorism Act 2000 (Information) Order 2002, online: http://www.homeoffice.gov.uk/atoz/pax_and_goods.pdf (date accessed: 8 November 2002) at point 12: "The measure will enable the police to build an intelligence picture which will allow them to target and track terrorists in a way that has become essential in the aftermath of September 11 and the subsequent ongoing campaign against the threat of global terrorism".

[387] Regulatory Impact Assessment: Introduction to Extended Powers of Information Collection On Passenger and Goods, Schedule 7 to the Terrorism Act 2000 (Information) Order 2002, at point 39: "We are confident that the enforcement agencies would apply the legislation fairly, proportionately and appropriately requesting the information and the police utilizing it. This approach has been confirmed by representatives of the police at meetings with the carriers".

D. The Passenger Name Record 145

in early March 2002,[388] it informed the Office of Regulations and Rulings of the United States of its support towards the API system and its compliance by a Memorandum of Understanding [hereinafter referred to as: MOU] with the US Customs of 1998. This MOU consisted of voluntary release of passenger information to the US. According to this new *Aviation and Transportation and Security Act*, British Airways believes that an automated version should be considered in order for the collection of data such as PNR information so long as it is not stored manually and complies with the machine readable information.

British Airways also considers this as the best method of transferring passenger information from the airline system to the government. It considered the automatic PNR release and ruled that it would not affect the *UK Data Protection Act*.[389] BA's pretension is that the carrier should not be obliged to change any reservation system, which could encounter additional costs or in the least reduce them to a strict minimum.[390] The WCO also considered this position to be viable as it announced in its recommendation that information should be kept to a strict minimum or it otherwise, such operations become time and cost consuming:

> 8.2.1 Perhaps the most critical aspect of API is the means by which the data to be transmitted to the Border Control Agencies in the destination country is captured. Data capture can be costly, time consuming, labour intensive and error prone. The capture of data concerning departing passengers at the airport of departure introduces a delay in the check in process that could, if not managed properly, offset the potential advantage to passengers provided by efficient API applications. If the check-in process in unduly prolonged, then API will simply shift much of the delays and congestion away from the arrival area to the departure area. It is vital therefore that the effect of API on the check-in process is kept to the absolute minimum.[391]

However, the WCO also claims that API can also reduce staff costs because of this automated process that can therefore bring some form of saving for the air carrier.[392]

Furthermore, such information should not permit access to any other passenger that is not on a flight bound to the US.[393]

---

[388] British Airways letter dated 1 March 2002 (not published).

[389] *UK Data Protection Act*, online: http://www.legislation.hmso.gov.uk/acts1998/19980029.htm (date accessed: 10 January 2003).

[390] Refer to supra note 9 at clause 6.5.1 and 6.5.2.

[391] *Ibid.* at clause 8.2.1.

[392] *Ibid.* at clause 6.9.3. and 6.3.

[393] Refer to supra note 59 at attachment A:

> (3) A general request to oblige the carrier to give access only to passenger name record information relating to passengers whose itineraries include at least one flight operated to or from or within the United States. In the event that carrier's systems are not designed or configured so as to allow such access without also giving access to information about other passengers, the Customs Service shall adopt procedures or take other appropriate measures to ensure that its officers do not access information relating to such other passengers. In

This air carrier also had specified its concerns to the US Customs in a previous correspondence whereas the American legislation gave no assurance that other information pertaining to that were not bound to the US would be transmitted.[394]

Virgin Atlantic also expressed grave concerns over PNR transmission as some private information relating to passenger's file are private and would contravene the UK Data Protection Act, according to local management.[395] According to authorities, it is now necessary under the UK Data Protection Act that recording of personal data of passengers do not leave the territory of the European Economic Area unless enough protection can be assured. In order to achieve this, the US Customs Service Agency would have to adopt the *Safe Harbour Principles*, set forth by the European Commission, under EU Directive (94/46/EC).[396] In this Directive, European companies can only send out information to any foreign country outside of the community if it should so correspond to a reasonable protection of sharing of information.

4. Safe Harbour Principles

In order to fully comply with the EU Directive (95/46EC), it introduced seven principles, otherwise referred to as Safe Harbour Principles:

- Notice must be given to individuals informing them of the purposes for which their data has been collected and how it will be used;
- Choice must be offered to individuals, allowing them to choose (opt out) whether and how their personal information is disclosed to third parties or used for purposes which differ from the ones which were originally notified;
- Onward transfer of personal data by organizations to third parties must be consistent with the principles of notice and choice;
- Security of personal data must be maintained using reasonable precautions;

---

addition, prior to implementing any online processes, the Customs Service will agree to appropriate security protocols with the carrier.

(4) No carrier shall be obliged to change or modify its computer systems (hardware or software) in order to comply with a general or specific request, unless the changes or modifications and the allocation of the cost of making them are agreed in advance between the carrier and the Customs Service.

[394]British Airways letter date 26 August 2002 (not published): "There appears to be nothing in the Interim Rule to protect the security and integrity of the carrier's systems. This is essential for British Airways to have confidence that cooperation will protect the integrity of its departure control systems and the legitimate rights and interests of its passengers. The Rule should provide such protection and British Airways respectfully requests the Customs Service agree to a security protocol prior to any direct systems access [...] British Airways requests that the agreements be finalized before access is activated".

[395]Unofficial letter by Virgin Atlantic Airways Ltd. dated 30 August 2002.

[396]Frashfields Bruckhaus Deringer. "Data Protection," online: http://www/freshfields.com/practice/ipit/publications/22367.pdf (date accessed: 6 February 2003).

D. The Passenger Name Record   147

- Data integrity must be ensured so that personal data is relevant for the purposes for which it used, not processed in ways which are incompatible with the purposes for which it has been collected and steps taken to ensure that it remains accurate;
- Access to personal data must be maintained so that individuals can ensure that it is corrected or deleted where inaccurate;
- Enforcement should be available through independent recourse mechanisms to deal with complaints, disputes and remedies, and provide sufficiently rigorous sanctions to ensure compliance.[397]

BA's charter counterpart, Britannia, had mentioned to the American authorities that it was not able to comply with the interim rule on passenger name records requirements because it did not process the necessary computerized reservation system.[398] Furthermore, in regards to API, it urged the US Customs Agency to waive applicability of data transmission on flights that are not bound for the US. It should also consider reducing the penalties imposed on air carriers if compliance cannot be performed on time. The time frame allotted to airlines in regards to changing their reservation systems is also an aspect that should be considered for API transmission.

5. The Australian Position

The Australia Immigration and Customs have already implemented API systems in order to accelerate the process and enhance border control.[399] In order to achieve these goals, Australia implemented the Advance Passenger Processing, hereinafter referred to as APP. This system provides a rapid clearance by the participating carriers. In essence, at foreign check-in points, the passengers passport is read and a magnetic card is then given if authorization to travel is granted with the individual's details enabling him or her to use the "Express Lane" upon arrival in Australia.[400] The government has also amended their national legislation in order to permit capture of data of API and PNR without infringing on privacy rights. This APP system works in cooperation with the Electronic Travel Authority (ETA), which is a communications network. When data is captured, it is sent through the ETA system that verifies the validity of the visa for those passengers who require such travel documents as well as the status of Australian and New Zealand passports.

---

[397] Frashfields Bruckhaus Deringer. "Data Protection," at page 1.

[398] Unofficial letter by Air 2000 Limited (26 August 2002).

[399] Manning, J. (Australian Delegate), "Facilitation Panel Fourth Meeting Information Paper," (Montreal, 2–5 April 2002) ICAO Doc FAL/4-IP/8.

[400] *Ibid.* at clause 3.2.5 and 3.2.6: "At check-in, the airline prints the passenger's bio data and flight number on a special Australian Incoming Passenger Card with the word 'EXPRESS' indicated. The card also has a magnetic strip that is coded with an identifier to retrieve that data on arrival in Australia. On arrival in Australia, the passenger will be directed to the appropriate processing lanes by use of dynamic signage and Customs marshals who are on-hand. APP passengers using the Express lanes are expected to be cleared in about half the time of other passengers who are not APP".

The APP/ETA system has been accepted by air carriers as it meets the individual need of each and every one of them. As some carriers have voluntarily participated to this plan, it is of the government wish to implement mandatory procedures for APP.[401] According to Australian Customs Service, APP gives a quantifiable reduction in undocumented travels and therefore reducing the possibility of being imposed fines by other Contracting States.

Qantas Airways, Australia's leading air carrier, expressed a similar concern to the one of British Airways but asked the US Customs Agency for further precisions on the interim rule pertaining to PNR information. It has concerns pertaining to the legislation when it permits sharing of all relevant information to the different Federal Agencies.[402] Furthermore, Qantas has asked the authorities to sign an agreement in order to prevent the US Customs Service[403] to send such data to an undetermined amount of agencies.

6. The German and Swiss Positions

According to present privacy laws, certain data is protected by the *Federal Data Protection Act ("Bundesdatenschutzgesetz," BDSG)*[404] and requires special permission from each airobia before permitting access to such information by other States. Furthermore, all data must be deleted from any banks after a certain amount of time. In response to this legislation, Germany's leading air carrier, die Deutsche Lufthansa Aktiengesellschaft, had informed the US Customs Service of these legislative impediments and was awaiting assistance in order to comply. According to the in-house legal counsel department, it appears that this could be implemented throughout the year of 2003.[405] Unless clear amendments can be made, violations of the Federal Data Protection Act can be of substantial financial and legal consequence.[406]

---

[401] Permanent Technical Committee, "Review of the WCO/IATA Guidelines on Advance Passenger Information" WCO Doc PW0045E1 (Brussels, 20 August 2001).

[402] Permanent Technical Committee, "Review of the WCO/IATA Guidelines on Advance Passenger Information" WCO Doc PW0045E1 (Brussels, 20 August 2001) at clause 4.2.

[403] Qantas Airways letter dated 22 August 2002 (not published): "Prima facie, Qantas has not identified any incompatibility between USCS Passenger Name Record (PNR) requirements and Australia's national protection laws. However the statement in the CFR that 'PNR information that is made available to Customs electronically may, upon request, be shared with other Federal Agencies,' requires further clarification. Specifically, whether or not carriers will be notified when and with whom this information is being shared and how the integrity of the data will be maintained during this process".

[404] *Bundesdatenschutz*, online: http://www.datenschutz-berlin.de/recht/de/bdsg/bdsg1.htm#absch1 (date accessed: 17 January 2003).

[405] Unofficial letter dated 30 August 2002 by the Deutsche Lufthansa Aktiengesellschaft: "Implementation by Lufthansa in the first quarter of 2003 appears feasible, provided that the present legal issues can be resolved".

[406] *Ibid.* at page 2: "Administrative offences are applicable and punishable by fines up to Euros 250,000.00 to anyone who, whether intentionally or through negligence, collects or processes

D. The Passenger Name Record   149

As for the Swiss Government, personal data is regulated by the *Data Protection Act(DSG)*,[407] all transmissions must be transferred in good faith and must be done in a secure manner. As the API transmission is currently used is relevant to current border inspections, it therefore does not go against these legal previsions.

As for Swiss International, Switzerland's main air carrier, API does not cause any infringement on the DSG, however the compulsory PNR unless certain conditions are meet could cause legal consequences to the carrier:

> The unlimited access by a third party in a foreign jurisdiction to the entire PNR data of a Swiss air carrier, without legal safeguards described above, turns out to cause major legal problems for the carrier concerned.

However, provided that data can be restricted to PNR data on in- and outbound US-flights, SWISS might be able to comply with national data protection laws when providing PNR access to US Customs. Compliance with Swiss and European Data Protection law could be achieved, if (a) the air carrier receives permission from the Swiss National Data Protection Officer and (b) obtains the required guarantees from the US authorities (see Point 2.2 above), eventually by applying the "Safe Harbour" principles. Furthermore (c), the air carrier would have to change its booking procedures by asking the passenger for additional data and an explicit consent to make this data available to U.S. Customs and other explicitly named U.S. authorities.[408]

Swiss International AirLines also raised the matter of implementing a filter system in order to protect a leakage of information to other authorities in the US. This filter concern was also brought up by the Bundeskriminalamt but is currently being resolved by the creation of new biometric procedures slowly being introduced in Frankfurt's airport that facilitates the creation of such a filtering data processing that would not infringe on any federal data protection legislation.[409]

7. The Mexican Approach

As of the present time, Mexico submits all API information on passengers and crew on all international flights and intended as of July 2002 to submit to the United States to a minimum of 95% sufficiency all information in the UN-EDIFACT messaging format. As a counterpart, Mexico also plans to fully request API information and penalize air carriers that are either late or not submitting such

---

personal data which are not generally accessible without authorization (Section 43 BDSG); additionally, certain violations of this law can also carry criminal penalties of up to 2-years imprisonment and/or fines up to Euros 250,000.00 per offense (Section 44 BDSG)".

[407]*Swiss Federal law On Data Protection*, online: http://www.datenschutz-berlin.de/recht/de/bdsg/bdsg1.htm#absch1 (date accessed: 5 March 2003).

[408]Unofficial letter by Swiss International Air Lines dated 26 August 2002 (not published).

[409]Unofficial interview with Dr. Edgar Friedrich, Bundeskriminalamt, Wiesbaden Germany in February of 2003.

information. Contrary to the US, it does not plan to request any passenger name record other than the Record Locator reference to be used upon request.[410]

Mexico also signed an agreement with the United States, the *Smart Border 22 Point Agreement*,[411] stipulating that on a voluntary basis, Mexico would exchange some information with the United States on a mutual level in order to prevent illegal migration and detection of high potential risk passengers. According to the US Customs Service, there is presently exchange of information on international flights bound for Mexico even though the port of arrival is not the Unites States. For example, at this time, a flight from Frankfurt to Mexico City non-stop may have to submit to the US API and PNR information on its passengers.

The Instituto Nacional De Migracion proposed an electronic database collecting information of passengers when making a reservation and together with ICAO/IATA and the Simplifying Travel Procedures established a data processing that will be later discussed regarding the understanding of biometric procedures.[412]

Other airlines have as well expresses grave concerns at the new Final Rule RIN 1515-A06 on Passenger Name Record Information Required for Passengers on Flights in Foreign Air Transportation To or From the United States.[413] In fact, according to VARIG, Brazil's leading air carrier, PNR violates the Brazilian Constitution where unless express authorization is given by the airobia or by other competent authority, it cannot comply with this new legislation.[414]

IATA, the International Air Transport Association, which represents 274 member airlines noted that considerable discussions should continue to be held with the US Bureau of Customs and Border Protection in order to assure its carriers that privacy laws are being complied with. It founded its remarks on the *EC Directive (95/46EC)*[415] that regulates the processing of personal data for all countries falling under the European Union. According to IATA, with the EC Directive, if the United States would adopt the Safe Harbour Principles, it would give sufficient protection for other States and their air carriers to comply without being held liable for data

---

[410]Unknown sheet of paper?

[411]Refer to supra note 53.

[412]Secretaria De Gobernacion, Instituto Nacional De Migracion, "Technical Specifications INM Fast-Track" Confidential INM Presentation (not published).

[413]Passenger Name Record Information Required For Passengers On Flights In Foreign Air Transportation To Or From The United States, 67 Fed. Reg. 42710 (25 June 2002).

[414]Unofficial letter by Varig's legal counsel, Mrs. Constance O'Keefe dated 18 September 2002 (not published): "Due to Constitutional provision, information contained in air travel reservations, which is of a confidential nature, can only be disclosed upon written request by competent public authorities, by public administrative agencies, by an individual passenger – with proper identification – or by a legal representative duly authorized by the passenger".

[415]*EC Data Protection Directive (95/46EC), Protection of the individuals in relation to the processing of personal data*, online http://wwwdb.europarl.eu.int/oeil/oeil4.Res213 (date accessed: 5 march 2003).

transmission only if all agencies of the US receiving such data also adopt such principles. Furthermore, the US Customs Service should:

- Self-certify under the Department of Commerce "Safe Harbour" Principles or develop and implement self-regulatory data privacy policies that conform to those Principles
- Communicate that self-certification or privacy policy development to all governments having data privacy legislation adopted in accordance with the EU Directive
- Provide guarantees that limit sharing of data obtained through access to airline systems only to those agencies that have self-certified under, or fully adopted the "Safe Harbour" principles
- Limit its access to "read only" capability and provides assistance in blocking illegal outside access
- Provide assurances to governments and to carriers alike that it will limit access to information pertaining only to those flights touching U.S. territory[416]

In conclusion, although many air carriers deem that potential liability could be foreseeable, it is to be noted that under the Chicago Convention, a State has the right to request information in order for proper border control to be established.[417] Therefore, as national policy of a member State has the right to infringe upon others requesting clearance into their sovereign State, it can request or infringe upon another its principles for proper border control.[418]

E. Machine Readable Travel Documents

Border controls and identity management, while being traditionally a law enforcement matter, have direct implications to the modern security agenda worldwide. While effective border and identity controls cannot eliminate terrorism and related trans-border crime per se, they can reduce the threats dramatically and offer a powerful tool to mitigate and counter such threats. The use of false identities and fraudulent travel documents remains an important modus operandi that tends to be exploited by trans-border criminal networks. The terrorist attacks, including the 9/11, Bali atrocities, London and Madrid bombings, further highlighted the

---

[416]Unofficial letter by IATA dated 26 August 2002 (not published).

[417]Chicago Convention, Art.13: "The laws and regulations of a contracting State as to the admission to or departure from its territory of passengers, crew or cargo of aircraft, such as the regulations relating to entry, clearance, immigration, passports, customs, and quarantine shall be complied with by or on behalf of such passengers, crew or cargo upon entrance into or departure from, or while within the territory of that State".

[418]Chicago Convention, Art.1: "The contracting States recognize that every State has complete and exclusive sovereignty over the airspace above its territory".

importance of travel document security and identity management for preventing and combating these acts.

ICAO develops standards, specifications and recommended practices for the issuing and reading machine-readable passports and other travel documents to promote efficient facilitation, increase the reliability of travel documents, and contribute to national and international security.[419] While ICAO Contracting States have an obligation to implement the standards, it has been recognized that some States may lack the capacity to comply and require technical assistance from the international community. In this context, ICAO and partner agencies have been deploying outreach and capacity building efforts to promote the adoption of ICAO standards, specifications and best practices, and to assist States in need of help and guidance.

The Implementation and Capacity Building Working Group (ICBWG) of the TAG/MRTD[420] has been created to help develop and implementation of these technical assistance initiatives in priority areas. In particular, recognizing that States that do not yet issue machine-readable passports may lack the necessary technical knowledge or financial resources, ICAO has launched a plan of action called the "Universal Implementation of Machine Readable Travel Documents" to help them meet the new standard. Support offered under this plan includes: technical assistance to apply the ICAO.

ICAO's capacity-building work worldwide is closely linked with the technical cooperation facilitation efforts by the UN Security Council's Counter-Terrorism Committee with particular reference to the implementation of the UN Security Council's Resolution 1373 (2001). Resolution 1373 constitutes a landmark in international efforts to fight terrorism by creating formal obligations to all 191 UN Member States to join efforts in building their capacity to prevent and combat terrorism. In particular, Resolution 1373 creates an obligation to member States to "prevent the movement of terrorists or terrorist groups by effective border controls and controls on issuance of identity papers and travel documents, and through measures for preventing counterfeiting, forgery or fraudulent use of identity papers and travel documents."[421] In addition, it also calls upon all States to "find ways of intensifying and accelerating the exchange of operational information, especially regarding actions or movements of terrorist persons or networks" and "forged or falsified travel documents." Those obligations and ongoing counter-terrorist capacity-building efforts have been reinforced by Resolution 1624 (2005) which calls further upon all States to cooperate, inter alia, to strengthen the security of their international borders, including by combating fraudulent travel documents and, to

---

[419]Identifying current border security and identity management capacity gaps and vulnerabilities in ICAO Contracting States through questionnaires, consultations, assessment missions, etc.

[420]Formulating capacity-building interventions through technical cooperation projects or programmes based on international best practices and time-proven responses in other Contracting States, in partnership with other international agencies where needed.

[421]Securing donor funding for specific technical cooperation projects and launching project implementation, in partnership with other international agencies where needed.

E. Machine Readable Travel Documents

the extent available, by enhancing terrorist screening and passenger security procedures with a view to preventing those guilty of [incitement to commit a terrorist act] from entering their territory.

The ICBWG, in particular, has been working on a global Environmental Scan that collects and analyses information about known border security and identity management weaknesses worldwide. While the main focus of the Environmental Scan is MRTD compliance to ICAO document security standards, it also addresses to some extent the broader counter-terrorism capacity building agenda with reference to border security including: the production of secure and integral breeder documents; current levels of national staff's technical knowledge and vocational expertise; automated migrant processing systems including passport readers and alert lists; and assessing the security of travel document issuing procedures in cases where secure MRTDs might be compromised by lack of integrity and abuse of the system. The Environmental Scan is meant to be an evolving document, effectively a database that is being enhanced and updated, and provides knowledge about main capacity gaps globally that could be addressed through international technical cooperation assistance.

ICAO's global capacity-building strategy is based on the following logical sequence:

1. Identifying current border security and identity management capacity gaps and vulnerabilities in ICAO Contracting States through questionnaires, consultations, assessment missions, etc.
2. Formulating capacity-building interventions through technical cooperation projects or programmes based on international best practices and time-proven responses in other Contracting States, in partnership with other international agencies where needed.
3. Analysing the counter-terrorist, international development and capacity building priorities of international donor agencies, to match the needs of beneficiary States, project objectives and donor funding priorities.
4. Securing donor funding for specific technical cooperation projects and launching project implementation, in partnership with other international agencies where needed.
5. Implementing the projects, achieving their objectives and purposes, eliminating the original border security gaps and vulnerabilities in beneficiary States, and enhancing their administrative and operational capacity to further the objectives of the UNSC Resolution 1373 (2001) with reference to border security and identity management.
6. Using lessons learnt and project implementation experience and expertise acquired in further capacity-building project cycles replicating the technical cooperation solutions on a broader regional or global scale.

It is recognized that many developing countries are unlikely to have sufficient funds and technical expertise to deal with the challenge effectively on its own. Hence the need for international technical assistance to implement the project that would enhance beneficiary States' good governance, effective administrative and operational capacity to manage and control their borders, and their national regional security.

ICAO has developed the only global MRTD programme which offers technical specifications on machine readable passports, visas, identity cards and other related travel documents. The ICAO MRTD Programme has gained much momentum since its inception and continues to be the global leader in establishing and maintaining specifications on MRTDs (contained in Doc 9303). In addition, the MRTD Programme has been proactive in enhancing its non-Regular Programme revenue generating activities through advocacy, outreach and technical consultations with States.

While the MRTD Programme continues with its core activities, the evolving international environment and extended globalization present new challenges and opportunities that affect its future work. Two main emerging factors that shape and influence the MRTD Programme's future activities are the need to render assistance to needy States in the light of tighter security standards and growing technical complexity, and the increasing focus on security and counterterrorism on the international agenda.

Traditionally, MRTDs were perceived in the context of facilitation, with an emphasis on the speed and efficiency of passenger processing, partly due to their relevance to Annex 9 – *Facilitation* and partly due to strong links with the aviation industry. While facilitation and convenience aspects remain important, MRTDs have emerged as a powerful tool to prevent terrorism and transborder crime. The current international environment requires a stronger focus on these areas, in order to keep the MRTD Programme at the heart of global developments and priorities.

The MRTD Programme remains uniquely placed to address current terrorism and transborder crime concerns through the establishment of specifications and expanded capacity-building projects, and ever closer cooperation with other international agencies. Some of the latest achievements of the MRTD programme are the establishment of the New Technology Working Group (NTWG) of the Technical Advisory Group on Machine Readable Travel Documents (TAG/MRTD) which assists the Secretariat in developing and maintaining the currency of the specifications contained in *Machine Readable Travel Documents*[422] as well as in issuing related guidance material. Regular updates are provided through Supplements, which serve as maintenance vehicles and are based on States' implementation experiences or other related technical developments. Much of the contents of the Supplements are eventually incorporated into Doc 9303, a Technical Report, or both. To date, seven Supplements have been issued, the last of which was completed in November 2008.

Additionally, the ICAO Secretariat has provided assistance to several States and international organizations on matters related to MRTDs, on request. To coordinate the efforts and resources deployed by several international organizations and Contracting States in this direction, the TAG/MRTD, during its 18th meeting

---

[422]ICAO (Doc 9303).

E. Machine Readable Travel Documents

from 5 to 8 May 2008, supported the creation of the Implementation and Capacity Building Working Group (ICBWG).

Advances in the provision of assistance to States were made in 2008. Such advances were possible through collaboration with other UN and non-UN bodies, including the UN Counter-Terrorism Committee Executive Directorate (CTED), INTERPOL, the International Organization for Standardization (ISO), Airports Council International (ACI), the International Air Transport Association (IATA), the Organization of American States' (OAS) Secretariat of the Inter-American Committee Against Terrorism (CICTE), the Organization for Security and Co-operation in Europe (OSCE) and the International Organization for Migration (IOM). A total of 46 States were so assisted in 2008 and 2009.[423]

ICAO has adopted a strategy to develop and implement an MRTD training initiative. Vocational training in MRTD-related matters remains an area in increasing demand as currently, there are no credible training providers in this field beyond ad hoc seminars or conferences. The Secretariat has started planning and designing a comprehensive vocational training programme, in close cooperation with the Technical Cooperation Bureau Trainair Programme, and relying on their instructional design and delivery methodology. Subject to the availability of funds, this programme is envisaged to be fully deployed by the second quarter of 2010.

Global efforts in capacity building in combating terrorism and trans-border crime have gained considerable momentum during the recent years, bringing together numerous agencies actively involved in this field. While the UN Security Council's Counter-Terrorism Committee provides the lead and overall coordination of global counter-terrorism technical cooperation efforts, other international agencies such as the International Centre for Migration Policy Development (ICMPD), International Organization for Migration (IOM), Interpol, Organisation for Security and Cooperation in Europe (OSCE), Organization of American States' Inter-American Committee Against Terrorism (OAS/CICTE), UN Development Programme (UNDP) and UN Office on Drugs and Crime (UNODC), to name just a few, have been actively contributing to the cause through their projects and programmes in a broad range of areas. In addition, regional forums and agencies (e.g., the ASEAN Secretariat, the European Commission, FRONTEX, Pacific Islands Forum, etc.) and donor States have been also actively contributing to security and development needs in beneficiary States.

Given the complexity of the task and a broad range of actors involved, coordination of such efforts remains of the utmost importance. While presenting a detailed compendium of ongoing projects and programmes worldwide is outside the scope of this paper, ICAO has been taking into account and tracking other relevant capacity-building activities in order to ensure that the current and future projects do not duplicate but complement and reinforce each other. Also, where possible, creating synergies, economies of scale and adding to sustainability of such projects.

---

[423] A table of coordinated assistance activities may be found in Volume 4, Number 1 of the MRTD Report, available at http://www2.icao.int/en/MRTD/Pages/ICAOMRTDReport.aspx.

Another important aspect is to ensure that such capacity-building work capitalises on each international agency's strengths and areas of expertise. Inter-agency MOUs are of particular help in this respect as they assist in outlining the roles and focus of each actor.

I. Some Problem Areas

(a) Some States do not issue machine-readable passports compliant with ICAO standards and specifications, and are unable to meet the 1 April 2010 deadline for their introduction in accordance to their international obligations under the Chicago Convention. In addition, their old-style passports present insufficient security features, use glued-in passport photos that can be easily altered including photo substitution on the bio-data page and exploited by transborder criminals.[424]

(b) In a number of countries, including those that already have ICAO-compliant MRPs, breeder documents (birth certificates, national ID cards, etc.) required to apply for a passport generally have no security features and can be easily faked or altered. When combined with weaknesses in their current Civil Registry system (e.g., no matches between birth and death records or name changes), this situations presents numerous opportunities for identity fraud where a "legitimately-issued" passport could be obtained on the basis of fake or doctored breeder documents. As modern MRPs are very difficult to forge or alter effectively, the focus of document fraud worldwide has been increasingly shifting to breeder documents which often have insufficient security features and can be exploited for criminal purposes.

(c) In some countries, including those that already have ICAO-compliant MRPs, immigration staff process passport controls manually, without the use of an automated database with passport readers that would store information which could be retrieved and analyzed as required. In addition, there is often no watch list functionality which makes it impossible to detect known cases of document fraud and persons of interest. Such lack of processing infrastructure eliminates the facilitation advantages that MRTDs offer and constitute vulnerability in preventing terrorism and trans-border crime.

(d) In a number of States, immigration and passport officials appear to have very limited knowledge of migrant processing and passport examination, and generally have not undergone any specialized vocational training relevant to their daily duties. They cannot take advantage of modern secure MRTDs and migrant processing techniques, often reducing their work to a mechanical stamping of passports, with obvious implications to border security and integrity.

---

[424]While a laser-engraved passport photo is not part of the ICAO MRTD standards yet, it is a recommended good practice as photo substitution of glued-in passport photos remains perhaps the most common *modus operandi* in travel document fraud.

Based on the Environmental Scan data, these are the main capacity gaps that present security vulnerabilities which can be exploited by criminals and terrorists. While implications to those specific States' national security are obvious, the current situation presents security risks regionally and even globally. Trans-border crime is international by definition, and border security and identity management weaknesses have the potential of spilling over across international borders. Most developing nations reliant on international aid are unlikely to be able to solve the identified security challenges without international technical assistance that would provide funding and expertise to tackle the problem in a coordinated manner.

F. Unmanned Aerial Vehicles

An Unmanned Aerial Vehicle (UAV)[425] is a self piloted or remotely piloted aircraft[426] that can carry cameras, sensors, communications equipment or other payloads. The United States Department of Defence defines a UAV as "a powered aerial vehicle that does not carry a human operator, uses aerodynamic forces to provide vehicle lift, can fly autonomously or be piloted remotely, can be expendable or recoverable, and carry a lethal or non-lethal payload."[427] Ballistic or semi-ballistic vehicles, cruise missiles and artillery projectiles are not considered UAVs by this definition.[428] UAVs have been used to conduct reconnaissance and intelligence-gathering for nearly 60 years (since the 1950s). The future role of the UAV is a more challenging one which, in addition to its current uses will include involvement in combat missions.[429] The issues and challenges that UAVs bring to civil aviation can be bifurcated into two main areas. The first concerns airworthiness regulations which are required to ensure that a UAV is built, maintained and operated at high standards that ensure the safety of all involved including crew and passengers of manned civilian and military aircraft with which UAVs will share

---

[425] For more details of UAV operations and their nature visit http://www.uvs-info.com.

[426] An aircraft is defined as "any machine that can derive support in the atmosphere from the reactions of the air other than the reactions of the air against the earth's surface." This definition appears in Annexes 1, 2, 3, 7, 8, 11, 13, 16 and 17 to the Convention on International Civil Aviation, signed at Chicago on 7 December 1944. See ICAO Doc 7300/9 Ninth Edition, 2006.

[427] *Unmanned Aerial Vehicles: Background and Issues for Congress*, Report for Congress written by Elizabeth Bone and Christopher Bolkcom, Congressional Research Service: The Library of Congress, 25 April 2003 CRS 1.

[428] *Unmanned Aerial Vehicles: Background and Issues for Congress*, Report for Congress written by Elizabeth Bone and Christopher Bolkcom, Congressional Research Service: The Library of Congress, 25 April 2003 CRS 1.

[429] Since 1964 the US Defense Department has developed 11 different UAVs, though due to acquisition and development problems only 3 entered production. The US Navy has studied the feasibility of operating Vertical Take off and Landing (VTOL) UAVs since the early 1960s, the QH-50 Gyrodyne torpedo-delivery drone being an early example. However, high cost and technological immaturity have precluded acquiring and fielding operational VTOL UAV systems.

de-segregated airspace as well as persons and property on the ground.[430] There is currently no international Standards and Recommended Practices (SARPs) adopted under the auspices of the International Civil Aviation Organization[431] applicable to the UAV and the Unmanned Aircraft System (UAS)[432] although UAVs are increasingly requiring access to all categories of airspace including non segregated airspace.

The second challenge is more far reaching and concerns the possibility of the UAV encroaching on air traffic control (ATC) functions in non segregated airspace. In doing so, UAVs should not place an added burden and demands on airspace management and the flow of general air traffic within the en-route air space structure which must not be impeded by the presence of UAVs. In this context, the priority would lie in collision avoidance, primarily through effective separation of aircraft by which aircraft could be kept apart by the application of appropriate separation minima. The two key players in this exercise would be the pilot of the manned aircraft involved and the air navigation service provider who would be jointly or severally liable if a separation minima were compromised.

Although there are international regulations in place that address the operation of UAVs in non segregated airspace, there is provision under ICAO regulations for the appropriate procedure to be followed. Annex 11 to the Chicago Convention,[433] which deals with the subject of air traffic services, lays down requirements for coordination of activities that are potentially hazardous to civil aircraft. Standard 2.17.1 stipulates that arrangements for activities potentially hazardous to civil aircraft, whether over the territory of a State or over the high seas, shall be coordinated with the appropriate air traffic services authorities, such coordination to be effected early enough to permit timely promulgation of information regarding the activities in accordance with the provisions of Annex 15 to the

---

[430]The main concern of the International Civil Aviation Organization in its role as regulator in this context is with international civil UAV operations and those standards that affect such operations. ICAO should therefore, not be expected to take on a leading role in the development of aircraft performance specifications.

[431]An ICAO Exploratory Meeting on Unmanned Aerial Vehicles (UAVs) was held at ICAO Headquarters in Montreal from 23 to 24 May 2006. The primary objective of the meeting was to explore the current state of affairs with respect to development of regulatory material related to UAVs and to discuss the possible role of ICAO in the regulatory process. The meeting was informed that the ICAO Secretariat would use the results of the meeting as the basis for developing a report to the ICAO Air Navigation Commission (ANC) along with recommendations on an ICAO work programme.

[432]At least four States: Australia; France; South Africa; and the United States are known to have commenced a programme developing standards for UAV operations. See Alexander ter Kuille, UASD and the ATM Community: The CANSO Policy of Engagement, *UAV Systems, The Global Perspective, 2006/2007* Blyenburgh & Co.: France, p. 24 at 25.

[433]*Convention on International Civil Aviation, Supra* note 3. Air traffic Services: Annex 11 to the Convention on International Civil Aviation, 13th Edition, July 2001.

F. Unmanned Aerial Vehicles

Chicago Convention.[434] Standard 2.17.2 of Annex 11 explains that the objective of the coordination referred to in the earlier provision shall be to achieve the best arrangements that are calculated to avoid hazards to civil aircraft and minimize interference with the normal operations of aircraft.

The Chicago Convention[435] is focused on civil aviation, and applies to civil aircraft. The Convention does not apply to State aircraft, which are identified as aircraft engaged in police, military an customs services.[436] Therefore, principles of the Convention will apply only to UAVs not engaged in such activities as are excluded. One of the provisions which may have a bearing on UAVs in the Convention is Article 8 which stipulates that no aircraft capable of being flown without a pilot shall be flown without a pilot over the territory of a contracting State without special authorization by that State. Furthermore states allowing the operation of aircraft that do not have a pilot in air space open to civil aircraft are required to ensure that they are so controlled as to obviate danger to civil aircraft. One of the common usages of UAVs – aerial photography – is affected by Article 36 of the Chicago Convention which empowers contracting states to prohibit or regulate the use of photographic apparatus in aircraft over its territory. Presumably this provision can be tagged on to Article 1 of the Convention whereby every State has complete and exclusive sovereignty over the airspace above its territory. Another important consideration could lie in Finally Annex 17[437] to the Chicago Convention, on the subject of aviation security where. Article 2.1.2 of the Annex states that each Contracting State shall establish an organization and develop and implement regulations, practices and procedures to safeguard civil aviation against acts of unlawful interference taking into account the safety, regularity and efficiency of flights. This could impel States to develop regulations and practices addressing the interference of control signals or even the hostile takeover of the command of an UAV which is a very common hazard to the operation of UAVs.

Another challenge in the operation of UAVs is licensing of personnel in charge of the operation of the vehicle and certification of the UAV. Article 31 of the Convention provides that every aircraft engaged in international navigation shall be provided with a certificate of airworthiness issued or rendered valid by the State in which it is registered. The Standards and Recommended Practices (SARP) for the issuance of an airworthiness certificate are laid down in Annex 8[438] to the Chicago

---

[434] Annex 15 contains Standards and Recommended Practices relating to Aeronautical Information Services.

[435] Convention on International Civil Aviation, signed at Chicago on 7 December 1944. ICAO Doc 7300/9, Ninth Edition, 2006.

[436] Convention on International Civil Aviation, signed at Chicago on 7 December 1944. ICAO Doc 7300/9, Ninth Edition, 2006, Article 3.

[437] Annex 2 to the Convention on International Civil Aviation (note 385), "Safety – Safeguarding International Civil Aviation Against Acts of Unlawful Interference," 8th edition, April 2006.

[438] Annex 8 to the Convention on International Civil Aviation (note 385), "Airworthiness of Aircraft," 10th edition, April 2005.

Convention. Annex 8 (in its 9th Edition) only addresses aeroplanes[439] over 5,700 kg certificated take-off mass and helicopters[440] without a limitation on the mass of an aircraft which is intended for the carriage of passengers or cargo or mail in international air navigation[441] This might provoke the argument that Annex 8 would not usually apply to UAVs since only large UAVs exceed the weight of 5,700 kg. The lack of internationally recommended and accepted standards and practices for smaller aeroplanes is a challenge for the operation of UAVs as well as for aeroplanes with a pilot on board. This point is covered in the 10th edition of Annex 8 which, in addition to the provisions in part VI on helicopters has been amended to be applicable for helicopters with a certificated take-off mass over 750kg only. In terms of licensing it has to be noted that Annex 1[442] to the Chicago convention, defines SARPs for personnel licensing, in that a person shall not act as an air crew member unless a valid license is held[443] by that person. Pilots are considered not only flight crew but as well flight navigators, flight engineers and radiotelephone operators.[444] Implicitly, this means that not only is the remote pilot of UAVs subject to licensing, but also personnel who are involved in the navigation and technical operation of UAVs should be licensed as well. Furthermore mechanics of UAVs be should also be licensed according the provisions in chapter 4.1 and 4.2 of Annex 1 to the Chicago Convention. Article 29 of the Chicago Convention requires the carriage of documents in aircraft such certificates of registration and airworthiness but also the appropriate licenses for each member of the crew. Although certificates of airworthiness can be carries in an aircraft in the manner required, the carriage of other documents may pose difficulties as some UAV are designed to operate over extended periods of time, up to several months, and the specific operators who would operate for such long periods may not be known at the initial stage of the flight. One potential solution could be to electronically store the data and electronic licenses (be it in the form of scanned documents or other forms) of the current crew on board of the vehicle, but this would need in depth assessment in regards to the legal validity of such a form.

Annex 2 to the Chicago Convention, detailing the rules of the air referred to in Article 12 of the Convention, states *inter alia* that the rules of the air shall apply to

---

[439]"A power-driven heavier-than-air aircraft, deriving its lift in flight chiefly from aerodynamic reactions on surfaces which remain fixed under given conditions of flight," see definitions in note 388.

[440]"A heavier-than-air aircraft supported in flight chiefly by the reactions of the air on one or more power driven rotors on substantially vertical axes," see definitions in note 388.

[441]Supra note 15, part IV, Article 1.1.2 (wording identical to 9th edition).

[442]Annex 1 to the Convention on International Civil Aviation (note 385), "Personnel Licensing," 10th edition, July 2006.

[443]*Supra* note 15, Article 1.2.1.

[444]*Supra* note 3925, Chapter 3.

F. Unmanned Aerial Vehicles

aircraft bearing the nationality and registration marks of a Contracting State.[445] These rules applicable to UAVs as well. Two main categories of rules of the air exist: visual flight rules and instrument flight rules.[446] The note to article 2.2. of Annex 2 states *inter alia* that a pilot may elect to fly in accordance with instrument flight rules in visual meteorological conditions. The rules of the air adhered to are thus distinct and separate from the metrological conditions prevailing in the area of operation, except for instrument metrological conditions, requiring instrument flight rules to be applied. Chapter 3.1 of Annex 2 contains an article on unmanned free balloons, stating that they shall be operated in such a manner as to minimize hazards to persons, property or other aircraft and in accordance with the conditions specified in Appendix 4. Appendix 4 states inter alia that heavy balloons[447] need to comply with similar provisions like normal aeroplanes, inter alia minimum height over "congested areas of cities, towns or settlements or an open-air assembly of persons not associated with the operation,"[448] SSR equipment,[449] and lightening.[450] Article 3.3 of Appendix 4 to the Annex 2 to the Chicago convention contains a remarkable requirement to unmanned balloons. Such vehicles shall be equipped with at least two payload flight termination devices or systems. It may well be argued that such devises or systems are required for UAVs as well. An analogy to the operation of UAVs exists in Annex 2 which requires obliges pilots-in-command to take action as will best avert collision. The Annex also requires that vigilance for the purpose of detecting potential collisions be exercised on board an aircraft, regardless of the type of flight or the class of airspace in which the aircraft is operating. Therefore, it can be concluded that pilots flying according instrument flight rules are required to scan the environment visually in order to detect potentially conflicting traffic. This task may prove difficult in the case of UAVs in that although many UAV are equipped with video cameras, it would be difficult for UAV operators to detect vehicles nearby, to assess the potential for conflict and to initiate appropriate actions. This inability might result in infringement of Article 3.2.1 of Annex 2, which provides that an aircraft shall not be operated in such proximity to other aircraft as to create a collision hazard. A potential solution to this problem could be that movement sensors, based on radar or ultrasound devices, similar to parking assistants for cars, are built into UAVs. The drawback of such a measure would be the cost involved and the additional weight that has to be carried by the UAV.

---

[445]*Supra* note, Article 2.1.1.
[446]*Supra* note, Article 2.2.
[447]*Supra*, note 14, Appendix 4, Article 1(c).
[448]*Id*, Article 3.2.
[449]*Id*, Article 3.4.
[450]*Id*, Article 3.6.

I. Legal and Regulatory Issues

The conduct of operations of UAVs are essentially State based and therefore becomes an issue of State Responsibility.[451] State responsibility in turn is founded on the basic legal principle of sovereignty and the rights and liabilities of States. International responsibility relates both to breaches of treaty provisions and other breaches of legal duty. In the *Spanish Zone of Morocco Claims* case, Justice Huber observed:

> [R]esponsibility is the necessary corollary of a right. All rights of an international character involve international responsibility. If the obligation in question is not met, responsibility entails the duty to make reparation.[452]

The principle of State sovereignty in airspace is embodied in Article 1 of the Chicago Convention which recognizes that every State has sovereignty over the air space above its territory, the latter being defined in Article 2 as land situated within and water adjacent to the State concerned. As for rights over airspace over the high seas, Article 87 of the *United Nations Convention on the Law of the Sea* of 1982[453] awards freedom for the aircraft of all States to fly over the high seas. An important consideration in delineating territorial sovereignty lies in the expansion of Flight Information Regions (FIR) and the provision of air traffic management services by States particularly when such measures are influenced by the revenue generating capabilities that are inherent in such an expansion of scope. The Chicago Convention, in its vision and wisdom, incorporates various provisions regarding the provision of air navigation services by States to aircraft flying over their territories. Firstly, the Convention guarantees, through provisions included in Chapter XV, that States which are unable to provide air navigation services to aircraft will be assisted.[454] Secondly, Article 15 of the Convention assures airlines that every airport in a Contracting State that is open to public use by its national aircraft shall also be open under uniform conditions to the aircraft of all the other Contracting States. The conditions are deemed to apply to the use, by aircraft, of every Contracting State of all air navigation facilities, including radio and meteorological services, which may be provided for public use for the safety and expedition of air navigation. Charges levied for such services are deemed by Article 15 to be anti-discriminatory whereby aircraft are not to be charged for airports and air navigation services provided to them at a rate higher than those levied on the national carrier of the State which provides

---

[451] The International Court of Justice in the *Barcelona Traction Case* held: [A]n essential distinction should be drawn between the obligations of a State towards the international community as a whole, and those arising *vis-a-vis* another State in the field of diplomatic protection. By their very nature, the former are the concerns of all States. In view of the importance of the rights involved, all States can be held to have a legal interest in their protection; they are obligations *erga omnes*. See *Barcelona Traction, Light and Power Company Limited, I.C.J. Reports, 1974*, 253 at 269–270.

[452] *1925 RIAA ii* 615 at 641.

[453] The Law of the Sea, Original Text of the *United Nations Convention on the Law of the Sea*, all Annexes and Index, United Nations: New York, 1983.

[454] Articles 69 and 70.

the service. To this end, Article 28 of the Convention obligates Contracting States to provide, as far as practicable in their territories, airports, radio services, meteorological services and other air navigation facilities to facilitate international air navigation according to Standards established pursuant to the Convention.

The tightly-set legal parameters of the Chicago Convention, particularly the assurance of air navigation services on an equal and non-discriminatory basis, are relevant in the twenty-first century, where service providers and airline operators have to collaborate in ensuring a seamless global air navigation system. Modern technology offers sophisticated air-ground data communications by VHF (very high frequency) and satellite, assisted by precise navigation by inertial/GNSS and computing in air traffic services. These will be used in the negotiation of dynamic user preferred routes offering various alternatives to airline operators which provide fuel and time savings. However, such preferences for flight profiles and uses thereof will be subject to growing air traffic demands which have to be cautiously assessed. This imposes an added burden on both the service provider and airline operator. Judgment and interpretation will be critical factors in this process, an inevitable corollary of which will be the need to examine legal aspects of the modern seamless air traffic management system.

As already stated, responsibility of States for the provision of air navigation services in their territories is founded in principles contained in Article 28 of the Chicago Convention of 1944. It must be noted that this is not an absolute obligation as the State is called upon to provide such services only in so far as it finds practicable to do so. In order to cover an eventuality of a State not being able to provide adequate air navigation services, the Convention imposes an overall obligation on the Council of ICAO in Article 69 to the effect that the Council shall consult with a State which is not in a position to provide reasonably adequate air navigation services for the safe, regular, efficient and economical operations of aircraft. Such consultations will be with a view to finding means by which the situation may be remedied. Article 70 of the Chicago Convention even allows for a State to conclude an arrangement with the Council regarding the financing of air navigation facilities and the Council is given the option in Article 71 of agreeing to provide, man, maintain and administer such services at the request of a State.

II. Operations Over the High Seas

Article 12 of the Chicago Convention unambiguously states that over the high seas, the rules in force shall be those under the Convention and each Contracting State undertakes to insure the prosecution of all persons violating the regulations applicable. This peremptory principle,[455] of adherence by States and aircraft bearing

---

[455] Bin Cheng confirms that over the high seas there is absolutely no option for States to deviate from rules established under the Chicago Convention for the manoeuvre and operations of aircraft. See Cheng (1962, p. 148).

their nationality to any Standards and Recommended Practices (SARPs) adopted in regard to the high seas, effectively precludes any possible reliance by States on Article 38 of the Convention which allows States to deviate from SARPs in general. In other words, Annex 2 on Rules of the Air, which contains provisions relating to the operation of aircraft over the high seas, is sacrosanct and inviolable. The first legal issue that would emerge from this clear principle is the question of applicability of Annexes (other than Annex 2) to the high seas and whether their provisions, if directly related to the principles of manoeuvre and navigation of aircraft over the high seas, would be binding with no flexibility offered by Article 38 of the Convention. Kaiser offers the opinion:

> Over the high seas, the rules of the air have binding effect under Article 12, Sentence 3 of the Chicago Convention. It should be clarified that rules of the air have a broader meaning than Annex 2 and encompass the Standards and Recommended Practices of all other Annexes as far as their application makes sense over the high seas.[456]

Kaiser is of course referring mainly to Annexes 10 and 11 to the Chicago Convention relating to air traffic services and air traffic management, while at the same time drawing the example of Annex 16 (on environmental) protection being applicable in a future date if extended beyond noise and engine emissions to the high seas under Article 12 of the Chicago Convention.[457] This argument, which would ascribe to the ICAO Council wider control over larger spans of the world's air space, would be acceptable only if provisions of other Annexes (other than those of Annex 2) would directly have a bearing on the manoeuvre and navigation of aircraft over the high seas, as exclusively provided for by Article 12 of the Chicago Convention.

The provision of air navigation services are mainly regulated by three Annexes to the Chicago Convention, namely Annex 2 (Rules of the Air), Annex 3 (Meteorological Service for International Air Navigation) and Annex 11 (Air Traffic Services).[458] Of these, compliance with Annex 2 is mandatory[459] and does not give the States the

---

[456]Kaiser (1995, p. 455). Bin Cheng states that contracting States are expected to be able to exercise control over all that takes place within their territories, but outside their respective territories only over aircraft bearing their nationality. Cheng (1962, p. 110).

[457]*Ibid.*

[458]Article 54(l) of the Chicago Convention stipulates as a mandatory function of the Council the act of adopting, in accordance with Chapter VI of the Convention, international standards and recommended practices (SARPs) and for convenience designate them as Annexes to the Convention. Article 37 of the Convention reflects the areas in which SARPs should be developed and Annexes formed. Article 38 obliges contracting States to notify ICAO of any differences between their own regulations and practices and those established by international standards or procedures. The notification of differences however, does not absolve States from their continuing obligation under Article 37 to collaborate in securing the highest practicable degree of uniformity in international regulations, standards, and procedures.

[459]In October 1945, the Rules of the Air and Air Traffic Control (RAC) Division at its first session made recommendations for Standards, Practices and Procedures for the Rules of the Air. These were reviewed by the then Air Navigation Committee and approved by the Council on 25 February 1946. They were published as *Recommendations for Standards, Practices and Procedures – Rules*

F. Unmanned Aerial Vehicles

flexibility provided in Article 38 of the Chicago Convention to register differences from any provisions of the Annex.

With regard to maritime navigation, the *United Nations Convention on the Law of the Sea (UNCLOS)*, Article 39, lays down the duties of ships and aircraft involved in transit navigation to the effect that ships and aircraft, while exercising the right of transit passage, should: proceed without delay through or over the strait; refrain from any threat or use of force against the sovereignty, territorial integrity or political independence of States bordering the strait, or in any other manner in violation of the principles of international law embodied in the Charter of the United Nations; refrain from any activities other than those incident to their normal modes of continuous and expeditious transit unless rendered necessary by force majeure or by distress; and comply with the relevant provisions of the Convention. Article 39 (3) explicitly states that aircraft in transit passage shall observe the Rules of the Air established by ICAO as they apply to civil aircraft and that state aircraft will normally comply with such safety measures and will at all times operate with due regard for the safety of navigation. The provision further states that at all times aircraft shall monitor the radio frequency assigned by the competent internationally designated air traffic control authority or the appropriate international distress radio frequency.

Standard 2.1.1 of Annex 2 to the Chicago Convention provides that the rules of the air shall apply to aircraft bearing the nationality and registration marks of a Contracting State, wherever they may be, to the extent that they do not conflict with the rules published by the State having jurisdiction over the territory over-flown.[460] The operation of an aircraft either in flight or on the movement area of an aerodrome shall be in compliance with the general rules and, in addition, when in flight, either with: visual flight rules (VFR); or the instrument flight rules (IFR).[461] Standard 2.3.1 further provides that the pilot-in-command of an aircraft shall, whether manipulating the controls or not, be responsible for the operation of the aircraft in accordance with the rules of the air, except that the pilot-in-command

---

of the Air in the first part of Doc 2010, published in February 1946. The RAC Division, at its second session in December 1946 – January 1947, reviewed Doc 2010 and proposed Standards and Recommended Practices for the Rules of the Air. These were adopted by the Council as Standards and Recommended Practices relating to Rules of the Air on 15 April 1948, pursuant to Article 37 of the Convention on International Civil Aviation (Chicago, 1944) and designated as Annex 2 to the Convention with the title *International Standards and Recommended Practices – Rules of the Air*. They became effective on 15 September 1948. On 27 November 1951, the Council adopted a complete new text of the Annex, which no longer contained Recommended Practices. The Standards of the amended Annex 2 (Amendment 1) became effective on 1 April 1952 and applicable on 1 September 1952.

[460]The Council of the International Civil Aviation Organization resolved, in adopting Annex 2 in April 1948 and Amendment 1 to the said Annex in November 1951, that the Annex constitutes Rules relating to the flight and manoeuvre of aircraft within the meaning of Article 12 of the Convention. Over the high seas, therefore, these rules apply without exception.

[461]Information relevant to the services provided to aircraft operating in accordance with both visual flight rules and instrument flight rules in the seven ATS airspace classes is contained in 2.6.1 and 2.6.3 of Annex 11. A pilot may elect to fly in accordance with instrument flight rules in visual meteorological conditions or may be required to do so by the appropriate ATS authority.

may depart from these rules in circumstances that render such departure absolutely necessary in the interests of safety.

III. Air Traffic Services

The provision of air traffic services[462] is addressed in Annex 11 to the Chicago Convention which provides *in limine* that Contracting States shall determine, in accordance with the provisions of the Annex and for the territories over which they have jurisdiction, those portions of the airspace and those aerodromes where air traffic services will be provided. They shall thereafter arrange for such services to be established and provided in accordance with the provisions of this Annex, except that, by mutual agreement, a State may delegate to another State the responsibility for establishing and providing air traffic services in flight information regions, control areas or control zones extending over the territories of the former.[463]

The Standards and Recommended Practices in Annex 11, together with the Standards in Annex 2, govern the application of the *Procedures for Air Navigation Services – Air Traffic Management*[464] and the *Regional Supplementary Procedures – Rules of the Air and Air Traffic Services*, contained in Doc 7030, Annex 11 pertains to the establishment of airspace, units and services necessary to promote a safe, orderly and expeditious flow of air traffic. A clear distinction is made between air traffic control service, flight information service and alerting service. Its purpose, together with Annex 2, is to ensure that flying on international air routes is carried out under uniform conditions designed to improve the safety and efficiency of air operation.

---

[462] According to Paragraph 2.2 of the Annex, The objectives of the air traffic services shall be to (a) prevent collisions between aircraft; (b) prevent collisions between aircraft on the manoeuvring area and obstructions on that area; (c) expedite and maintain an orderly flow of air traffic; (d) provide advice and information useful for the safe and efficient conduct of flights; (e) notify appropriate organizations regarding aircraft in need of search and rescue aid, and assist such organizations as required.

[463] Standard 2.1.1. It is also provided in the Annex that if one State delegates to another State the responsibility for the provision of air traffic services over its territory, it does so without derogation of its national sovereignty. Similarly, the providing State's responsibility is limited to technical and operational considerations and does not extend beyond those pertaining to the safety and expedition of aircraft using the concerned airspace. Furthermore, the providing State in providing air traffic services within the territory of the delegating State will do so in accordance with the requirements of the latter which is expected to establish such facilities and services for the use of the providing State as are jointly agreed to be necessary. It is further expected that the delegating State would not withdraw or modify such facilities and services without prior consultation with the providing State. Both the delegating and providing States may terminate the agreement between them at any time.

[464] Doc 4444, PANS-ATM.

F. Unmanned Aerial Vehicles

The Standards and Recommended Practices in Annex 11 apply in those parts of the airspace under the jurisdiction of a Contracting State wherein air traffic services are provided and also wherever a Contracting State accepts the responsibility of providing air traffic services over the high seas or in airspace of undetermined sovereignty. A Contracting State accepting such responsibility may apply the Standards and Recommended Practices in a manner consistent with that adopted for airspace under its jurisdiction.

Standard 2.1.2 of the Annex stipulates that those portions of the airspace over the high seas or in airspace of undetermined sovereignty where air traffic services will be provided shall be determined on the basis of regional air navigation agreements. A Contracting State having accepted the responsibility to provide air traffic services in such portions of airspace shall thereafter arrange for the services to be established and provided in accordance with the provisions of the Annex.[465] The Annex goes on to say that when it has been determined that air traffic services will be provided, the States concerned shall designate the authority[466] responsible for providing such services.[467] Situations which arise in respect of the establishment and provision of air traffic services to either part or whole of an international flight are as follows:

Situation 1: A route, or portion of a route, contained within airspace under the sovereignty of a State establishing and providing its own air traffic services.

Situation 2: A route, or portion of a route, contained within airspace under the sovereignty of a State which has, by mutual agreement, delegated to another State, responsibility for the establishment and provision of air traffic services.

Situation 3: A portion of a route contained within airspace over the high seas or in airspace of undetermined sovereignty for which a State has accepted the responsibility for the establishment and provision of air traffic services.

For the purpose of the Annex, the State which designates the authority responsible for establishing and providing the air traffic services is:

In *Situation 1*: the State having sovereignty over the relevant portion of the airspace

In *Situation 2*: the State to whom responsibility for the establishment and provision of air traffic services has been delegated

In *Situation 3*: the State which has accepted the responsibility for the establishment and provision of air traffic services

---

[465] The phrase "regional air navigation agreements" refers to the agreements approved by the Council of ICAO normally on the advice of Regional Air Navigation Meetings. The Council, when approving the Foreword to this Annex, indicated that a Contracting State accepting the responsibility for providing air traffic services over the high seas or in airspace of undetermined sovereignty may apply the Standards and Recommended Practices in a manner consistent with that adopted for airspace under its jurisdiction.

[466] The authority responsible for establishing and providing the services may be a State or a suitable Agency.

[467] Standard 2.1.3.

IV. UAVs as State Aircraft

One of the legal issues that has to be considered is that UAVs are usually State aircraft and as such might not come within the purview of the Chicago Convention in the context of regulation through an Annex to the Convention unless such an Annex were to address issues affecting civil aircraft. Article 3 (a) of the Chicago Convention provides that the Convention will be applicable only to civil aircraft and not to state aircraft. It is an inclusionary provision which identifies military, customs and police service aircraft as being included in an undisclosed list of state aircraft. The Convention contradicts itself in Article 3 (c), where it says that no state aircraft of a contracting State shall fly over the territory of another State or land thereon without authorization by special agreement or otherwise, and in accordance with the terms thereof. The question arises as to how an international treaty, which on the one hand prescribes that it applies only to civil aircraft, turns around and prescribes a rule for state aircraft. Article 3 (c) effectively precludes relief flights over the territory of a State by state aircraft if the State flown over or landed upon does not give authorization for the aircraft to do so.

The distinction between civil and state aircraft is unclear as the Chicago Convention does not go to any length in defining or specifying as to how the two categories have to be distinguished. The ICAO Assembly, at its 14th Session held in Rome from 21 August to 15 September 1962, adopted Resolution A14-25 (Coordination of Civil and Military Air Traffic) which was on the subject addressed in Article 3(d) – that the Contracting States undertake, when issuing regulations for their state aircraft, that they will have due regard to the safety of navigation of civil aircraft. In A14-25, the Assembly directed the Council to develop guidance material for the joint civil and military use of airspace, taking into account the various policies, practices and means already employed by States to promote the satisfactory coordination or integration of their civil and military air traffic services.

At its 21st Session of the Assembly, Held in Montreal from 21 September to 15 October 1974, ICAO saw the adoption of Resolution A21-21 (Consolidated Statement of Continuing Policies and Associated Practices Related Specifically to Air Navigation) where, at Appendix O, on the subject of coordination of civil and military air traffic, the Assembly resolved that the common use by civil and military aviation of airspace and of certain facilities and services shall be arranged so as to ensure safety, regularity and efficiency of international civil air traffic, and that States would ensure that procedures and regulations pertaining to their state aircraft will not adversely affect or compromise the regularity and efficiency of international civil air traffic. In order to effectively implement the proposals of the Resolution, Contracting States were requested to initiate and improve the coordination between their civil and military air traffic services and the ICAO Council was required to ensure that the matter of civil and military coordination in the use of airspace is included, when appropriate, in the agenda of divisional and regional meetings.

F. Unmanned Aerial Vehicles

At its 35th Session, held in Montreal in September/October 2004, the ICAO Assembly adopted Resolution A 35-14[468] (Consolidated statement of continuing ICAO policies and associated practices related specifically to air navigation) which, in Appendix P (Coordination of civil and military traffic) recognized that the airspace as well as many facilities and services should be used in common by both civil and military aviation. The resolution also went on to note that full integration of the control of civil and military air traffic may be regarded as the ultimate goal, and that coordination between States in achieving this goal should be the ultimate aim in resolving current difficulties. The Assembly resolved that the common use by civil and military aviation of airspace and of certain facilities and services shall be arranged so as to ensure the safety, regularity and efficiency of international air traffic and that regulations and procedures established by contracting States to govern the operation of their State aircraft over the high seas shall ensure that such operations do not compromise the safety, regularity and efficiency of international civil air traffic. The Resolution concludes that, to the extent practicable, these operations should comply with provisions of Annex 2 to the Chicago Convention on Rules of the Air.[469]

At the 36th Session of the ICAO Assembly held in Montreal from 18 to 28 September 2007, the Assembly adopted Resolution A 36-13 (Consolidated statement of continuing ICAO policies and associated practices related specifically to air navigation) which superseded Resolution A 35-14 which noted *inter alia* that the airspace as well as many facilities and services should be used in common by civil aviation and military aviation and resolved that the common use by civil and military aviation of airspace and of certain facilities and services shall be arranged so as to ensure the safety, regularity and efficiency of international traffic.[470]

With regard to Conventions other than the Chicago Convention, one can see some provisions which are relevant to the discussion on the distinction between civil and military aircraft, the latter of which, by implication, includes UAVs. The Convention on the International Recognition of Rights in Aircraft (Geneva, 1948),[471] the Convention on Offences and Certain Other Acts Committed on Board Aircraft (Tokyo, 1963),[472] the Convention for the Suppression of Unlawful Seizure of Aircraft (The Hague, 1970)[473] and the Convention for the Suppression of Unlawful Acts Against the Safety of Civil Aviation (Montreal, 1971),[474] all contain

---

[468]*Assembly Resolutions in Force (as of 8 October 2004)*, ICAO Doc 9848, ICAO Montreal, at II-2.

[469]*Assembly Resolutions in Force (as of 8 October 2004)*, ICAO Doc 9848, ICAO Montreal, at II-12.

[470]Assembly 36th Session, A 36-TE, Report of the Technical Commission (Report Folder), Resolution A 36-13, *Appendix O* at 36-19.

[471]The Convention entered into force on 17 September 1953. See http://www.mcgill.ca/files/iasl/geneva1948.pdf.

[472]Signed at Tokyo on 14 September 1963. See ICAO Doc. 8364.

[473]Signed at The Hague on 16 December 1970. See ICAO Doc. 8920.

[474]Signed at Montreal on 23 September 1971. See ICAO Doc. 8966.

a provision that "this Convention shall not apply to aircraft used in military, customs or police services." This appears to be a more simple way to indicate the scope of applicability of these Conventions than the provisions of Article 3 (a) and (b) of the Chicago Convention, although the end result seems to be the same. Furthermore, the clear implication is that all aircraft not so used would be subject to the provisions of the respective Conventions.

The Convention on Damage Caused by Foreign Aircraft to Third Parties on the Surface (Rome, 1952)[475] states in Article 26 that, "this Convention shall not apply to damage caused by military, customs or police aircraft." It should be noted that a "military, customs or police aircraft" is not necessarily the same thing as an "aircraft used in military, customs and police services" although again the expression "military, customs or police aircraft" was left undefined. Similarly, other "state" aircraft fall within the scope of the Convention. However, the 1978 Protocol to amend this Convention reverts to more familiar language; it would amend Article 26 by replacing it with, "this Convention shall not apply to damage caused by aircraft used in military, customs and police services."

The Convention for the Unification of Certain Rules Relating to the Precautionary Attachment of Aircraft (Rome, 1933) provides that certain categories of aircraft are exempt from precautionary attachment, including aircraft assigned exclusively to a government service, including postal services, but not commercial aircraft. On the other hand, the Convention for the Unification of Certain Rules Relating to Assistance and Salvage of Aircraft or by Aircraft at Sea (Brussels), 1938 "apply to government vessels and aircraft, with the exception of military, customs and police vessels or aircraft . . ."

The Convention for the Unification of Certain Rules Relating to International Carriage By Air (Warsaw, 1929)[476] applies, *inter alia* to all international carriage of persons, luggage or goods performed by aircraft for reward, regardless of the classification of the aircraft. Article 2 specifically provides that the Convention applies to carriage performed by the State or by legally constituted public bodies, but by virtue of the Additional Protocol, Parties may make a declaration at the time of ratification or accession that Article 2 (1) shall not apply to international carriage performed directly by the State. The Hague Protocol of 1955 to amend this Convention, in Article XXVI allows a State to declare that the Convention as amended by the Protocol shall not apply to the carriage of persons, cargo and baggage for its military authorities on aircraft, registered in that State, the whole capacity of which has been reserved by or on behalf of such authorities. Identical provisions are contained, *mutatis mutandis,* in the Guatemala City Protocol of 1971 (Article XXIII) the *1975* Additional Protocol No. 2 (Montreal), the *1975* Additional

---

[475] Signed at Rome on 7 October 1952. See ICAO Doc. 7364.

[476] Signed at Warsaw on 12 October 1929. The authentic French text of this Convention can be referred to in II Conférence Internationale de Droit Privé Aérien (4–12 Octobre 1929). The English translation is at the Schedule to the United Kingdom Carriage by Air Act, 1932; 22 & 23 Geo.5, Chap. 36.

F. Unmanned Aerial Vehicles

Protocol No. 3 (Montreal) and in Montreal Protocol No. 4 of 1975. It is submitted that Article 3 (b) of the Chicago Convention has no bearing on the applicability of these instruments of the "Warsaw System" which specify their own scope of applicability.

The Montreal Convention of 1999[477] which replaced the Warsaw Convention of 1929 also stipulates in its Article 1 that the Convention applies to all international carriage of persons, baggage or cargo performed by aircraft for reward. Like its predecessor, the Montreal Convention does not distinguish between civil and military or other State aircraft.

This analysis of some international air law instruments illustrates that many post-Chicago air law instruments (Geneva 1948, Tokyo 1963, The Hague 1970, Montreal 1971 and Rome 1952 and as amended in 1978) all have broadly similar provisions to Article 3 (a) and (b) of the Chicago Convention. The private air law instruments of the Warsaw System on the other hand, because of their nature, have adopted different formulae.

The provisions of the Chicago Convention and Annexes would not apply in a case where a state aircraft is (mistakenly or otherwise) operated on the basis that it is a civil aircraft. Similarly, the Geneva Convention of 1948, the Tokyo Convention of 1963, The Hague Convention of 1970, the Montreal Convention of 1971 and the Rome Convention *(1952)* as amended in 1978, will also not be applicable where it is determined that the aircraft was "used in military, customs or police services." The converse, of a civil aircraft being operated on the basis that it is a state aircraft, would theoretically raise the same problems (i.e., legal regimes thought to be inapplicable are in fact applicable). Concern is not often expressed in this regard.

Another frequently mentioned difficulty is claimed to be the loss of insurance coverage in respect of the aircraft (hull), operator, crew and passengers or other parties where the aircraft is in fact state aircraft. The question whether a particular insurance coverage is rendered invalid in such situations is primarily a private law matter of the construction and interpretation of the insurance contract. Unless the contract has an exclusion clause which specifically makes reference to the classification in Article 3 of the Chicago Convention (e.g., loss of coverage where the operation is of a state (or civil) aircraft as defined in the Chicago Convention), then the Convention will have no bearing on the contract, and this issue of the loss of insurance coverage is not germane to this study. Frequently, the policy will exclude usage of the aircraft for any purpose other than those stated in a Schedule; among the exclusions would be any use involving abnormal hazards. Nearly every aviation hull and liability policy now excludes losses due to war, invasion, hostilities, rebellion, etc., although insurance to cover such losses can usually be obtained by the payment of a higher premium. However, the instances mentioned do not require a determination of whether the aircraft is considered to be state or civil under the Chicago Convention.

---

[477]*Convention for the Unification of Certain Rules for International Carriage by Air*, signed at Montreal on 28 May 1999. ICAO Doc 9740.

A question sometimes asked is whether national civil laws and regulations would apply to civilian flight crews operating what is a state aircraft under the Chicago Convention. Would civil or military investigative and judicial processes be applied, for example, in the case of an accident? The answer would depend largely on the domestic laws of the State concerned. The fundamental principle is stated in Article 1 of the Convention: every State has complete and exclusive sovereignty over the airspace above its territory. Furthermore, subject to the provisions of the Convention, the laws and regulations of a contracting State relating to the admission to or departure from its territory of aircraft engaged in international air navigation, or the operation and navigation of such aircraft within its territory, shall be complied with by (civil) aircraft of other contracting States, upon entering or departing from or while in the territory of that State. *A fortiori,* state aircraft are also subject to the laws of the subjacent State.

In the case of an accident involving state aircraft, States are not bound by Article 26 of the Chicago Convention and Annex 13. However, they can voluntarily (through their legislation) apply' these provisions. Sometimes, the legislation specifies a different procedure in relation to military aircraft only; all other aircraft, including those used in customs or police services, are treated as civilian in this regard. In the case of other incidents, where for example the requisite over-flight permission has not been obtained by a state aircraft, which is then forced to land and charges brought against the crew, again the answer would depend on the domestic laws of the over-flown State and the factual circumstances. The classification of an aircraft as "state" aircraft under the Convention does not necessarily mean that military laws and procedures of a State would apply to that aircraft or its crew. The current or any different classification of aircraft under the Convention would not be determinative whether a particular State, in the exercise of its sovereignty, would make that aircraft and/or its crew subject to civil or military laws and regulations. As a matter of practice States usually apply military rules and processes to military aircraft and personnel only. At the international level, attempts to arrive at a common, acceptable definition of military aircraft have met with a singular lack of success.

Even though there are no international regulations applicable to UAVs, it is clear that there are certain rules that States are required to adhere to in order to ensure that UAVs operated under their control do not adversely affect civil air transport. The various provisions of the Chicago Convention and its Annexes cited in this article as well as the numerous ICAO Assembly resolutions quoted leave no room for doubt that there is an existing regime that addresses the safety of de-segregated air space when it comes to the operation of civil and State aircraft. This regime derives its legal legitimacy from the principles of State responsibility which are now accepted as binding on States. Article 1 of the *Articles of Responsibility* of the International Law Commission (ILC) expressly stipulates that every internationally wrongful act entails the international responsibility of a State.[478]

---

[478]See Crawford (2002, p. 77).

F. Unmanned Aerial Vehicles

Paul Stephen Dempsey[479] sums it up well, when he says that the issue of air traffic management has two critical considerations, one relating to legal issues and the other impacting public policy. Dempsey states correctly that the skies belong to the public and the sovereign is but the trustee in this regard. Therefore, inasmuch as States cannot abdicate or pass on their responsibility and accountability of their traditional function and fiduciary responsibility, ICAO too has responsibility under Chapter XV of the Chicago Convention to assist States needing help with regard to the provision of air navigation services.

However, the bottom line with regard to legal challenges posed by the operation of UAVs lies in the issue of liability and the responsibility of States. It is also now recognized as a principle of international law that the breach of a duty involves an obligation to make reparation appropriately and adequately. This reparation is regarded as the indispensable complement of a failure to apply a convention and is applied as an inarticulate premise that need not be stated in the breached convention itself.[480] The ICJ affirmed this principle in 1949 in the *Corfu Channel Case*[481] by holding that Albania was responsible under international law to pay compensation to the United Kingdom for not warning that Albania had laid mines in Albanian waters which caused explosions, damaging ships belonging to the United Kingdom. Since the treaty law provisions of liability and the general principles of international law as discussed complement each other in endorsing the liability of States to compensate for damage caused by space objects, there is no contention as to whether in the use of nuclear power sources in outer space, damage caused by the uses of space objects or use thereof would not go uncompensated. Furthermore, under the principles of international law, moral damages based on pain, suffering and humiliation, as well as on other considerations, are considered recoverable.[482]

The sense of international responsibility that the United Nations ascribed to itself had reached a heady stage at this point, where the role of international law in international human conduct was perceived to be primary and above the authority of States. In its Report to the General Assembly, the International Law Commission recommended a draft provision which required:

> Every State has the duty to conduct its relations with other States in accordance with international law and with the principle that the sovereignty of each State is subject to the supremacy of international law.[483]

This principle, which forms a cornerstone of international conduct by States, provides the basis for strengthening international comity and regulating the conduct of States both internally – within their territories – and externally, towards other

---

[479]See Dempsey (2003, pp. 118–119).

[480]*In Re. Chorzow Factory (Jurisdiction) Case, (1927) PCIJ, Ser. A, no. 9* at 21.

[481]*ICJ Reports (1949)*, 4 at 23.

[482]Christol (1991, p. 231).

[483]*Report of the International Law Commission to the General Assembly on the Work of the 1st Session, A/CN.4/13*, 9 June 1949, at 21.

States. States are effectively precluded by this principle of pursuing their own interests untrammelled and with disregard to principles established by international law.

These obligations are *erga omnes* affecting all States and thus cannot be made inapplicable to a State or group of States by an exclusive clause in a treaty or other document reflecting legal obligations without the consent of the international community as a whole. Besides, holding governments responsible will ensure proper quality control in the provision of air navigation services.

References

Abeyratne RIR (1992) The development of the machine readable passport and visa and the legal rights of the data subject. Annals of Air and Space Law 17(Part II):1–31
Abeyratne RIR (1998) Aviation and regulatory issues. Ashgate, London
Abeyratne RIR (1999) Mental injury caused in accidents during international air carriage – a point of view. The Aviation Quarterly 1999(4):193–205
Abeyratne RIR (2000) Mental distress in aviation claims – emergent trends. Journal of Air Law and Commerce 65(2):225–261
Abeyratne RIR (2001a) Aviation trends in the new millennium. Ashgate, London
Abeyratne RIR (2001b) The exchange of airline passenger information – issues of privacy. Communication Law 6(5):153–162
Abeyratne RIR (2002a) Attacks on America – privacy implications of heightened security measures in the United States, Europe, and Canada. Journal of Air Law and Commerce 67(1):83–115
Abeyratne RIR (2002b) Intellectual property rights and privacy issues: the aviation experience in API and biometric identification. Journal of World Intellectual Property 5(4):631–650
Abeyratne RIR (2003) Profiling of passengers at airports – imperatives and discretions. European Transport Law 38(3):297–311
Abeyratne RIR (2006) Theoretical justification for modernizing the Rome Convention of 1952. Annals of Air and Space Law 31:185–212
Becker T (2006) Terrorism and the state: rethinking the rules of state responsibility. Hart Monographs in Transnational and International Law. Hart, Portland, OR
Brownlie I (1990) Principles of public international law, 4th edn. Oxford University Press, Oxford
Cheng B (1962) The law of international air transport. Oceania, London
Christol CQ (1991) Space law past, present and future. Kluwer Law and Taxation, Deventer
Crawford J (2002) The International Law Commission's Articles on State Responsibility: Introduction, Text and Commentaries. Cambridge University press, Cambridge
Dempsey PS (2003) Privatization of the air: governmental liability for privatized air traffic services. Annals of Air and Space Law 27:118–119
Heitmeyer R (2000) Biometric ID and airport facilitation. Airport World (ACI) 5(1):18–20
Honig JP (1956) The legal status of aircraft. Martinus Nijhoff, The Hague
Kaiser SA (1995) Infrastructure, airspace and automation – air navigation issues for the twenty-first century. Annals of Air and Space Law 20(Part I):447
Lazareff F (1971) Status of military forces under current international law. Sijthoff, Leiden
Mankiewicz RH (1979) The application of Article 17 of the Warsaw Convention to mental suffering not related to physical injury. Annals of Air and Space Law 4:187–211
McMunn MK (1996) Aviation security and facilitation programmes are distinct but closely intertwined. ICAO Journal 51:9

References

Morganthau H, Thompson KW (1950) Principles and problems of international politics. Knopf, New York

Petras CM (2007) An alternative proposal to modernize the liability regime for surface damage caused by aircraft to address damage resulting from hijackings or other unlawful interference. Gonzaga Journal of International Law 10:316. http://www.gonzagajil.org/content/view/164/1/

Shaw MN (2003) International law, 5th edn. Cambridge University Press, Cambridge

Tompkins GN Jr (2009) Who bears the costs of terrorism? Allocating the risk under draft ICAO Unlawful Interference Compensation Convention. Royal Aeronautical Society, 3502679.7

Turack DC (1972) The passport in international law. DC Heath & Co., Lexington, MA

Weishaupt G (ed) (1979) Selected international agreements relating to air law. Butterworths, London, Publications for the Association for the Development of the Academic Institute for Air Transport Education and Research

Woodliffe J (1992) The peacetime use of foreign military installations under modern international law. Martinus Nijhoff, Dordrecht

Chapter 4
Narco-terrorism

A. Introduction

The trafficking and illicit use of drugs has been a serious problem globally over the years, affecting peoples' lives and the aviation industry has not escaped the scourge. The illegal carriage by air of narcotics and other psychotropic substances and its various corollaries of violence or Narco-terrorism as it is popularly known, has shown that the activity can cause various other forms of unlawful interference with civil aviation such as the unlawful seizure of aircraft and the causing of damage to persons and property related to international civil aviation. Furthermore, in addition to the disturbing problem of drug use by air crews, the threat of drug trafficking has already had a tremendous adverse impact on aviation and the financial community which supports it Narco-terrorism is considered an offence on two grounds: the fact that the illicit trafficking of drugs is an offence against public health; and the illicit carriage by air of these substances threatens to sabotage the legitimate carriage by air of passengers and freight. Narco-terrorism involves two facets: the transportation of drugs and narcotics by aircraft and across national boundaries by air; and the act of loading and unloading them at aerodromes and airports. The two acts are claimed to be integrally linked to one another in that the essential elements of the unlawful act, i.e., "transport by air" and "trafficking" are inseparable.

The problem has blown into unmanageable proportions owing to the rapid proliferation of air travel as a means of communication. It is not difficult therefore to figure out the tremendous encouragement given to the drug trafficking trade by the numerous aircraft movements that keep increasing exponentially. The Annual Report of the ICAO Council for the year 2007 has recorded that the total scheduled traffic carried by the airlines of the 190 member States amounted to approximately 2,260 million passengers and some 41 million tonnes of freight. They were carried in 29.3 million scheduled flights in 2007 (of which 16.5 million were international flights and the rest were domestic). 3,730 airports served these flights.

Throughout known history, human society has used substances to alter moods and alleviate physical and mental suffering. These substances, although proving indispensable for the policy of pain and suffering, also proved to be addictive and destructive when misused or abused. As a result, early society made rules which allocate the use of these substances for medicinal or religious purposes and entrusted them to priests, leaders and doctors.[484]

The abuse of drugs has been proliferating as a corrosion of social intercourse from the mid-nineteenth century due to the increased availability of products, the expansion of connections, the necessities brought about by changing socio-economic factors, rapid urbanization and changes in attitudes and values. These factors have contributed to a rapid increase in criminality in human society, leading to the exploitation of human society by insidious criminal elements.

As a response to this problem, global control mechanisms have been introduced by the international community – one of which is a regulatory system for the control of illicit transport of narcotic drugs by air. Not only does illicit transportation of narcotics by air *per se* constitute an offence, but it also leads to other criminal acts related to terrorism such as the destruction of airports by those involved in the carriage of narcotic drugs by air, destruction of property and aircraft resulting from attempts to transport narcotic drugs and the threats posed to traffic installations.

The need for an international drug control system was first felt in 1909 when representatives of 13 States met in Shanghai to discuss the proliferation of instances relating to the transportation of narcotics for non-medical consumption. Furthermore, the conference was considered necessary, as by the end of the nineteenth century, opium smoking had become rampant in China, affecting a third of the adult male population. As early as 1729, Emperor Yong Cheng forbade opium smoking in China, which resulted in a decrease in trafficking. However, this attempt was to little avail, as opium was being smuggled at that time to China through India by the Portuguese and later by the English. The amount of opium that had been smuggled into China had increased from 13 metric tons in 1729 to 64 metric tons in 1767. During the decade 1820–1830 the quantity of drugs brought into China had taken an upward turn to 2,500 metric tons. As a result, China had 20 million opium smokers in 1838. By 1773 the East India Company of England had established a monopoly in the drug trade, thus inaugurating the first recorded enterprise involved in legal drug trafficking on a large scale.[485]

Today, a wide variety of illicit drugs are traded around the world, the most prolific being cannabis, which involves an estimated 160 million people. Cannabis production was estimated at 47,000 metric tons in the year 2006. The most important supplier to the European hashish market is Morocco, which delivers 80% of the supply. In addition, Amphetamine type drugs are being abused by 26 million people. It is estimated that the global production of amphetamine and methamphetamine is 300 metric tons. The production of Opium occurs mainly in Afghanistan (notably in all its 34 provinces) and opiate abusers number 16 million,

---

[484]Report of the International Conference on Day Abuse and Illicit Trafficking, Vienna, 17–26 June 1987. United Nations: New York 1987 at 7.
[485]Bell (1991, p. 2).

and 10 million of them are abusing heroin. The sophistication of heroin production in Europe has grown, making the product purer and the supply richer and more prolific. It is encouraging that Southeast Asia has, however, reduced its production by 78% since 1996. The United Nations Office of Drug Control opines, in a somewhat conservative manner that it is not unrealistic to believe that Southeast Asia may become virtually free of opium production in a few years. This is heartening in the context that the abuse of cocaine involves 13 million people, with a concentration in North America and Western Europe.

It is an understatement to say that the problem of illicit carriage of narcotics is acute today. The illicit carriage of drugs and psychotropic substances, which is a crime *per se*, invariably gives rise to Narco-terrorism.[486] The Oxford English Dictionary has the overarching definition of Narco-terrorism as *Terrorism associated with the trade in illicit drugs*. Criminals who perpetrate offences in dealing in drugs and associated crimes often copy methods from political assailants to influence the politics of a country by causing terror and obstructing justice. The problem is further compounded by the fact that, in the modern world of terrorist warfare and civil war, terrorist organizations have taken to the illegal drug trade as a source of income funding their ideology driven organizations. Over the years, several definitions of "Narco-terrorism" have been used.

Illegal returns from the illicit trade in drugs is phenomenal. One estimate claims the FARC cartel of Colombia has a net profit from drug related crime (including the "taxation" and "protection" of the illegal cocaine trade) of at least 300 million USD every year.[487] The United Nation has estimated that the annual total income from the drug trade for movements such as al-Qaeda is 2.4 billion US dollars. It is also claimed that 12 of the 28 organizations, have been listed as terrorist organizations by the U.S. State Department, are stated to be involved in the illegal drug trade, ranging from Sendero Luminoso of Peru to the "Tamil Tigers" of Sri Lanka.[488]

B. United Nations Initiatives

In its resolution 40/122 dated 13 December 1985, related to the International Conference on Drug Abuse and Illicit Trafficking, the General Assembly expressed its grave concern and that of nations of the world regarding the awesome and

---

[486]Narco-terrorism is a term coined by former President Fernando Belaunde Terry of Peru in 1983. President Terry considered Narco-terrorism to be attacks of a terrorist nature against his nation's anti-narcotics police. In the original context, Narco-terrorism is understood to mean the attempts of narcotics traffickers to influence the policies of a government or a society through violence and intimidation, and to hinder the enforcement of the law and the administration of justice by the systematic threat or use of such violence.

[487]Hartelius (2008, p. III).

[488]Hartelius (2008, p. III).

vicious effects of drug abuse and illicit trafficking, which threaten the stability of nations and the well-being of mankind and which therefore constitute a grave threat to the security and development of many countries, focused on the dangers posed for producer, consumer and transit countries alike by the illegal cultivation, production and manufacture of and demand for drugs and by their illicit traffic.

Recalling its earlier resolutions and relevant resolutions and decisions of the Economic and Social Council and the Commission on Narcotic Drugs in the international campaign against traffic in and abuse of narcotic drugs and psychotropic substances, the Assembly recognized special responsibilities of the United Nations and the international community to seek viable solutions to the growing scourge of drug abuse and illicit trafficking. It also noted the work of the Commission on Narcotic Drugs towards the preparation of a draft convention against illicit traffic in narcotic drugs and psychotropic substances,

The Assembly also noted with appreciation the statement made by the Secretary-General before the Economic and Social Council on 24 May 1985,[489] referred to in Council decision 1985/131 of 28 May 1985, which drew attention to the gravity, magnitude and complexities of the international drug problem and in response proposed a world-wide conference at the ministerial level in 1987 to consider all aspects of the problem. Recognizing that the interregional meeting of heads of national drug law enforcement agencies, which be convened at Vienna in 1986, could make a significant contribution to the deliberations of the conference at the ministerial level proposed by the Secretary-General, and taking into account the various reviews of the activities of the United Nations agencies in the narcotics field that have already been undertaken and noting with satisfaction the Secretary-General's designation of the Under-Secretary-General for Political and General Assembly Affairs as the overall co-ordinator of all United Nations activities related to drug control, the Assembly strongly urged[490] all States to summon the utmost political will to combat drug abuse and illicit trafficking by generating increased political, cultural and social awareness, it called upon the United Nations, the specialized agencies and other organizations of the United Nations system to give the highest attention and priority possible to international measures to combat illicit production of, trafficking in and demand for drugs.

Further, the Assembly called upon all States that have not already done so to become parties to the Single Convention of Narcotic Drugs of 1961[491] and the 1972 Protocol Amending the Single Convention on Narcotic Drugs of 1961[492] and to the Convention on Psychotropic Substances of 1971, and, in the meantime, to make serious efforts to comply with the provisions of these instruments. The General Assembly also took the step of addressing strategy and policies for drug control in its Resolution 40/129. The Assembly recalled *in limine* its resolution 32/124 of

---

[489] A/C.3/40/8, Annex.

[490] A/C.3/40/8.

[491] United Nations, Treaty Series, vol. 520, No. 7515, p. 204.

[492] United Nations, Treaty Series, vol. 976, No. 14151, p. 4.

B. United Nations Initiatives

16 December 1977, in which it requested the Commission on Narcotic Drugs to study the possibility of launching a meaningful programme of international drug abuse control strategy and policies.

It also recalled its resolution 36/168 of 16 December 1981, by which it adopted the International Drug Abuse Control Strategy and the basic five-year programme of action[493] proposed by the Commission on Narcotic Drugs in its resolution 1 (XXIX) of 11 February 1981[494] as well as its resolution 38/98 of 16 December 1983, in which it decided that, beginning with its eighth special session, the Commission on Narcotic Drugs, meeting in plenary during its sessions and in the presence of all interested observers, would constitute the task force envisaged in General Assembly Resolution 36/168 to review, monitor and co-ordinate the implementation of the International Drug Abuse Control Strategy and the basic five-year programme of action.

Taking note of resolution 2 (XXXI) of 21 February 1985 of the Commission on Narcotic Drugs[495] and Economic and Social Council decision 1985/130 of 28 may 1985 the Assembly approved the programme of action for 1986, the fifth year of the United Nations basic five-year programme of the International Drug Abuse Control Strategy, reviewed by the Commission on Narcotic Drugs at its 31st session.[496]

In a separate exercise, the Assembly taking into consideration that in response to the threat posed by the drug problem the international community had adopted numerous declarations and initiatives, interregional and regional, multilateral and bilateral, in order to condemn and combat the problem and to achieve its total eradication,[497] commended the valuable contributions of the Secretary-General of

---

[493] *Official Records of the Economic and Social Council*, 1981, Supplement No. 4 (E/1981/24), annex II.

[494] See *Official Records of the Economic and Social Council*, 1981, Supplement No. 4 (E/1981/24), Chap. XI, sect. A.

[495] See *Official Records of the Economic and Social Council*, 1985, Supplement No. 3 (E/1985/23 and Corr. 1), Chap. IX, sect. A.

[496] See A/40/773, annex.

[497] As illustrated, by, *inter alia*, the following meetings and initiatives:

The Inter-American Programme of Action against the Illicit Use and Production of Narcotic Drugs and Psychotropic Substances and Traffic Therein, adopted by the Inter-American Specialized Conference on Traffic in Narcotic Drugs, held at Rio de Janeiro, Brazil, from 22 to 26 April 1986;

The Tokyo Declaration entitled "Looking forward to a better future," issued at the Tokyo Economic Summit, held from 4 to 6 May 1986 (see A/41/354, annex I, para. 5);

The 19th Ministerial Meeting of the Association of South-East Asian Nations, held at Manila on 23 and 24 June 1986;

The recommendations of the first Interregional Meeting of Heads of National Drug Law Enforcement Agencies, held at Vienna from 28 July to 1 August 1986 (see A/41/559, para 10);

The Economic Declaration of the Eighth Conference of Heads of State or Government of Non-Aligned Countries, held at Harare from 1 to 6 September 1986;

The Puerto Vallarta Declaration, adopted at the Regional Meeting of Ministers of Justice and Attorneys-General, held at Puerto Vallarta, Mexico, from 8 to 10 October 1986 (A/C.3/41/5, annex);

the United Nations and the Secretary-General of the International Conference on Drug Abuse and Illicit Trafficking to the preparatory work for the Conference and noting the continuing efforts of the Commission on Narcotic Drugs, the United Nations Fund for Drug Abuse Control, the International Narcotics Control Board, the Division of Narcotic Drugs of the Secretariat, intergovernmental and non-governmental organizations and the regional commissions in this regard.

The Assembly also welcomed the commitment of the Secretary-General to cover the cost of holding the Conference through absorption within the regular budget for the biennium 1986–1987, without prejudice to ongoing initiatives, programmes and work of the United Nations in the field of drug control.

In December 1988, the General Assembly, as a further measure towards developing its preparation of a draft convention against illicit traffic in narcotics drugs and psychotropic substances considered the report of the Secretary-General[498] on the progress achieved in the preparation of the draft convention, and welcomed with appreciation Economic and Social Council resolution 1988/8, in which it decided to convene the Review Group in mid-June 1988 at Vienna, with the mandate of continuing the preparation of the draft convention and preparing the organizational aspects of the Conference of plenipotentiaries for the adoption of a convention against illicit traffic in narcotic drugs and psychotropic substances. It took note with satisfaction of the report of the Secretary-General, and the report of the Commission on Narcotic Drugs on its tenth special session[499] and the recommendations therein, approved by the Economic and Social Council in its resolution 1988/8 and decisions 1988/118 and 1988/120 of 25 May 1988 and 1988/159 of 26 July 1988, in which it decided, *inter alia*, to convene the Conference of plenipotentiaries to adopt the convention, and to extend to ten working days the 33rd session of the Commission on Narcotic Drugs in order to allow it to consider suitable measures to be taken prior to the entry into force of the convention; and requested the Commission on Narcotic Drugs, as the principal United Nations policy-making body on drug abuse control, to identify suitable measures to be taken prior to the entry into force of the convention. The Assembly also urged all States to adopt a constructive approach with a view to resolving any outstanding differences over the text of the convention; and requested all States, while reaffirming their commitment to the Declaration of the International Conference on Drug Abuse and Illicit Trafficking[500] as an expression of the political will of nations to combat the drug problem, to assign the highest

---

The Meeting of Ministers of Interior and Justice of the 12 member States of the European Community, held in London on 20 October 1986;

The recommendations of the Interregional Conference on the Involvement of Non-Governmental Organizations in Prevention and Reduction of the Demand for Drugs, held at Stockholm from 15 to 19 September 1986 (A/C.3/41/7, annex, paragraph 84).

[498] A/43/678.

[499] See Official Records of the Economic and Social Council, 1987, Supplement No. 3 (E/1988/13).

[500] *Report of the International Conference on Drug Abuse and Illicit Trafficking*, Vienna, 17–26 June 1987 (United Nations publication, Sales No. E.87.I.18), Chap. I, Sect. B.

B. United Nations Initiatives

priority to the Conference of plenipotentiaries and to participate actively in it, at the highest possible level, for the adoption of the Convention.

Whilst expressing its appreciation to the Secretary-General, the Commission on Narcotic Drugs and all related organs established by the Commission, for their effectiveness in responding to its request to prepare the draft Convention, the Assembly once again urged all States that had not yet done so to ratify or to accede to the Single Convention on Narcotic Drugs of 1961, as amended by the 1972 Protocol Amending the Single Convention on Narcotic Drugs of 1961, and the Convention on Psychotropic Substances of 1971; and requested the Secretary-General to report to the General Assembly at its 44th session on the implementation of the present resolution, particularly on the conclusions of the Conference of plenipotentiaries for the adoption of a convention against illicit traffic in narcotic drugs and psychotropic substances.

At its 43rd Plenary Meeting, in November 1989, the United Nations summoned a special session of the General Assembly to consider the question of international co-operation against illicit production, supply, demand, trafficking and distribution of narcotic drugs, with a view to expanding the scope and increasing the effectiveness of such co-operation. At this session, the General Assembly reiterated its concern about the serious problem of the illicit production, supply, demand, trafficking and distribution of narcotic drugs and about the devastating effect of drug abuse on individuals and society. Noting statements delivered before the Assembly in plenary meeting during its 44th session, including the address given by the President of the Republic of Colombia on 29 September 1989[501] and, in particular, his call for a special session of the General Assembly, the Assembly decided to hold a special session, at a high political level, to consider as a matter of urgency the question of international co-operation against illicit production, supply, demand, trafficking and distribution of narcotic drugs, with a view to expanding the scope and increasing the effectiveness of such co-operation. It requested the Secretary-General to make the necessary administrative arrangements for the convening of the special session.

In December 1989 at its 82nd Plenary meeting, the General Assembly, on the subject of implementation of the United Nations Convention against Illicit traffic in Narcotic Drugs and Psychotropic Substances, expressed its appreciation to the Secretary-General for the report on the conclusions of the conference of plenipotentiaries that adopted the United Nations Convention against Illicit Traffic in Narcotic Drugs and Psychotropic Substances at Vienna. It also expressed its appreciation to the States that participated in the preparation and adoption of the Convention; and urged States that have not yet done so to proceed rapidly to sign and to ratify the Convention, so that it may enter into force as early as possible. The Assembly also urged States to establish the necessary legislative and administrative measures to that their internal juridical regulations may be compatible with the

---

[501] See *Official Records of the General Assembly*, 44th Session, Plenary Meetings, 13th meeting (A/44/PV.13).

spirit and scope of the Convention; and invited States, to the extent that they are able to do so, to apply provisionally the measures set forth in the Convention, pending its entry into force for each of them.

The Secretary-General was requested to modify the section of the annual reports questionnaire regarding the implementation of international treaties to that the Commission on Narcotic Drugs, at its regular and special sessions, may review that steps that States have taken to ratify, accept, approve or formally confirm the Convention. The Assembly invited the Commission on Narcotic Drugs, as the principal United Nations policy-making body on the subject, to identify suitable measures to be taken prior to the entry into force of the Convention. It also requested the Secretary-General to assign the appropriate priority to providing the Division of Narcotic Drugs of the Secretariat ant he secretariat of the International Narcotics Control Board with the necessary financial, technical and human resources that would enable them to carry out the additional responsibilities under the Convention for the biennium 1990–1991.

The Secretary-General was further urged to provide assistance to States, at their request, to enable them to establish the legislative and administrative measures necessary for the implementation of the Convention. All States that had not yet done so were requested to ratify or to accede to the Single Convention on Narcotic Drugs of 1961, as amended by the 1972 Protocol Amending the Single Convention on Narcotic Drugs of 1961, and the Convention on Psychotropic Substances of 1971.

The Assembly also requested the Secretary-General, within existing resources and drawing, in particular, on funds available to the Department of Public Information of the Secretariat, to provide for, facilitate and encourage public information activities relating to the Convention and also to disseminate the text of the Convention in the official languages of the United Nations.

In December 1989, the Assembly considered measures to prevent international terrorism which endangers or takes innocent human lives or jeopardizes fundamental freedoms and study of the underlying causes of those forms of terrorism and acts of violence which lie in misery, frustration, grievance and despair and which cause some people to sacrifice human lives, including their own, in an attempt to effect radical changes.

The General Assembly expressed the view that it was convinced that a policy of firmness and effective measures should be taken in accordance with international law in order that all acts, methods and practices of international terrorism may be brought to an end. In this context, the Assembly noted the ongoing work within ICAO regarding research as to the detection of plastic or sheet explosives and the devising of an international regime for the marking of such explosives for the purposes of detection, and taking note of Security Council Resolution 635 (1989) of 14 June 1989 relating thereto, expressed its confidence that the ICAO work would contribute significantly towards curbing acts of unlawful interference with civil aviation.

The Assembly also expressed deep concern at the world-wide persistence of acts of international terrorism in all its forms, including those in which States are directly or indirectly involved, which endanger or take innocent lives, have a deleterious effect on international relations and may jeopardize the territorial integrity and security of States. The Assembly called attention to the growing connection between terrorist groups and drug traffickers. Its was also convinced of the importance of the observance by States of their obligations under the relevant international conventions to ensure that appropriate law-enforcement measures are taken in connection with the offenses addressed in those conventions, and the importance of expanding and improving international co-operation among Stats, on a bilateral, regional and multilateral basis.

The Assembly was convinced further that international co-operation in combating and preventing terrorism will contribute to the strengthening of confidence among States, reduce tensions and create a better climate among them, and was mindful of the need to enhance the role of the United Nations and the relevant specialized agencies in combating international terrorism.

The necessity of maintaining and protecting the basic rights of, and guarantees for, the individual in accordance with the relevant international human rights instruments and generally accepted international standards was recognized to the extent that the Assembly reaffirmed the principle of self-determination of peoples as enshrined in the Charter of the United Nations.

The Assembly noted the efforts and important achievements of the International Civil Aviation Organization and the International Maritime Organization in promoting the security of international air and sea transport against acts of terrorism, and recognized that the effectiveness of the struggle against terrorism could be enhanced by the establishment of a generally agreed definition of international terrorism. The Assembly, while condemning unequivocally as criminal and unjustifiable, all acts, methods and practices of terrorism wherever and by whomever committed, including those which jeopardize friendly relations among States and their security, deeply deplored the loss of human lives which results from such acts of terrorism, as well as the pernicious impact of these acts on relations of co-operation among States. It called upon all States to fulfil their obligations under international law to refrain from organizing, instigating, assisting or participating in terrorist acts in other States, or acquiescing in or encouraging activities within their territory directed towards the commission of such acts. States were also urged to fulfil their obligations under international law and take effective and resolute measures for the speedy and final elimination of international terrorism and to that end, in particular, to prevent the preparation and organization in their respective territories, for commission within or outside their territories, of terrorist and subversive acts directed against other States and their citizens.

The Assembly also called upon States to ensure the apprehension and prosecution or extradition of perpetrators of terrorist acts; and to endeavour to conclude special agreements to that effect on a bilateral, regional and multilateral basis.

Co-operation of States with one another in exchanging relevant information concerning the prevention and combating of terrorism was considered vital, together with the adoption of steps necessary to implement the existing international conventions on this subject to which they are parties, including the harmonization of their domestic legislation with those conventions. The Assembly appealed to all States that have not yet done so to consider becoming party to the international conventions relating to various aspects of international terrorism referred to in the preamble to the present resolution; and urged all States, unilaterally and in co-operation with other States, as well as relevant United Nations organs, to contribute to the progressive elimination of the causes underlying international terrorism and to pay special attention to all situations, including colonialism, racism and situations involving mass and flagrant violation of human rights and fundamental freedoms and those involving alien domination and foreign occupation, that may give rise to international terrorism and may endanger international peace and security. Finally, it firmly called for the immediate and safe release of hostages and abducted persons, wherever and by whomever they are being held.

I. The United Nations Convention Against Illicit Traffic in Narcotic Drugs and Psychotropic Substances

On 19 December 1988 the United Nations adopted its Convention Against Illicit Traffic in Narcotic Drugs and Psychotropic Substances. The Convention brought to bear the deep concern of the United Nations regarding the magnitude of and rising trend in the illicit production of, demand for and traffic in narcotic drugs and psychotropic substances, which pose a serious threat to the health and welfare of human beings and adversely affect the economic, cultural and political foundations of society. Concern was also raised at the steadily increasing inroads into various social groups made by illicit traffic in narcotic drugs and psychotropic substances, and particularly by the fact that children are used in many parts of the world as an illicit drug consumers market and for purposes of illicit production, distribution and trade in narcotic drugs and psychotropic substances, which entails a danger of incalculable gravity.

The Convention recognized the links between illicit traffic and other related organized criminal activities which undermine the legitimate economies and threaten the stability, security and sovereignty of States and that illicit traffic is an international criminal activity, the suppression of which demands urgent attention and the highest priority.

The Convention also provides that subject to its constitutional principles and the basic concepts of its legal system, the acquisition, possession or use of property, knowing, at the time of receipt, that such property was derived from an offence or offences, the possession of equipment or materials or substances listed in the Convention, knowing that they are being or are to be used in or for the illicit cultivation, production or manufacture of narcotic drugs or psychotropic substances, are offences. Also, those who publicly incite or induce others, by any means, to commit any of the offences established in accordance with the Convention or use narcotic drugs or psychotropic substances illicitly and participate in, associate or conspire to commit, attempts to commit and aid, abet, facilitate and counsel the commission of any of the offences established in accordance with its definitive provisions shall be guilty of an offence under the Convention.

B. United Nations Initiatives 187

Subject to its constitutional principles and the basic concepts of its legal system, each Party is required to adopt such measures as may be necessary to establish as a criminal offence under its domestic law, when committed intentionally, the possession, purchase or cultivation of narcotic drugs or psychotropic substances for personal consumption. Knowledge, intent or purpose required as an element of an offence set forth in paragraph 1 of this article may be inferred from objective factual circumstances.

II. Some Recent Efforts of the United Nations

One of the significant steps in the international campaign against traffic in drugs of the General Assembly was its pronouncement in December 1994 where, at its 101st Plenary meeting, the Assembly recalled *inter alia,* its Resolutions 36/132 of 14 December 1981 and 38/93 of 16 December 1983, in which it specifically acknowledged the economic and technical constraints impeding many developing countries from combating the illegal production of and illicit traffic in drugs and drug abuse. The Assembly noted the concern expressed by the Secretary-General in his report on the work of the Organization,[502] in which he recognized the need for greater efforts to reduce the traffic in and illicit use of drugs. The Assembly also considered the activities of the Commission on Narcotic Drugs and the International Narcotic Control Board, and appreciated the action being taken by the United Nations Fund for Drug Abuse Control in providing financial resources and support or integrated development programmes, including the replacement of illicit crops in affected areas. It reaffirmed the need to improve and maintain regional and interregional co-operation and co-ordination, particularly in law enforcement, in order to eliminate drug trafficking and drug abuse, and noted the growing interest in regional and interregional co-ordination.

Concern was expressed that, despite the significant national efforts deployed for this purpose, including those of a number of Latin American and Caribbean and Asian countries, the illicit traffic in narcotic drugs and psychotropic substances had increased noticeably. The Assembly was aware of the serious impact on the life and health of peoples and on the stability of democratic institutions resulting from the illicit production, marketing, distribution and use of drugs, and recognized that, to root out this evil, integrated action was required for simultaneously reducing and controlling illicit demand production, distribution and marketing.

It was considered that action to eliminate the illegal cultivation of and traffic in drugs must be accompanied by economic and social development programmes for the affected areas, programming activities for replacing illegal crops in such a manner as to conserve the environment and improve the quality of life of he social sectors concerned.

---

[502]Official Records of the General Assembly, 39th Session, Supplement No. 1 (A/39/1).

The Assembly recognized the dilemma of transit States which are seriously affected, both domestically and internationally, by drug trafficking, which was stimulated by demand for and production and use of illicit drugs and psychotropic substances in other countries. The Assembly expressed its awareness of the need to mobilize a co-ordinated strategy at the national, regional, and international levels, which would cover countries with illegal users and producers and countries used for transit in the world-wide distribution and marketing circuit, in order to eliminate drug trafficking and drug abuse.

In December 1993, at its 85th Plenary Meeting, the Assembly presented a global programme of action on international action to combat drug abuse and illicit production and trafficking. The General Assembly commenced its presentation by reiterating its grave concern that the illicit demand for, production of and traffic in narcotic drugs and psychotropic substances continue to threaten seriously the socio-economic and political systems and the stability, national security and sovereignty of an increasing number of States.

The Assembly was fully aware that the international community was confronted with the dramatic problem of drug abuse and the illicit cultivation, production, demand, processing, distribution and trafficking of narcotic drugs and psychotropic substances and that States needed to work at the international and national levels to deal with this scourge, which has a strong potential to undermine development, economic and political stability and democratic institutions. The Assembly emphasized that the problem of drug abuse and illicit trafficking has to be considered within the broader economic and social context, and also the need for an analysis of transit routes used by drug traffickers, which are constantly changing and expanding to include a growing number of countries and regions in all parts of the world. It was alarmed by the growing connection between drug trafficking and terrorism in various parts of the world, and recognized the efforts of countries that produce narcotic drugs for scientific, medicinal and therapeutic uses to prevent the channelling of such substances to illicit markets and to maintain production at a level consistent with illicit demand. It stressed the important role of the United Nations and its specialized agencies in supporting concerted action in the fight against drug abuse at the national, regional and international levels, and recognized the role of the Commission on Narcotic Drugs as the principal United Nations policy-making body on drug control issues, as a positive one.

Reaffirming the importance of the role of the United Nations International Drug Control Programme as the main focus for concerted international action for drug abuse control and commending its performance of the functions entrusted to it, the Assembly affirmed the importance of the role of the United Nations International Drug Control Programme as the main focus for concerted international action for drug abuse control and commended its performance of the functions entrusted to it. It also affirmed the proposals set out in the United Nations System-Wide Action Plan on Drug Abuse Control and recognizing that further efforts are needed to implement and update it, and invited the relevant agencies of the United Nations system to make greater progress in incorporating within their programmes and activities action aimed at dealing with drug-related problems.

C. ICAO Initiatives

The Air Navigation Commission at the 11th Meeting of its 24th Session on 25 February 1957 considered a request of ICAO of inviting comments on the carriage of opiates and derivatives in first-aid kits on board aircraft on international flights.[503] Special reference in this letter had been made on the following points:

(a) Is it believed necessary to carry opiates or drugs containing opiates and their respective derivatives in the first-aid kits of aircraft for use in case of emergency or in the kits carried by airlines for the relief of passengers suffering from certain diseases?
(b) Do the regulations in a country prohibit the carriage of drugs containing opiates or their derivatives in limited quantities in first-aid kits on board aircraft on international flights? If so, under what safeguards would you allow such drugs to be carried?
(c) Do the regulations of that country prohibit a qualified crew member from administering subcutaneous or intravenous injections in case of emergency on international flights?

Substantive replies were received from 28 Contracting States, Belgian Congo and Netherlands Antilles. In addition three States (Burma, Guatemala, Mexico) acknowledged receipt or had no comment to offer on the State Letter. Comments were also received from the United Nations European Office, Division on Narcotic Drugs, IFALPA, the Aero Medical Association and the ATA (through the United States government).

I. Basic Principles of Aeronautics on International Narcotic Control

The Commission noted in 1957 that any use or carriage of narcotics was subject to control under international laws on narcotic drugs. The ECOSOC Division of Narcotic Drugs, referring to the ICAO State Letter, had indicated that any discussions ICAO has on the subject or any decisions or regulations they adopt would be of great importance for the work of its Commission on Narcotic Drugs in connection with their preparation of a new Single Convention intended to codify all international treaties relating to narcotic drugs.

The Commission also noted that the international regulations on narcotic drugs consist of a complex system of several Conventions under the supervision of ECOSOC and its Committee on Narcotic Drugs. They were directed at establishing international control of all drugs causing addiction and the application of certain

---

[503] See C-WP/2372, 7/3/57 at p. 1.

principles as regards the manufacture, prescription, sale and traffic of narcotic drugs. These had generally been introduced into national regulations to prevent any abuse and illicit traffic. Here was nevertheless some lack of uniformity in the detailed application as not all States are parties to all Conventions. It was believed that the codification of all existing treaties into a single Convention, planned by ECOSOC, in the near future, would promote uniformity. In principle, there was no intent in the international regulations to restrict the legitimate use of opiates or narcotic drugs for medical purposes and in case of emergencies. A number of States had adopted specific legal provisions for the carriage of such drugs in aircraft first-aid kits. Other States believed that effective control of their contents was difficult so as to prevent any possibility of abuse.

The Commission was of the view that opiates or other restricted substances carried on board were generally subject to control and the laws applicable in the State of Registry. For instance, the supplies could only be obtained for legitimate, medical or scientific purposes in limited quantities under control, by medical prescription and from an authorized source. A record had to be kept and any use and replacement of such drugs must be accounted for. It is therefore desirable that first-aid kits be protected against misuse by being kept under lock and key or sealed. Preferably opiates or similar drugs might be kept in a sealed container within the first-aid kit. It was also necessary that a record be kept of the quantity, name of product and manufacturer, date of prescription, dispensing agency and signature of person responsible for the control and its use.

One of the significant achievements of the Commission was its conclusion that the terms "opiates and derivatives," "narcotics" and "narcotic drug" should be considered to include opium, coca leaves, their alkaloids and preparations or derivatives therefrom, whether prepared from substances of vegetable origin or by means of synthesis, or their combinations, and other synthetic drugs liable to produce addiction and controlled by international narcotic regulation.

The Commission concurred with the conclusion that the difficulties reported by some countries might have been caused by a lack of uniformity of national laws and practices developed under a complex system of international narcotic Conventions and the detailed conditions under which opiates and similar drugs had been carried on board. It appeared possible to eliminate existing differences – and potential difficulties – if certain principles for effective control and safeguards against abuse, such as quantity limitations, could be more generally agreed upon by all Contracting States.

As recommended by the Air Navigation Commission in its 545th Report,[504] the Council decided:

(a) That States should be informed of the Commission's study on the carriage of opiates and derivatives in first-aid kits on board aircraft on international flights and of its conclusion that the Recommendation of Annex 6 continues to be

---

[504] See C-WP/2372.

C. ICAO Initiatives

satisfactory and that the carriage of opiates and derivatives in aircraft first-aid kits is considered desirable; and

(b) That States should be invited to note the implication in this Recommendation that foreign carriers should be permitted to carry first-aid kits with contents as described in Annex 6 on international flights, under control, according to international narcotic laws and subject to satisfactory safeguards against abuse, and to notify ICAO when they do not permit this.[505]

It agreed that ECOSOC and WHO should be invited to study the related medical and legal problems, in particular the application of efficient safeguards against abuse and of uniform principles under which opiates or other drugs might be used and carried in first-aid kits on board aircraft, in an effort to promote uniformity under existing laws and to avoid difficulties. It was also agreed that such studies should take into account any factors affecting international civil aviation, such as the safety of persons on board and relief in the case of emergencies in flight or of aircraft accidents, and the ICAO Secretariat should call the attention of the bodies making the studies to these factors.

At its 30th Session in April 1957, Council considered further the question of the carriage of opiates in aircraft first aid kits and recognizing that there were three points for decision. The first was: "Is the carriage of opiates in aircraft first-aid kits desirable?" The Commission's conclusion was that it was desirable – there was a Recommended Practice in paragraph 6.2(a) of Annex 6 which listed narcotics and analgesics among the contents of first-aid kits, and the Commission believed that it should stand. The second point was: "Should their carriage be permitted?" From the ICAO standpoint the answer was obviously "Yes" – if the carriage of opiates and derivatives was desirable it should be permitted – but the issue was complicated by the fact that the international movement of drugs was subject to international narcotics control. The Commission was of the view that all that could be done in the circumstances was to inform contracting States of the present study, to ask them to note the implication in Annex 6 that foreign operators should be permitted to carry first-aid kits with the contents described in that paragraph, and to request them to notify the Organization of differences between their national regulations or practices and this provision. It recommended that the Council decide accordingly. States, ECOSOC and WHO might also be invited to study the related legal and medical problems, in particular the application of efficient safeguards against abuse and of uniform principles under which opiates or other drugs might be used and carried in first-aid kits on board aircraft, in an effort to promote uniformity in existing laws and avoid difficulty. The third point was whether qualified crew members should be permitted to administer narcotic drugs in cases of emergency. On this again national regulations and practices differed, and the Commission's conclusion was that States might be asked to accept the practice of the State of Registry of the aircraft.

---

[505] See AN-WP/1984, 1/12/58 at 1.

The Council in its Annual Report to the Assembly for 1957[506] advised that in the light of the replies received from States to the letter inviting comments on the carriage of opiates and derivatives in first-aid kits on board aircraft on international flights, the Air Navigation Commission had concluded that the recommendation in Annex 6 (that first-aid kits should contain analgesics and narcotic) was adequate but that there was a need for more uniformity in its practical application.

Consideration was also given to what action ICAO might take on a number of other medical and health problems in aviation such as the carriage of sick persons, pregnant women, live animals and dead bodies, pollution of food and drinking water on board and removal of refuse from aircraft. A majority of Council members felt that these problems were of much more concern to other organizations (particularly WHO and IATA) than to ICAO and that ICAO could not take any useful action on them. The Secretary General was instructed to keep in touch with developments in connection with them to ensure that any action taken by other organizations would not unnecessarily interfere with ICAO's Facilitation Programme.

The Economic and Social Council, acting on the request of the ICAO Council, to consider the problem of the carriage of narcotic drugs in first-aid kits of aircraft engaged in international flight, noted that the Council of the International Civil Aviation Organization at the eighth meeting of its 30th session, on 1 April 1957, invited the World Health Organization to study the medical aspects of this question.

The Air Navigation Commission of ICAO later noted that in 1958, the United Nations had recommended to Governments,[507] to take all necessary measures to prevent the misuse and diversion for illicit purposes of narcotics drugs carried in first-aid kits of aircraft engaged in international flight, in particular by ensuring that such drugs are kept in sealed or locked containers to which only authorized persons have access, that adequate records of supply and use, and of stocks, of narcotic drugs are maintained by the airline companies concerned, and that such records and stocks are subject to regular inspection.

In this context, the Economic and Social Council (ECOSOC) had also requested the Secretary-General to invite the views of the International Criminal Police Organization on the safeguards which should be taken to prevent the diversion of such drugs for illicit purpose and invited the Commission on Narcotic Drugs to consider the report referred to in the preceding paragraph, at its 14th session if possible, and to advise the Economic and Social Council whether further measures should be recommended to governments for application.

In an overview of action taken by ICAO on the carriage of opiates in first-aid kits in aircraft it was observed that the Air Navigation Commission first dealt with the subject when considering how to overcome the difficulties experienced by States in the application of Annex 6, dealing with the carriage of narcotic drugs in aircraft

---

[506]Doc 7866, A11-P/3 at 21.
[507]E/RES/689 (XXVI) 29 July 1958.

C. ICAO Initiatives

first-aid kits[508] These discussions when reported to Council resulted in council's invitation to ECOSOC and to the World Health Organization (WHO) to study various aspects of the problem.

The Air Navigation Commission was then informed on progress on this subject and at that time the Commission had noted that the problem of the carriage of opiates in aircraft first-aid kits on international flight was being studied further by the World Health Organization and the United Nations' Commission on Narcotic Drugs. The Secretary was requested to keep the Commission informed or later developments as appropriate.

It was also observed that the Economic and Social Council in Resolution 689F adopted at its 26th Session, July 1958, had requested the Secretary-General of the United Nations, to invite the views of the International Criminal Police Organization (ICPO or Interpol), on the safeguards which should be taken to prevent diversion of such drugs for illicit purposes.[509] And to prepare, if the World Health Organization study supports the carriage of narcotic drugs in first-aid kits of aircraft engaged in international flight, a report on the legal problems, in particular concerning the application of efficient safeguards against abuse and of uniform principles under which opiates or similar drugs might be used and carried in first-aid kits on board aircraft in an effort to promote uniformity under existing laws. This report was to be prepared in consultation with the Secretariats of the International Civil Aviation Organization and the World Health Organization.

Pursuant to this request, the World Health Organization, with the assistance of a consultant with wide experience in aviation medicine and air carrier problems had prepared a study which concluded that it was desirable to carry a limited amount of narcotics in aircraft first-aid kits and commented on a number of medical aspects which might be taken into consideration to prevent abuse. The United Nations Legal Office concluded that while drugs carried in first-aid kits are not exempted from other relevant provisions of the narcotic treaties, the import certificate and export authorization system then prevalent did not apply to drugs carried under appropriate safeguards in first-aid kits for emergency cases as long as they do not cross the customs lines at points of transit or destination.

The World Health Organization report, the United Nations' Secretariat legal opinion and the administrative measures proposed by the International Criminal Police Organization to prevent diversion of drugs for illicit purposes were considered by the United Nations Commission on Narcotic Drugs at its 14th Session in April/May 1959. On the basis of the Narcotic Commission's recommendation, the ECOSOC at its 28th Session in July 1959 unanimously adopted Resolution 730C (XXVIII). In this Resolution the UN Secretary-General had been invited "in co-operation with ICAO and WHO, and in consultation with ICPO, to prepare and to distribute to Governments in sufficient time for consideration at the 15th Session of the Commission on Narcotic Drugs, a set of requirements essential to

---

[508] AN-WP/1604; XXIV-11, 25/2/57.
[509] See AN-WP/1984 Appendix B.

ensure proper use of narcotic drugs and to prevent their abuse and diversion for illicit purposes, such requirements to be recommended to Governments as a basis for the control of the carriage of narcotic drugs in first-aid kits on board aircraft engaged in international flight". Accordingly a joint Secretariat Working Group of the UN Division of Narcotic Drugs, WHO, ICAO and ICPO (Interpol) had met in Geneva in January 1960 and prepared the set of requirements referred to above under which opiates or similar drugs may be carried in aircraft first-aid kits on international flight and used in emergency.

The UN Commission on Narcotic Drugs had later discussed and approved the Inter-Secretariat report at its 15th Session, May 1960. Pursuant to this, the ECOSOC at its 30th Session, July 1960, unanimously adopted the Resolution 770E (XXX) prepared by the Commission on Narcotic Drugs which included recommendations to States on the carriage of narcotic drugs in first-aid kits of aircraft engaged in international flights, and safeguards to prevent abuse.

In view of the ECOSOC recommendations aimed at eliminating the difficulties experienced by States, the Air Navigation Commission of ICAO considered whether any further action by ICAO was necessary or whether the subject should be deleted from the Work Programme.[510]

At its 39th Session the General Assembly of the United Nations adopted Resolution 39/143 on "International campaign against traffic in drugs" which, inter alia, called upon the specialized agencies to participate actively in its implementation. In December 1984, the Secretary General of the United Nations also addressed a letter to the Executive Heads of the specialized agencies requesting their co-operation in the efforts directed to control the abuse and illicit trafficking of narcotic drugs. Believing that a very large percentage of illicit narcotic drugs and psychotropic substances was carried by air, the United Nations Division of Narcotic Drugs requested the co-operation of ICAO in this field.

ICAO's activities in narcotics control became significant in 1984 when, in November of that year, ICAO was represented at the 11th meeting of Operational Heads of National Narcotics Law Enforcement Agencies, Far East Region (HONLEA). An ICAO observer also attended the Third and Fourth Sessions of the Enforcement Committee of the Customs Co-operation Council in February and September 1985 respectively, where matters of narcotics law enforcement were discussed. On the same two occasions, the ICAO observer attended the United Nations Ad Hoc Inter-agency meetings on Co-ordination in matters of International Drug Abuse Control.

At the informal meeting of the Council on 11 June 1985, a preliminary consideration was given to the constitutional mandate of ICAO in the field of suppression of illicit transport of narcotic drugs and psychotropic substances by air. In this connection it should be noted that:

---

[510] The Economic and Social Council (ECOSOC) in Resolution 770E (XXX) approved certain recommendations to Governments on safeguards to prevent abuse for the carriage of narcotic drugs in first-aid kits of aircraft engaged in international flights.

C. ICAO Initiatives

(a) Under the Agreement between the United Nations and the International Civil Aviation Organization,[511] the Organization is obliged to co-operate in establishing effective co-ordination of the activities of specialized agencies and those of the UN; in particular, ICAO is obliged to consider formal recommendations made by the United Nations and to furnish to the United Nations required information. In this context, it was therefore noted that UN General Assembly Resolution 39/143 called upon the specialized agencies of the United Nations system to participate actively in the implementation of that Resolution, entitled "International campaign against traffic in drugs". Consequently, the Organization had a responsibility to undertake a study of the problem of suppression of illicit transport on narcotic drugs and psychotropic substances.

(b) The Chicago Convention[512] contains several provisions referring to elements of International air law which may be relevant for the control and suppression of drug trafficking.

Article 10 – If all aircraft coming from abroad land only at designated customs airports and depart only from such airports, the control of illicit transport would be greatly facilitated.

Article 13 – Clearance and departure of cargo are subject to the regulations of the contracting States whose territories are involved; the movement of any specific cargo is subject to the legal regulations and effective control of the States concerned.

Article 16 – Contracting States have the right to search aircraft of other contracting States on landing or departure; obviously, the same right is applicable for the search of aircraft of their own registry.

Article 23 – Customs and immigration procedures affecting international air navigation should be in accordance with the practices established or recommended from time to time pursuant to the Convention; that clearly is reflected in the procedures established in Annex 9 to the Convention; however, the predominant provisions in this field are enacted by States through their immigration and customs legislation, the latter being co-ordinated internationally through the Customs Co-operation Council. Annex 9 deals with customs and immigration procedures but mainly in order to ensure that the procedures used are efficient and do not interfere with the speedy clearance of aircraft and their loads.

Article 35(b) – Each contracting State has the right, for reasons of public order and safety, to prohibit the carriage in or above its territory of certain articles; it is within the legislative power of the contracting States to adopt an unconditional interdiction on the carriage of narcotic drugs and psychotropic substances into or over their territory and to establish measures for enforcing such legislation.

---

[511] Doc 7970.

[512] Convention on International Civil aviation, signed at Chicago on 7 December 1944. See ICAO Doc 7300/9 Ninth Edition, 2006.

Article 37(j) – It is within the legislative function of the ICAO Council to adopt standards, recommended practices and procedures dealing with customs and immigration, import and export of specific articles.

These Articles of the Chicago Convention should be read in conjunction with Article 22 of the Convention under the terms of which each contracting State agrees to adopt all practical measures to facilitate and expedite air navigation and to prevent unnecessary delays to aircraft, crews, passengers and cargo. These requirements are further specified in Annex 9 to the Chicago Convention on Facilitation of Air Transport.

Although it might perhaps be suggested that Annex 9 could contain provisions urging aeronautical authorities to extend every possible assistance in the suppression of drugs trafficking, more detailed instructions or guidance material would seem out of context with the character of Annex 9, which is essentially aimed at overcoming obstacles to the rapid clearance of aircraft and their loads. This objective should be preserved and not diluted by inclusion in the Annex of provisions which would appear to run counter to the prime objective. On the other hand, recognition of ICAO's co-operation in drug trafficking control could be made by inclusion of text in the Annex, to provide for the following:

(a) That any special inspection procedures required in the control of narcotics traffic be accomplished speedily, with efficient equipment, without inconvenience to passengers and in such a way as to ensure the timely clearance of aircraft and their loads on arrival and departure; and
(b) Point out that most seizures of narcotics and related substances in illicit traffic result from police and drug enforcement intelligence rather than from routine inspection and consequently that sampling rather than exhaustive methods of inspection are to be preferred.[513]

Another programme of ICAO which has a bearing on or which would be affected by any ICAO activities related to the campaign against illicit trafficking in narcotic drugs and psychotropic substances is Aviation Security (Annex 17 to the Chicago Convention and the Security Manual, Doc 8973); however the protected interest in the aviation security programme is aviation itself.

Two other international organizations, i.e., ICPO-INTERPOL and the World Customs Organization (WCO), have a direct responsibility in the suppression of illicit traffic in narcotic drugs. INTERPOL's role has been involved with influencing national legislation and co-ordinating the work of police services world-wide

---

[513] A recommendation of the Customs Co-operation Council dated 17 June 1985, recognizes 'that the proper balance between the needs of customs enforcement and the facilitation of legitimate trade and travel can best be achieved if customs enforcement is selective and intelligence based and that it is therefore essential that every effort be made to develop and exploit the best possible intelligence.' (Recommendation on the Development of Co-ordinated Enforcement and Intelligence Operations aimed at Identifying and Interrupting Concealed Illicit Drugs).

C. ICAO Initiatives

to eradicate the sources of raw material, processing and refining facilities and the apprehension and prosecution of offenders.

The WCO has, for many years, been concerned with the suppression of smuggling narcotics and psychotropic substances within its general competence to deal with customs matters. Its main instrument in this field is the International Convention on mutual administrative assistance for the prevention, investigation and repression of customs offenses, signed in Nairobi in 1977. The "Nairobi Convention," as it is called, contains specific references to illicit traffic and an Annex (Annex X) dealing with assistance in action against the smuggling of narcotic drugs and psychotropic substances. These provisions are designed to complement those of prevailing treaties on narcotic drugs. The WCO is also developing with the International Air Transport Association a Memorandum of Understanding between the two organizations containing guideline for both customs authorities and airlines to follow to help prevent illicit traffic of drugs on board aircraft.

The definition of "smuggling" contained in the Nairobi Convention applies to all modes of transport. Pertinent exchanges of information undertaken by Customs Administrations pursuant to the Convention extend to all means of transport used or suspected of being used for the smuggling of narcotic drugs or psychotropic substances or that seem likely to give rise to such operations. Assistance, on request, relating to surveillance extends over particular vehicles, ships, aircraft or other means of transport reasonably believed to be used for smuggling narcotic drugs or psychotropic substances into the territory of the requesting Contracting Party.

The WCO's main deliberative organ in these matters is its Committee on Customs Enforcement. The current work programme of this Committee in the area of narcotics smuggling includes exchanges of information on couriers, their routes and pertinent traffic trends, development of catalogues of enforcement aids and places of concealment (in co-operation with INTERPOL and the UN Division of Narcotic Drugs), investigative techniques (undercover work), seminars and training programmes an action to monitor and pre-empt financial transactions relating to narcotics smuggling. A recommendation of the Customs Co-operation Council in 17 June 1985 adopted on the proposal of its Enforcement Committee seeks 'to secure the fullest co-operation of airline and shipping companies and others involved in the international transport and travel industries to assist the international Customs community in suppression the illicit traffic in narcotic drugs and psychotropic substances'.

In the light of the above, it appears that in this field and in accordance with its constitutional responsibilities, ICAO can play the following role:

(a) Monitor the adherence of States of the convention Against the Illicit Traffic in Narcotic Drugs and Psychotropic Substances in order to ensure *inter alia* that international civil aviation interests are not penalized by objective liability or responsibility unless there is a specific criminal involvement of the carrier or his staff
(b) Formulate and adopt as required technical specifications related to civil flight operations

(c) Develop as required guidance materials
(d) Co-operate with the United Nations Division of Narcotic Drugs and other international organizations through consultation and attendance at meetings
(e) Ensure that facilitation measures and measures directed to control the illicit traffic of drugs do not have an unnecessarily negative impact on each other so as to maintain the separate thrusts of these programmes

D. Other Regulatory Provisions

I. Article 4 of the Convention on International Civil Aviation

Article 4 of the Chicago Convention is the only provision in the Convention explicitly using the words "misuse of civil aviation"; even there, however, the expression is used only in the heading (in fact, in the margin in the original signature copy) and not in the substantive text of the Article. The first paragraph of the Preamble to the Convention refers to "abuse" of international civil aviation without any attempt at a definition of that term.

Article 4 of the Convention has never been the subject of nor involved in a decision or interpretation either by the Assembly or the Council. Therefore, that Article 4 is of no relevance to the problem since it refers only to the obligations of States and to the acts of States. The drafting history of this Article indicates that the underlying intent of Article 4 was to prevent the use of civil aviation by States for purposes which might create a threat to the security of other nations. The intent of Article 4 originated in the Canadian "Preliminary Draft" which stated as one of the purposes of ICAO (or PICAO, as was then envisaged), the future organization "to avert the possibility of the misuse of civil aviation creating a threat to the security of nations and to make the most effective contribution to the establishment and maintenance of a permanent system of general security". In the further drafting development ("Tripartite Proposal" presented to the Conference by the Delegations of the United States, United Kingdom and Canada) the wording was changed to read: "Each member State rejects the use of civil air transport as an instrument of national policy in international relations". This wording practically repeated the text of the Treaty for the Renunciation of War of 27 August 1928 (commonly known as the Briand–Kellogg Pact) in which the signatories renounced war "as an instrument of national policy in their mutual relations". The words "purposes inconsistent with the aims of this Convention" in Article 4 therefore essentially mean "threats to the general security".

Article 4 does not offer any solution to the problem of "misuse of civil aviation" within the scope of paragraph 2 above, namely, the status of an aircraft which is used for criminal purposes or other unlawful purposes.

The Chicago Convention in general does not contain any provisions which would foresee the specific situations when an aircraft is used for or involved in criminal activities or other activities violating the law and public order of the State. However, there are numerous provisions in the Convention which offer effective safeguards to States that their applicable laws and public order are observed by foreign aircraft (with respect to aircraft of its own registry, the State concerned has unrestricted jurisdiction). Articles 11, 12 and 13 of the Convention in essence confirm the rule of general international law that foreign aircraft, its crew, passengers and cargo do not enjoy any "extraterritorial" status while in the airspace or on the ground of another State; such aircraft are fully subject to the applicable laws of the State concerned. Under the Convention, the State may require the landing of a foreign aircraft involved in non-scheduled flight (Article 5), may prohibit or restrict foreign aircraft from flying over certain parts of its territory or over the whole territory (Article 9), may require landing of foreign aircraft at a designated customs airport (Article 10), may search the foreign aircraft (Article 16) and may regulate or prohibit the carriage of certain articles in or above its territory (Article 35(b)).

It is submitted that all States possess within the existing framework of the Chicago Convention full jurisdiction in the application of their respective laws to prevent or prohibit the use of civil aircraft for unlawful purposes. The practical problem therefore does not appear to arise in the field of the applicability of particular laws but in the field of practical enforcement of such laws with respect to aircraft, particularly aircraft in flight.

II. Article 3 bis

Problems of interception of and other enforcement measures with respect to a civil aircraft in flight are directly addressed in Article 3 bis adopted by unanimous consensus on 10 May 1984 by the 25th Session (Extraordinary) of the ICAO Assembly. The drafting history of this Article supports the conclusion that Article 3 bis is declaratory of the existing general international law with respect to the following elements:

(a) Obligation of States to refrain from resorting to the use of weapons against civil aircraft in flight
(b) Obligation, in case of interception, not to endanger the lives of persons on board and the safety of aircraft
(c) Right of States to require landing at a designated airport of a civil aircraft flying above its territory without authority or if there are reasonable grounds to conclude that it is being used for any purpose inconsistent with the aims of the Convention

While Article 3 *bis* accepted in paragraphs (b) and (d) the terminology "for any purpose inconsistent with the aims of the Convention" exactly as it is used in Article 4 of the Convention, the drafting history indicates conclusively that the scope of the phrase is different in Article 3 *bis* than in Article 4. At the 25th Session (Extraordinary) of the Assembly, convened in Montreal in 24 April to 10 May 1984, this phrase was meant to cover not only violations of the "aims" of the Chicago Convention as spelled out in the Preamble to the Convention and in its Article 44 (which deals with the aims and objectives of the Organization rather than the Convention), but any violation of the law and public order of the State concerned. In the Assembly discussions specific references were made to transport of illicit drugs, contraband, gun running, illegal transport of persons and any other common crimes.

It should be stressed that the scope of applicability of Article 3 *bis* is subject to significant restrictions; the protection of this Article is reserved only to:

(a) "Civil aircraft"; consequently, "state aircraft" would not enjoy the same protection.
(b) Civil aircraft "in flight"; while the Convention does not define the concept "in flight," it is likely that this phrase will be interpreted in harmony with the Rome Convention on Damage Caused by Foreign Aircraft to Third Parties on the Surface of 1952 (Article 1, paragraph 2) and the Tokyo Convention of 1963 (Article 1, paragraph 3). An aircraft shall be deemed to be in flight from the moment when power is applied for the purpose of take-off until the moment when the landing run ends. Consequently, aircraft which are not "in flight" do not enjoy the special protection of Article 3 *bis*.

It is also submitted that the protection of Article 3 bis is reserved to "foreign" aircraft and does not include aircraft of the State's own registration. After discussions in the Executive Committee of the Assembly, the reference to aircraft "of the other contracting State" was dropped for the specific reason that the protection was to be recognized as mandatory with respect of aircraft, whether belonging to contracting or non-contracting States. At no stage of the deliberations and drafting did the Assembly (in the Plenary, in the Executive Committee or in the Working Group) contemplate regulation of the status of an aircraft in relation to the State of its own registration; such regulation would have exceeded the scope of the Convention which deals with international civil aviation. Again, the purpose of the Chicago Convention is to establish conventional rules of conduct in the mutual relations of sovereign States but not to govern matters of their exclusive domestic jurisdiction. Consequently, Article 3 *bis* will not apply to the treatment of aircraft by the States of their registration. This conclusion does not imply that a State is free to treat aircraft of its own registration without regard to any rules of international law; other sources of international law (e.g., the International Covenants on Human Rights) may be relevant for the conduct of States (protection of the right to life, requirement of due legal process, presumption of innocence, etc.).

When requiring the landing of a civil aircraft flying above its territory or when issuing other instructions to the aircraft to put an end to a "violation," contracting States may resort to any appropriate means consistent with relevant rules of international law, including the Chicago Convention and, specifically, paragraph (a) of Article 3 *bis*. Consequently, Article 3 *bis* does not exclude enforcement against foreign aircraft in flight and does not rule out the use of adequate and proportionate force and does not rule out interception as such. Any act of interception

D. Other Regulatory Provisions

or other enforcement measure not involving the use of weapons against civil aircraft in flight and not endangering the lives of persons on board and the safety of flight is legitimate and acceptable. Any interception procedures consistent with the applicable Standards and Recommended Practices adopted by the Council of ICAO pursuant to Articles 37, 54(1) and 90 of the Chicago Convention would be "consistent with relevant rules of international law".

Two additional provisions of Article 3 *bis* are likely to deter the occurrences of "misuse" of civil aviation. Firstly, civil aircraft are unconditionally obliged to comply with an order to land or other instruction; contracting States are accepting, under paragraph (c) of Article 3 *bis*, an obligation to establish all necessary provisions in the national law or regulations to make such compliance mandatory for aircraft of their registration or operated by an operator having his principal place of business or permanent residence in that State. Contracting States are also accepting an obligation to make violation of such laws or regulations punishable by severe penalties and to submit the case to their competent authorities. This provision may offer a practical safeguard that no violators would go unpunished; even if they were to escape from the jurisdiction of the State where the unlawful act was committed, they should be prosecuted and punished by the State of the registration of the aircraft; in practical application this provision may be reinforced by existing or future arrangements for extradition of offenders; and secondly, all contracting States are accepting an unconditional obligation to take appropriate measures to prohibit any deliberate "misuse" of any civil aircraft of their registration or operated by an operator having his principal place of business or permanent residence in that State. Legislative implementation of such a prohibition will no doubt be accompanied by appropriate penalties.

III. Other Legal Aspects

States can exercise criminal jurisdiction over foreign aircraft in flight over their territory as well as over the territory not subject to Sovereignty of any State (e.g., the high seas) also under the conditions set forth in the Tokyo Convention of 1963.[514] Article 4 of that Convention permits "interference" with an aircraft in flight in order to exercise criminal jurisdiction over an offence committed on board in the following cases:

(a) The offence has effect on the territory of such State.
(b) The offence has been committed by or against a national or permanent resident of such State.
(c) The offence is against the security of such State.

---

[514]Convention on Offences and Certain Other Acts Committed on Board Aircraft, signed at Tokyo on 14 September 1963, ICAO Doc 8364.

(d) The offence consists of a breach of any rules or regulations relating to the flight or manoeuvre of aircraft in force in such State.
(e) The exercise of jurisdiction is necessary to ensure the observance of any obligation of such State under a multilateral international agreement.

Since the Tokyo Convention has been accepted by many of ICAO's Contracting States, this provision represents an important additional clarification to Article 3 *bis* of the Chicago Convention with respect to the interpretation of the term "any purpose inconsistent with the aims of this Convention". It is submitted that any offence foreseen in Article 4 of the Tokyo Convention gives right to the State concerned to "interfere," i.e., to require the landing or give the aircraft other instructions and to resort to proportionate and adequate use of force against such aircraft.

The United Nations Convention on the Law of the Sea does not foresee the right of hot pursuit of aircraft; the target of hot pursuit may be exclusively a ship but the procedures of hot pursuit may be effected by an aircraft (Article 111, paragraph 6).

IV. ICAO Assembly Resolution A 27-12

At its 27th Session, held in September/October 1989, the ICAO Assembly adopted the above Resolution which recognised the enormity of drug abuse and illicit trafficking in drugs and psychotropic substances and urged the ICAO Council to give the highest priority to adopt concrete measures in order to prevent and to eliminate possible use of illicit drugs by crew members, air traffic controllers and other staff of international civil aviation. The Resolution also urged the Council to continue its work in order to prevent illicit transport of narcotic drugs and psychotropic substances. *A fortiori*, the Assembly called upon Contracting States to continue their efforts to prevent the illicit trafficking of drugs by air, to take appropriate legislative measures to ensure that the crime of illicit transport of narcotic drugs and other psychotropic substances by air is punishable by severe penalties. Contracting States are also urged by this Resolution to become parties, as soon as possible to the United Nations Convention of 1988.

It is incontrovertible that the foregoing discussion brings to bear the concerted efforts of the international community both through the United Nations Organization (through its General Assembly) and through the International Civil Aviation Organization – towards controlling the problem of narcotic drug trafficking. A third force – the carriers themselves – have had considerable success in their anti-narcotic drug programme. One of the best examples of carrier action is reflected in the United States Customs Carrier Initiative Agreement Programme which was introduced in 1984. This programme is a purely voluntary arrangement between governments and carriers in which the government allows each airline to create a security programme that is approved by Customs. The Carrier Initiative Agreement Programmes are aimed at both prevention and interdiction, covering areas of training, prevention and co-operation.

D. Other Regulatory Provisions

The enormity of the offence of narcotic drug trafficking by air pervades multifarious delinquencies and criminal offences. These may lead to hijacking of aircraft, destruction of aircraft, interception of aircraft and several other offences leading to loss of life and destruction of buildings and installations. Therefore, in the overall perspective, the offence could be termed "misuse of civil aviation".

The problem of "misuse of civil aviation," "improper use of civil aviation," "undue use of civil aviation" or "criminal use of civil aviation" was raised by several delegations at the 25th Session (Extraordinary) of the ICAO Assembly in April/May 1984 and was mentioned by several Representatives on the Council in the context of discussions on the proposed amendment of Annex 2 to the Convention on International Civil Aviation with respect to interception of civil aircraft. In that context, several Representatives expressed concern that necessary procedures must be foreseen to prevent the use of civil aviation for unlawful purposes, in particular for drug trafficking which is more and more generally recognized to be a serious crime against humanity. The problem to be addressed is essentially how to reconcile the protection of civil aircraft in situations of interception with the protection of the law and order of States concerned and with the enforcement of such applicable laws. The scope of the problem encompasses in particular the concern of several States whether Article 3 *bis* of the Chicago Convention and the amended Annex 2 leave sufficient safeguards for States to prevent, prosecute and punish and deliberate use of civil aircraft for unlawful purposes.

Article 3 bis provides *inter alia* that every State must refrain from resorting to the use of weapons against civil aircraft in flight. The worlds "refrain from" do not provide the necessary strength to the provision as it does not explicitly prohibit the use of weapons against aircraft in flight. The study of the problem of "misuse" of civil aviation and of its consequences for law enforcement with respect to civil aircraft in flight leads to the following conclusions:

(a) Although the term "misuse of civil aviation" is a legally imprecise term which has no firm basis in the Convention on International Civil Aviation apart from the title of Article 4, it still reflects the overall threat posed by unlawful interference with civil aviation.
(b) The phrase "any purpose inconsistent with the aims of this Convention" has historically a different meaning in Article 4 of the Convention and in paragraphs (b) and (d) of Article 3 *bis*.
(c) The concept of "misuse of civil aircraft" should best be referred to as "deliberate use of civil aircraft for unlawful purposes."
(d) The Chicago Convention contains effective provisions safeguarding full jurisdiction of States to prevent or prohibit the use of foreign aircraft for unlawful purposes in their territory.

The above conclusions may be drawn upon to use the relevant provisions of the Chicago Convention as a base to formulate other legal documents that would enforce more stringent control over this offence.

From an administrative perspective, certain measures may be taken to address the problem, and to limit and control production and distribution of drugs. These measures could be coordinated under the broad umbrella of the United Nations through the International Criminal Police Organization (ICPO, Interpol) and the World Customs Organization (WCO). States have a lead role to play in the introduction and implementation of *Counterterrorist measures which* are directly aimed at disbanding terrorist organizations. They could also enact effective legislation against money laundering which could, through national financing oversight mechanisms track and freeze payments and assets of drug dealers, etc. This rests mainly with national financial oversight authorities. Domestically, States could introduce drug rehabilitation programmes and provide adequate counselling and treatment to drug abusers. Above all, a stringent police presence would be needed to arrest street crimes that are drug related and drug deals on the streets.

References

Bell R (1991) The history of drug prohibition and legislation. Interpol International Criminal Police Review, September – October 1991

Hartelius J (2008) Narcoterrorism, Policy Paper 3/2008. United Nations

Chapter 5
The Unlawful Interference Conventions

A. United Nations General Assembly Resolutions on Unlawful Interference with Civil Aviation

During its 24th session on December 1969, the United Nations General Assembly discussed the problem of "forcible diversion of civil aircraft" and adopted Resolution 2551 (XXIV),[515] in which the General Assembly stated its deep concern over acts of unlawful interference with international civil aviation. The General Assembly also called upon States to take every appropriate measure to see that their respective national legislation provides an adequate framework for effective legal measures against all kinds of acts of unlawful seizure of civil aircraft. It furthermore called upon States to ensure that persons on board who perpetrate such acts are prosecuted. The General Assembly urged that States give their fullest support to the International Civil Aviation Organisation in its endeavours towards the speedy preparation and adoption of a convention which would provide for appropriate measures which would make the offence of unlawful seizure of aircraft punishable. The commission of the offence would lead to the prosecution of persons who commit it. By this resolution, the General Assembly also invited States to ratify and accede to the Convention on Offenses and Certain Other Acts Committed On Board Aircraft, signed in Tokyo on 14 September 1963.[516]

On 25 November 1970 the General Assembly adopted Resolution 2645 (XXV)[517] which condemned without exception whatsoever all aerial hijacking or other interference with civil air travel caused through the threat or use of force. The Resolution also condemned all acts of violence which may be directed against passengers, crew and aircraft engaged in, and air navigation facilities and aeronautical communication used by civil air transport. The Assembly called upon States to

---

[515] Resolution 2551 (XXIV). The Resolution was adopted by a vote of 77 in favour, 2 against with 17 abstentions.

[516] The Tokyo Convention will be discussed in some detail later.

[517] Resolution 2645 was adopted by 105 in favour, none against and 8 abstentions.

take all appropriate measures to deter, prevent or suppress such acts within their jurisdiction, at every stage of the execution of those acts, and to provide for the prosecution and punishment of persons who perpetrate such acts, in a measure commensurate with the gravity of those crimes, or extradite such persons for the purpose of their prosecution and punishment. Furthermore, the Assembly condemned the exploitation of unlawful seizure of aircraft for the purpose of taking of hostages, calling upon States to take joint and separate action, in accordance with the United Nations Charter and in co-operation with the United Nations and International Civil Aviation Organisation so that passengers, crew and aircraft engaged in civil aviation are not used for purposes of extortion.

The international community thus condemned terrorism against air transport by giving official recognition to such condemnation and called upon all States to contribute to the eradication of the offence by taking effective, preventive and deterrent measures. Notwithstanding the weight of these resolutions the General Assembly has seemingly deprived itself of the opportunity of declaring the offence of hijacking an international crime under international law. The world condemnation of the offence has left the question open to States as to whether the international community would collectively respond in the face of a crisis related to unlawful interference with civil aviation. Another blatant weakness of the Resolution is that the provisions of the resolution regarding extradition are ambivalent. The Resolution has also remained silent as to whether political motive would be a valid ground against extradition or not. It is submitted that the General Assembly should have considered adopting the principle that political motive will not be a factor affecting the extradition of hijackers.

The Resolution, with all its lapses, has many advantages, such as its condemnation of the offence of unlawful interference and call for international action against the offence. The persuasive nature of Resolutions will facilitate nations in interacting with each other and assisting each other.

The United Nations has, over the past two decades extended an invitation to nations, to co-operate with each other in eradicating or controlling international terrorism. For instance Resolution 2645 (XXV) recognized that international civil aviation is a vital link in the promotion and preservation of friendly relations among States, and that the Assembly was gravely concerned over acts of aerial hijacking or other wrongful interference with civil air travel. The resolution condemned without exception, all acts of aerial hijacking or other interference with civil air travel and called upon States to take all appropriate measures to deter, prevent or suppress such acts within their jurisdiction.[518] Earlier, the Security Council had adopted Resolution 286 (1970) which expressed the Council's grave concern at the threat to innocent civilian lives from the hijacking of aircraft and any other interference in international travel. The Security Council appealed to all parties concerned for the

---

[518] A/RES/2645 (XXV) 30 November 1970. The Resolution was approved by the United Nations General Assembly on 25 November 1970 by a vote of 105 in favour, none against, and 8 abstentions.

A. United Nations General Assembly Resolutions

immediate release of all passengers and crews without exceptions, held as a result of hijackings and other interference in international travel, and called on States to take all possible legal steps to prevent further hijackings or any other interference with international civil air travel.[519]

On 18 December 1972, the United Nations General Assembly, at its 27th session adopted a resolution[520] expressing the deep concern of the Assembly over acts of international terrorism which are occurring with increasing frequency and recalled the declaration on principles of international law which called for friendly relations and co-operation among States in accordance with the Charter of the United Nations. The resolution urged States to devote their immediate attention to find quick and peaceful solutions to the underlying causes which give rise to such acts of violence.[521]

One of the salutary effects of this resolution was the sense of urgency it reflected in reaffirming the inalienable right to self-determination and independence of all people and the condemnation it issued on the continuation of repressive acts by colonial, racist and alien regimes in denying peoples their legitimate right to the enforcement of their human rights. The resolution followed up with the invitation to States to become parties to the existing international Conventions which relate to various aspects of the problem of international terrorism.[522]

On 21 January 1977, the General Assembly commenced drafting an international convention against the taking of hostages, which was authorized by Resolution A/RES/31/103 which broadly invoked the Universal Declaration of Human Rights; and the International Convention on Civil and Political Rights which provides that everyone has the right to life, liberty and security. The resolution established an *ad hoc* Committee on the drafting of an international convention against the taking of hostages. The Committee was mandated to draft, as early as possible, an international convention. The President of the General Assembly was requested by the Assembly to appoint the members of the *ad hoc* Committee on the basis of equitable geographical distribution and representing the principal legal systems of the world.[523] The resolution was adopted on 15 December 1976.

Three years later in December 1979, the General Assembly adopted a resolution[524] which revised the work of the *ad hoc* Committee and called for international co-operation dealing with acts of international terrorism. The resolution, while welcoming the results achieved by the Committee, called upon States to fulfil their obligations under international law to refrain from organizing, instigating, assisting or participating in civil strife or terrorist acts in another State, or

---

[519] S/RES/286 (1976) 9 September 1970.
[520] A/RES/3034 (XXVII), 18 December 1972.
[521] A/RES/3034 (XXVII), 18 December 1972.
[522] A/RES/3034 (XXVII), 18 December 1972.
[523] A/RES/31/103, 21 January 1977.
[524] A/RES/34/145, 22 January 1980.

acquiescing in organized activities within their territory directed towards consensus of such acts.[525]

A major contribution of this Resolution was its recognition that in order to contribute to the elimination of the causes and the problem of international terrorism, both the General Assembly and the Security Council should pay special attention to all situations, including, *inter alia*, colonialism, racism and situations involving alien occupation, that may give rise to international terrorism and may endanger international peace and security. The application, when feasible and necessary, of the relevant provisions of the Chapter of the United Nations, was also recommended. The resolution also requested the Secretary General of the United Nations to prepare a compilation on the basis of material provided by Mentor States of relevant provisions of material legislation dealing with the combating of international terrorism.

In December 1985 the United Nations General Assembly adopted Resolution 40/61 which unequivocally condemned as criminal, all acts, methods and practices of terrorism, whenever committed, including those which jeopardise international peace and security which affect States or their property.[526] The Resolution referred to the international conventions that had been adopted in relation to unlawful interference with civil aviation and called upon States to fulfil their obligations under international law to refrain from organizing, instigating, assisting or participating in any terrorist acts against other States, their people or property.

The Resolution, while citing the relevant conventions relating to unlawful interference with international civil aviation (a discussion of which will follow), once again appealed through the General Assembly to States that had not done so, to become parties to such conventions, including others which related to the suppression of international terrorism. While encouraging ICAO to continue its efforts aimed at promoting universal acceptance of and strict compliance with the international air services conventions, the Resolution also called upon all States to adhere to the ICAO conventions that provide for the suppression of terrorist attacks against civil aviation transport and other forms of public transport.[527]

Simultaneously, the Security Council, in December 1985 adopted Resolution S/RES/579 which expressed deep concern at the prevalence of incidents of hostage taking and abduction following terrorist acts. The Resolution appealed to all States to become parties *inter alia* to the ICAO Conventions. This resolution further urged the development of international co-operation among States according to international law, in the facilitation of prevention, prosecution, and punishment of all acts

---

[525] A/RES/34/145, 22 January 1980.

[526] A/RES/40/61, 14 January 1986. Djonovich DJ (ed) United Nations Resolutions (7 Series), Volume XXIV, 1985–1986, at 507.

[527] A/RES/40/61, 14 January 1986. Djonovich DJ (ed) United Nations Resolutions (7 Series), Volume XXIV, 1985–1986, at 507.

A. United Nations General Assembly Resolutions

of hostage taking and abduction which were identified as manifestations of international terrorism.[528]

The General Assembly, in December 1987, adopted another Resolution[529] which referred to the recommendations of the *ad hoc* Committee which had called for stringent measures of international co-operation in curbing international terrorism, which repeated the appeal of the previous resolutions for more participation by States in controlling the problem and welcomed the efforts of ICAO and IMO (International Maritime Organization) to curb unlawful interference with civil aviation and shipping respectively. The Resolution also called upon other specialized agencies and inter-governmental organizations, in particular, the Universal Postal Union, the World Tourism Organization and the International Atomic Energy Agency, within their respective spheres of competence, to consider what further measures could usefully be taken to combat and eliminate terrorism.[530] This resolution was followed by another, in December 1989, which called for a universal policy of firmness and effective measures to be taken in accordance with international law in order that all acts, methods and practices of international terrorism may be brought to an end.[531] The Resolution also expressed the grave concern of the United Nations Mentor States at the growing and dangerous link between terrorist groups, condemned traffickers of drugs and paramilitary gangs which had been known to perpetrate all types of violence, and thereby endanger the constitutional order of States and violating basic human rights.[532]

In 1991, the United Nations General Assembly once again unanimously condemned as criminal and unjustifiable all acts, methods and practices of terrorism; called firmly for the immediate and safe release of all hostages and abducted persons; and called upon all States to use their political influence in accordance with the Charter of the United Nations and the principles of international law to secure the safe release of all hostages and abducted persons and do their utmost to prevent commission of acts of hostage-taking and abduction.[533] The plea for international co-operation was reviewed by the General Assembly in December 1993 where the Assembly urged the international community to enhance co-operation in the fight against the threat of terrorism at national, regional and international levels.[534]

---

[528] S/RES/579 (1985), 18 December 1985.

[529] A/RES/42/159, 7 December 1987.

[530] A/RES/42/159, 7 December 1987.

[531] A/RES/44/29, 4 December 1989.

[532] A/RES/44/29, 4 December 1989, Clause 9.

[533] A/RES/46/51, 9 December 1991, Clauses 1 and 8.

[534] A/RES/48/122, 20 December 1993, Clause 2.

B. International Conventions

I. Convention for the Prevention and Punishment of Terrorism (1937)

Prior to the Tokyo Convention of 1963, most of the legal work relating to the security of international civil aviation was undertaken by the League of Nations or thereafter by the United Nations. The League of Nations, which was impelled to act in response to the increase of international terrorist activities following World War I, had already made several multilateral attempts to deal with the problem. Its initial efforts towards multilateral accord were directed towards the establishment of an International Convention for the Prevention and Punishment of Terrorism.[535] In spite of these attempts, governments took determined action against terrorism only after a major terrorist attack on 9 October 1934, which resulted in the assassination at Marseilles of King Alexander I of Yugoslavia, during his visit to France, and the murder of the French Foreign Minister, Mr. Louis Barthou, who was officially receiving the King in Marseilles.[536] The Yugoslav Government made a request of the Council of the League of Nations to investigate the incident.[537]

The Council of the League of Nations set up a Committee of Experts on 10 December 1934 to prepare a draft convention for the prevention and punishment of terrorism. The draft was submitted to an international conference in Geneva in November of 1937 and was adopted. Subsequent to approval of this convention, it was unfortunately precluded from entering into force owing to the outbreak of World War II.

II. Convention on International Civil Aviation (Chicago Convention of 1944)

At the time the Chicago Conference was held in 1944, and during the drafting of the Convention on International Civil Aviation,[538] although no explicit mention was made of the security of international civil aviation since such acts were unknown at that time, several States made reference to the significance of the Convention to security and safety of air travel. The Preamble to the Convention endorses its role at

---

[535]Opened for signature at Geneva on 16 November 1937. For the test see Hudson (1941, p. 862); U.N. Doc A/C.6/418 Annex 1, p. 1 (hereinafter 1937 U.N. Convention).

[536]McWhinney (1987, p. 128); see also, *A.J.I.L.*, Vol. 68, (1974), p. 69.

[537]McWhinney (1987, p. 129).

[538]*Convention on International Civil Aviation*, opened for signature at Chicago on 7 December 1944, entered into force on 4 April 1947. ICAO Doc 7300/6 (hereinafter Chicago Convention of 1944).

B. International Conventions

ensuing security and safety of international civil aviation in creating and preserving international civil aviation friendship and understanding among the nations and peoples of the world, and the necessity, therefore, to develop international civil aviation in a safe and orderly manner and to establish international air transport services on the basis of equal opportunity as well as sound and economic operation. Other provisions of the Convention also indicate clearly that safety of civil aviation is one of its main objectives. Article 25 of the Convention provides that:

> each Contracting State undertakes to provide such measures of assistance to aircraft in distress in its territory as it may find practicable, and to permit, subject to control by its own authorities, the owners of the aircraft or authorities of the State in which the aircraft is registered to provide such measures of assistance as may be necessitated by the circumstances. Each Contracting State, when undertaking search for missing aircraft, will collaborate in co-ordinated measures which may be recommended from time to time pursuant to this Convention.[539]

This principle gives effect to one of the oldest principles of customary international law,[540] which incorporates principles of humanitarian law, falling under the category of International Humanitarian Law. At the time of its incorporation into the Chicago Convention, however, it was deemed one of the lesser significant aspects of international law.

"Aircraft in distress" is not defined in the Chicago Convention or in other ICAO documents. In its report, the *Ad Hoc* Group of Experts on unlawful interference agreed that, regardless of the terminology used, the objective of assistance to aircraft in distress provided some sense of security for international civil aviation.[541]

The Convention defines its scope in Article 3:

(a) This Convention shall be applicable only to civil aircraft, and shall not be applicable to State aircraft;
(b) aircraft used in military, customs and police services shall be deemed to be State aircraft;
(c) no State aircraft of a Contracting State shall fly over the territory of another state or land thereon without authorization by special agreement or otherwise, and in accordance with the terms thereof; and
(d) the Contracting States undertake, when issuing regulations for their state aircraft, that they will have due regard for the safety of navigation of civil aircraft.

The Chicago Convention applies only to civil aircraft, to the exclusion of state aircraft.[542] While the Convention defines state aircraft to include aircraft used in

---

[539] Chicago Convention of 1944, ICAO Doc 7300/6.
[540] McWhinney (1987, p. 131).
[541] Report of the *Ad Hoc* Group of Experts – Unlawful Interference, Montreal, ICAO Doc. AH-UI/2, 14–18 July 1986.
[542] Art. 3(a) of Chicago Convention of 1944, ICAO Doc 7300/6.

military, custom, or police services,[543] it fails to define a civil aircraft. All aircraft not devoted to military, customs and police services may be deemed to be civil aircraft, although it would not be incorrect to apply definitive boundaries to Article 3 in the light of the ambivalence of the provision.

Article 3(d) requires Contracting States, when issuing regulations for their state aircraft, to have due regard for the safety of navigation of civil aircraft.

It should also be noted that a State cannot use civil aircraft in a manner that is incompatible with the purposes of the Chicago Convention. In other words, according to the Convention, abuse of civil aviation is prohibited.

Article 4 of the Convention states:

> Each Contracting State agrees not to use civil aviation for any purpose inconsistent with the aims of this Convention.

and therefore deals explicitly with the problem of misuse of civil aviation. It can therefore be assumed that the intent of the States parties to the Chicago Convention was to preclude any threat to the security of nations by adopting this provision.[544]

III. United Nations Charter

Although the Charter contains no provision which deals directly with the security of civil aviation, it is one of the most salutary international legal documents in the area of civil aviation security. The Preamble to the Charter stipulates that citizens of the member States of the United Nations will practice tolerance and live together in peace with one another as good neighbours. The principle of security is embodied in several articles of the Charter. Article 1(2) provides that the purpose of the United Nations is to pursue the development of friendly relations among nations based on respect for the principle of equal rights and self-determination of peoples, and to take other appropriate measures to strengthen universal peace.

As civil aircraft are by definition presumed to transport civilians, the principles of the Chicago Convention should ensure the protection of civilians and their property from dangers affecting civil aircraft in flight. The United Nations Charter can therefore be regarded as imputing to the international community a duty to protect the human being and his property in relation to flight:

> There is a mandatory obligation implied in article 55 of the Charter that the United Nations "shall promote respect for, and observance of, human rights and fundamental freedoms"; or, in terms of article 13, that the Assembly shall make recommendations for the purpose of assisting in the realization of human rights and freedoms. There is a distinct element of legal duty in the understanding expressed in article 56 in which all members pledge

---

[543] Art. 3(a) of Chicago Convention of 1944, ICAO Doc 7300/6.
[544] Milde (1986, p. 122).

themselves to take joint and separate action in co-operation with the organization for the achievement of the purpose set forth in article 55.[545]

A civil aircraft, when identified as such cannot be attacked.[546] The United Nations Charter opposes the use of force against civilian aircraft. Article 2(4) of the charter prohibits the use of force in any manner inconsistent with the purposes of the Charter. There is also provision for the settlement of disputes by peaceful means.[547]

An armed attack against an aircraft is a special kind of aggression[548] and is protected by the right of self-defence which is recognized against such an attack, by Article 51 of the Charter. This provision narrows the field of the exercise of self-defence to circumstances involving an armed attack. An unauthorized entry into the airspace of a State by an unarmed aircraft does not constitute an armed attack, even if such entry is effected for the purposes of espionage or provocation.[549] Although no authoritative definition of an armed attack has ever been adopted internationally, it is generally presumed that an armed attack would constitute belligerence endangering the safety of those affected by such attack when it is carried out by an offender(s) wielding weapons.

IV. The Geneva Convention on the High Seas (1958)

Transportation systems have often attracted terrorist attacks and the international community has come to terms with the vulnerability of modern aviation, taking sustained steps towards the protection of aviation.

The earliest forms of terrorism against international transportation was piracy. Pirates are considered by international law as common enemies of all mankind. The world has naturally an interest in the punishment of offenders and is justified in adopting international measures for the application of universal rules regarding the control of terrorism. The common understanding between States has been that pirates should be lawfully captured on the high seas by an armed vessel of any particular State, and brought within its territorial jurisdiction for trial and punishment. Lauterpacht recognized that:

> Before international law in the modern sense of the term was in existence, a pirate was already considered an outlaw, a hostis humani generis. According to the Law of Nations, the act of piracy makes the pirate lose the protection of his home State, and thereby his national character. Piracy is a so-called international crime, the pirate is considered enemy of all States and can be brought to justice anywhere.[550]

---

[545]Lauterpact (1950, p. 149).
[546]Vlasic (1982, p. 161).
[547]Art. 33 of the U.N. Charter.
[548]Kunz (1948, pp. 111, 115).
[549]Vlasic (1982, p. 275).
[550]Cited in Oppenheim (1958, p. 609).

It is worthy of note that under the rules of customary international law the international community had no difficulty in dealing with acts of terrorism which forms the offence of sea piracy. Due to the seriousness of the offence and the serious terroristic acts involved, the offence was met with the most severe punishment available – death. The universal condemnation of the offence is reflected in the statement:

> In the former times it was said to be a customary rule of international law that after the seizure, pirates could at once be hanged or drowned by the captor.

The laws dealing with the offence of piracy went through a sustained process of evolution. In 1956, while considering legal matters pertaining to the law of the sea, the International Law Association addressed the offence of piracy and recommended that the subject of piracy at sea be incorporated in the Draft Convention of the Law of the Sea. This was followed by the United Nations General Assembly Resolution (Resolution No. 1105 (XI) in 1957 which called for the convening of a diplomatic conference to further evaluate the Law of the Sea). Accordingly, the Convention of the High Seas was adopted in 1958 and came into force in September 1962.

The Geneva Convention of the High Seas of 1958[551] was the first attempt at international accord to harmonize the application of rules to both piracy at sea and in air. The Convention adopted authoritative legal statements on civil aviation security, as it touched on piracy over the high seas.[552]

Article 5 of the Convention inclusively defines piracy as follows:

> Piracy consists of any of the following acts:
> (1) Any illegal acts of violence, detention or any act of depredation, committed for private ends by the crew or the passenger of a private ship or a private aircraft, and directed:
>
>> (a) on the high seas, against another ship or aircraft, or against persons or property on board such ship or aircraft;
>> (b) against a ship, aircraft, persons, or property in a place outside the jurisdiction of any state;
>
> (2) Any act of voluntary participation in the operation of a ship or of an aircraft with knowledge of facts making it a pirate ship or aircraft;
> (3) Any act of inciting or of internationally facilitating an act described in sub-paragraph 1 or sub-paragraph 2 of this article.

As provided for by Article 14 of the Convention, there is incumbent on all States a general duty to "co-operate" to the fullest extent in the repression of piracy as defined by the Convention. One commentator has observed,

> The International Law Commission in its 1956 report, however, deemed it desirable to enjoin co-operation in the repression of piracy, to define the act to include piracy by

---

[551]The Geneva Convention was opened for signature at Geneva on 16 November 1937. See Hudson (1941, p. 862), U.N. Doc. A/C.6/418, Annex 1, at 1.

[552]League of Nations, *Official Journal*, 1934, at 1839.

B. International Conventions

aircraft, as set forth in the repressive measures that may justifiably be taken. The United Nations conference on the Law of the Sea in Geneva in 1958 accordingly incorporated these adjustments of the law to modern times in its convention on the High Seas.[553]

Article 14 seemingly makes it a duty incumbent upon every State to take necessary measures to combat piracy by either prosecuting the pirate or extraditing him to the State which might be in a better position to undertake such prosecution. The Convention, in Article 19, gives all States universal jurisdiction under which the person charged with the offence of aerial or sea piracy may be tried and punished by any State into whose jurisdiction he may come. This measure is a proactive one in that it eliminates any boundaries that a State may have which would preclude the extradition or trial in that State of an offender. Universal jurisdiction was conferred upon the States by the Convention also to solve the somewhat complex problem of jurisdiction which often arose under municipal law where the crime was committed outside the territorial jurisdiction of the particular State seeking to prosecute an offender. The underlying salutary effects of universal jurisdiction in cases of piracy and hijacking which was emphasized by the Convention, is discussed in the following manner:

> the absence of universal jurisdiction in relation to a given offence, means that, if a particular State has no jurisdiction either on the basis of territoriality or protection, or on the personality principle, whether passive or active, it will not be authorized to put the offender on trial, even if he is to be found within the territorial boundaries of the State.[554]

The inclusion of the offence of "piracy" in the Convention brings to bear the glaring fact that the crime is international in nature, giving the international community the right to take appropriate measures to combat or at least control the occurrence of the offence. The General Convention by its very nature and adoption has demonstrably conveyed the message that piracy is a heinous crime which requires severe punishment. The Convention also calls for solidarity and collectivity on the part of nations in combating the offence in the interests of all nations concerned.[555]

Notwithstanding the above, it is worthy of note that the phenomenon of hijacking as it exists today need not necessarily fall within the definition of piracy as referred in Article 15 of the High Seas Convention (1958). Although there exists a marked similarity between the offenses of unlawful seizure of aircraft and acts of piracy directed against ships on the high seas, in that in both cases, the mode of transportation is threatened and abused and the safety of the passengers, crew members and the craft itself is endangered by the unlawful use of force or threat, there may still be a subtle difference that may exist between the offence as applying to sea transport and to air transport.

---

[553]Reiff (1959, p 6).
[554]Feller (1972, p. 212).
[555]Feller (1972, p. 212).

Whilst admittedly, there are similarities between the acts of piracy against ships and those against aircraft, the legal differences that may exist should have to be determined in order to inquire whether aircraft hijacking amounts to piracy as defined by the Convention.

The essential features of definition of piracy as are incorporated in the Geneva Convention are as follows: (1) the pirate must be motivated by "private" as opposed to "public" ends; (2) the act of piracy involves action affecting a ship, an aircraft; (3) the acts of violence, detention, and depredation take place outside the jurisdiction of any State, meaning both territorial jurisdiction and airspace above the State; (4) acts committed on board a ship or aircraft, by the crew or passengers of such ship or aircraft and directed against the ship or aircraft itself, or against persons or property, do not constitute the offence of piracy.

Upon close examination, it appears that the definition of piracy does not apply to the phenomenon of aerial piracy or hijacking. Firstly it is a fact that most hijackings are not carried out in pursuance of private ends. INTERPOL[556] reported in 1977 that the percentage of cases in which political motives had impelled the offender was 64.4%. Hijacking of aircraft for political motives would thus not relate to Article 15(1) of the Convention on the High Seas (1958) since acts solely inspired by political motives are excluded from the notion *piracy jure gentium*. Sami Shubber has observed of the 1958 Convention that its inapplicability to the notion of aerial piracy may lie in the fact that private ends do not necessarily mean that they can affect private groups, acting either in pursuance of their political aims, or gain. The fact that it is not always possible to distinguish between private ends and public ends in defiance of the political regime of the flag State may be said to be covered by Article 15(1) of the Convention.[557] The reasons given by Shubber were that "private ends" do not necessarily mean private gain.

Under the definition, the act of illegal violence or detention must be directed on the high seas, against another ship or aircraft. It is obvious therefore that this interpretation does not apply to hijacking since the offence of hijacking is committed by the offender who travels in the aircraft. It is hard to imagine that an offender could enter an aircraft from outside while the aircraft is in flight. The Convention also excludes acts committed on board a ship by the crew or passenger and directed against the ship itself, or against persons or property on the ship, from the scope of piracy,[558] which will also make the definition inconsistent with the exigencies related to the offence of aerial piracy.

Although piracy, according to the Convention, must be committed on the "high seas," instances of hijacking may occur anywhere. Furthermore, piracy under Article 15 of the Convention must involve acts of violence, detention or depredation.

---

[556]INTERPOL had submitted to the Legal Committee of ICAO in 1977 that out of recorded hijackings up to that year, the percentage of instances of hijackings which were motivated politically was 6.2 at a ratio of 64:4. See *ICAO Doc 8877-LC/161* at 132.

[557]Shubber (1973a, p. 226).

[558]Aircraft Hijacking, *Harvard International Law Journal*, Vol. 12 (1971) at 65.

Most hijackings, however, have been carried out simply by the use of threats, and may even be carried out through a variety of means other than those involving violence or force.

It is therefore reasonable to conclude that hijacking does not necessarily and absolutely fall within the "aircraft piracy" as defined by the Geneva Convention on the High Seas.[559] The hopes of the international community to control the crime of hijacking through the application of Geneva Convention on the High Seas (1958) may therefore have been frustrated by the exclusivity of the nature of the two offenses of aerial piracy and piracy related to the high seas. The Convention remains therefore to be of mere academic interest for those addressing the issue of aerial piracy.

C. Concerted Action Under the Auspice of the International Civil Aviation Organization: The Tokyo Convention (1963)

Shocked by the rising trend of aircraft hijacking in the early 1960s and the failure of the Geneva Convention on the High seas to offer rules applicable to the offence of hijacking, the international community considered adopting the Tokyo Convention of 1963, which was adopted under the aegis of ICAO. This Convention attempted to provide certain rules that would address the offence.

Prior to 1960, most of the collective action to combat international terrorism was undertaken by the United Nations or its predecessor, the League of Nations. Although the League of Nations made cohesive efforts to create an international criminal court, to deal, among other things, with acts of international terrorism by drafting a Convention to Combat International Terrorism in 1937,[560] it was unfortunate that this Convention was signed only by 13 States and ratified by one State and was effectively precluded the Convention from coming in force.

At the end of 1950, a new crusader against international terrorism – particularly that which applied to aerial incidents of terrorism – appeared in the international scene to adopt necessary international measures to combat terrorism against air transport. This new entity was the International Civil Aviation Organization. In retrospect, it is noted that although the United Nations was unsuccessful in adopting sufficiently compelling measures of international co-operation to deal with aircraft hijacking, ICAO has made significant strides in the area of adoption of multilateral conventions. The primary aim of these Conventions has been to adopt measures, through international agreement, to control and arrest terrorist activities which are aimed against international air transport. It has been said of ICAO on its regulatory attempts in this field:

---

[559]Van Panhuys (1970, p. 13).
[560]This Convention was opened for signature at Geneva on 16 November 1937. See Hudson (1941, p. 862). See also *U.N Doc, A/C.6/418*, Annex 1 at 1.

these menacing incidents during the last few years have resulted in intense activities aimed at finding possible solutions on the basis of universally accepted international treaty and/or other technical remedies. The beginning of concerted international effort since the formation of ICAO in relation to the so-called problem of hijacking can be traced back to the formulation of certain provisions in the "Convention on Offenses and Certain Other Acts Committed on Board Aircraft Commonly Known as the Tokyo Convention 1963."[561]

The Tokyo Convention was the first substantial effort at dealing with terrorism in the air. It was followed by The Hague and Montreal Conventions.[562]

In 1950, the Legal Committee of ICAO, upon a proposal from the Mexican Representative on ICAO Council for study of the legal status of airports, referred the subject to the *ad hoc* Sub-Committee established by the Legal Committee.[563] After a survey had been made of all the problems relating to legal status of aircraft, it was decided by the Committee that the best course would be to confine the work to a detailed examination of some particularly important matters, namely crimes and offenses committed on board aircraft, jurisdiction relating to such crimes and the resolution of jurisdictional conflicts. The Sub-Committee thought that resolving these problems was of vital importance for the following reasons:[564]

(1) One characteristic of aviation is that aircraft fly over the high seas or over seas having no territorial sovereign. While national laws of some States confer jurisdiction on their courts to try offenses committed on aircraft during such flights, this was not the case in others, and there was no internationally agreed system which would co-ordinate the exercise of national jurisdiction in such cases. Further, with (the) high speed of modern aircraft and having regard to the great altitudes at which they fly as well as other factors, such as meteorological conditions, and, in certain parts of the world, the fact that several States may be overflown by aircraft within a small space of time, there could be occasions when it would be impossible to establish the territory in which the aircraft was at the time a crime was committed on board. There was, therefore, the possibility that in such a case, and in the absence of an internationally recognized system with regard to exercise of national jurisdiction, the offender may go unpunished.

(2) National jurisdictions in respect of criminal acts are based on criteria which are not uniform; for example, on nationality of the offender, or nationality of the victim, on the locality where the offence was committed, or on nationality of the aircraft on which the crime occurred. Thus, several States may claim jurisdiction over the same offence committed on board aircraft, in certain cases. Such conflict of jurisdictions could be avoided only by international agreement.

---

[561] See Sarkar (1972, p. 200).

[562] See *International Legal Materials* 1963, (II) at 1042.

[563] See Boyle (1964, pp. 305–328), for a detailed analysis of the Tokyo Convention.

[564] Report of the Sub-Committee, LC/SC "Legal Status," WD No. 23, 10 October 1956.

C. Concerted Action Under the Auspice of the ICAO

(3) The possibility that the same offence may be triable in different States might result in the offender being punished more than once for the same offence. This undesirable possibility could be avoided by a suitable provision in the Convention.

After sustained deliberation and contradiction, the Sub-Committee on Legal Status of Aircraft produced a draft convention which was submitted to the Legal Committee on 9 September 1958.[565] The Legal Committee in turn considered the draft convention at its 12th Session held in Munich in 1959,[566] undertaking a substantial revision of the draft. The revised text was submitted to the ICAO Council subsequently, who in turn submitted the draft to Member States and various international organizations for their comments. A new Sub-Committee was formed for the purpose of examining the Convention of State organization in 1961 to examine and prepare a report. This report was studied by the Legal Committee in its 14th Session held in Rome in 1962. A final text of a Convention was drawn up at this meeting and communicated to Member States with a view to convening a Diplomatic conference in Tokyo with a long-term prospect of adopting a Convention on aerial rights. This Convention was signed in Tokyo on 14 September 1963 by the representatives of 49 ICAO Member States, and entered into force after six years, on 4 December 1969.[567] This slow process of ratification of the Convention (5 years) was by no means due to the ineptitude of the Convention as has been claimed[568] but was solely due to the fact that the Convention was drafted prior to the series of hijacking in the late 1960s and was not implemented with due dispatch by most States. Another reason for the delayed process was the complicated legal and political issues facing many countries at the time of the adoption of the Convention.[569] A significant feature of the Tokyo Convention is that although at first, States were slow in acceding to or ratifying the Convention, 80 States ratified the convention within one year (1969–1970) presumably in response to the spate of hijackings that occurred during that period.

The main purpose of the Tokyo Convention was to secure the collaboration of States in restraining terrorist activity directed at air transport. It has therefore been said that:

> The first action taken by the international community to combat hijacking was the Tokyo Convention 1963. This Convention was originally designed to solve the problem of the commission of crimes on board aircraft while in flight where for any number of reasons the criminal might escape punishment.[570]

---

[565]Boyle (1964, p. 320).
[566]Boyle (1964, p. 321).
[567]Boyle (1972, p. 463).
[568]Boyle (1972, p. 463).
[569]Abramovsky (1974, p. 89).
[570]Boyle (1972, p. 463).

The objectives of the Tokyo convention may be subsumed into four principal areas:

(1) The Convention makes it clear that the State of registration of the aircraft has the authority to apply its laws. From the standpoint of States such as the United States, this is probably the most important aspect of the Convention, since it accords international recognition to the exercise of extraterritorial jurisdiction under the circumstances contemplated in the Convention;
(2) The Convention provides the aircraft commander with the necessary authority to deal with persons who have committed, or are about to commit, a crime or an act jeopardising safety on board his aircraft through use of reasonable force when required, and without fear of subsequent retaliation through civil suit or otherwise;
(3) The convention delineates the duties and responsibilities of the contracting State in which an aircraft lands after the commission of a crime on board, including its authority over, and responsibilities to, any offenders that may be either disembarked within territory of that State or delivered to its authorities;
(4) The crime of 'hijacking' has been addressed in some degree of depth.[571]

The Convention applies to any act that is an offence under the penal laws of a Contracting State, as well as to acts which, whether or not they are offenses, may jeopardise safety, good order and discipline on board. The Convention thus does not define the offence at the international level nor does it explicitly explain the nature of the offence. Alona E. Evans has observed:

> The offence is not made a crime under international law; its definition is to be determined by the municipal laws of the contracting State.[572]

Admittedly, there are some limitations placed upon the scope of the application of the Convention. Firstly, the Convention excluded from its operations aircraft used in military, customs or police services. It should be noted that reference is not made in the Convention to "State aircraft" as mentioned in Article 3 of the Chicago Convention, which does not apply to such aircraft. This difference in terminology is explained by the fact that State aircraft provide air transport that is usually provided by civil aircraft and civil transport in some cases. Secondly, offenses against penal laws of a political nature or those based on racial or religious discrimination are not covered by the Convention except to the extent that the Convention addressed such acts which jeopardise safety or good order and discipline on board. the reason for excluding those offenses from the scope of the Convention could be attributed to the view:

> Penal laws forbidding various forms of racial and religious discrimination take many and varied forms, and the views of the Courts of the Contracting States may differ on the issue of whether one or the other is within or without the Convention. Even more divergence of

---

[571]Boyle (1964, p. 329).
[572]Evans (1969, p. 708).

view can be expected in decisions which involve the question of whether a particular offence is of a "political nature."[573]

Although the Convention does not define the offence of hijacking, Article 11 specifies the circumstances that would constitute the offence as:

> When a person on board has unlawfully committed by force or threat thereof an act of interference, seizure or other wrongful exercise of control of an aircraft in flight or such an act is bound to be committed.

When the offence of hijacking is committed in the above manner, the State in which the aircraft lands has two obligations which it must satisfy according to the terms of the Convention. The first obligation is that the landing State "shall take all appropriate measures to restore control of the aircraft to its lawful commander or to preserve control of the aircraft and shall return the aircraft and its cargo to the person lawfully entitled to possession."

R.P. Boyle emphasized the above contention when he stated:

> The obligation assumed by a State under the Tokyo Convention with respect to the disposition of the hijacker... is to take all appropriate measures to restore control of the aircraft to its lawful commander and to permit the passengers and crew to continue their journey as soon as practicable and to return the aircraft and cargo to persons lawfully entitled.[574]

I. The Powers Given to Aircraft Commander and Others in Order to Combat Hijackings

The Convention gives wide powers to the aircraft commander to control the offence of hijacking. Article 6 enables the aircraft commander to use reasonable measures, including restraint, to protect the safety of the aircraft, and maintain good order and discipline, when he has a reasonable ground to believe that a person has committed an offence contemplated in Article I (1), *viz.*:

(a) Offences against penal laws
(b) Acts which, whether or not they are offenses, may jeopardise the safety of the aircraft or of persons or property therein or which jeopardise good order and discipline on board

An interesting observation may be made in respect to requirement in (a) above. The aircraft commander will have, according to that paragraph, the power to take measures and restrain a passenger even if his act did not amount to jeopardising the safety of the aircraft or the person or the property therein. This may lead to absurdity. If for example two passengers conspire, while on board the aircraft,

---

[573]Boyle (1964, p. 333).
[574]Boyle (1964, p. 331).

to commit some illegal act upon landing, or upon termination of the flight, according to sub-paragraph (a) above the commander can restrain them on the suspicion that the act they are conspiring to commit, is against penal law of a particular jurisdiction. This seems to be illogical when one recalls that the principal objective of the Convention is to assure the maintenance of safety and good order "on board" the aircraft.

The aircraft commander in discharging his duties according to the Convention can require or authorize the assistance of the crew and request the assistance of passengers for that purpose. Even passengers and crew members are authorized under Article 6(2) to take reasonable preventive measures without any authorization from the aircraft commander whenever they have reasonable grounds to believe that such action is immediately necessary for safety reasons. Although this clause has tried to give powers to other people beside the aircraft commander in order to tighten the measures that leads to the thwarting of acts of unlawful interference against civil aviation, some delegates at the Tokyo Conference attacked this approach on the ground that passengers normally would not be qualified to determine whether a particular act jeopardized the safety of the aircraft or persons and property therein. For this reason, it was unwise to give this authority to passengers.[575] However, this argument was rejected on the ground that this provision contemplated an emergency type of situation on which the danger of the aircraft or persons and property on board was clearly present, and in fact no special technical knowledge would be required to recognize the peril.[576]

The powers entrusted to the commander in order to suppress any unlawful act that threatens the safety of the aircraft go as far as requiring the disembarking of any person (who commits any of the acts referred to in Article 1(1) and discussed above) in the territory of any State in which he lands and delivering him to its competent authorities.[577] The State is under an obligation to allow the disembarkation and to take delivery of the person so apprehended by the aircraft commander, but such custody may only be continued for such time as is reasonably necessary to enable the criminal extradition proceedings (if any) to be instituted. In the meantime the State of landing should make a preliminary enquiry into the facts and notify the State of registration of the aircraft[578] (Articles 12 and 13).

In any event, the commander as well as the crew members and passengers are given immunity from suits by the alleged offender against whom they acted. Article 10 expressly provides:

> Neither the aircraft commander, any member of the crew, any passenger, the owner or operator of the aircraft, nor the person on whose behalf the flight is performed shall be held responsible in any proceedings on account of the treatment undergone by the person against whom the actions were taken.

---

[575]Boyle (1964, p. 340).

[576]Boyle (1964, p. 340).

[577]See Articles 8 and 9 of the Tokyo Convention.

[578]See Articles 12 and 13 of the Tokyo Convention.

This protection was given to the aircraft commander and other persons in order to encourage them to fight the wrongful acts contemplated by the Convention.

II. Jurisdiction to Punish the Terrorists

The major problem that States often face in the process of combating terrorism is the issue of jurisdiction. This is most evident in cases of hijacking where the crime often takes place outside the jurisdiction of the receiving State, although in most, if not all of the cases it could be argued that the offence is of a continuing nature. Under international law, State's jurisdiction to prosecute is founded upon two traditional concepts. First, there must exist a substantial link between the person or the act and the State claiming sovereign jurisdiction and second, this theoretical basis must be actualized through a sovereign act, i.e., legislation for implementation of this theoretical act. In an act of international nature, such as hijacking, two or more States involved may possess jurisdiction to prosecute. As a result, jurisdictional conflicts are eminent, since two or more of those States can claim the right to prosecute and press claims against each other through diplomatic channels. In order to eliminate these conflicts, the jurisdictional rules incorporated in the Tokyo Convention were preferred. The Tokyo Convention was adopted to grant powers to States to establish jurisdiction which would be uncomplicated by diplomatic claims over criminal acts committed on board aircraft.

Jurisdiction over offenses and acts committed on board appertain primarily to the State of registration of the aircraft (Article 3(1)). The adoption of this rule guarantees to the flights over the High Seas the assured presence of the criminal law. It provides a sound legal basis for extra-territorial exercise of criminal jurisdiction extending even to cases of flight within foreign airspace. A.I. Mendelssohn has observed:

> As a matter of international law, therefore, any crime abroad an international carrier, no matter where, by or against whom it is committed, can be punished by at least one sovereign – the State of registration of the carrier. All doubts are removed on the question whether the flag will henceforth follow the aircraft as it traditionally has followed a vessel.[579]

Article 3(2) of the Convention provides:

> Each Contracting State shall take measures as may be necessary to establish its jurisdiction as State of registration over offenses committed on board aircraft registered in such State.

It is clear that the fundamental objective of this sub-paragraph was to make the act of combating hijacking an international issue in which all States must take part when need arises.

Article 3(3) went further to give more grounds of jurisdiction in order to eliminate the gravity of the obstacles that hinders the prosecution of hijackers.

---

[579]Mendelsohn (1967, p. 515).

It provides that the Convention does not exclude criminal jurisdiction exercised in accordance with the national law. Mendelssohn has commented on this sub-paragraph:

> Its objectives are (a) to retain all existing jurisdiction presently asserted by the various States; (b) to enable them to enact further legislation providing for even more extensive jurisdiction; and most important, (c) to require the State of registration to extend at least some of its criminal laws to its aircraft and to provide an internationally accepted basis for the application and enforcement of these laws.[580]

The Convention also authorizes a contracting State which is not a State of registration to interfere with an aircraft in flight in five cases in which the offence (a) has an effect on the territory of State; (b) has been committed by or against a national or permanent resident of State; (c) is against the security of the State; (d) consists of a breach of any rules or regulations relating to the flight or manoeuvre of aircraft in force in such State; (e) that the exercise of such jurisdiction is necessary in order to ensure the observance of any obligation of such State under a multilateral international agreement.

As regards the geographic scope of the Convention for jurisdictional purposes, Article 1 provides that the Convention applies in respect of acts or offenses committed while the aircraft is "in flight" or on the surface of the High Seas or on another area which does not have a territorial sovereign. The term "in flight" is defined in Article 1(3) as "from the moment when the power is applied for the purpose of take-off until the moment when the landing run ends." Hence, hijacking attempts initiated during the time the aircraft is parked or taxing are not considered to be within the ambit of the Convention. As a consequence, the provisions of Tokyo convention fell short of curbing the crime of sabotage of air transport facilities. This shortcoming of the Tokyo Convention, *inter alia*, led to the adoption of Montreal Convention (1971).

III. Powers and Duties of States

It is a basic obligation of a State to co-operate with other States in order to ensure the safety of international civil aviation. Article 11 of the Tokyo Convention, which is referred to above, provides that Contracting States have certain obligations whenever a person on board an aircraft has unlawfully committed by force or threat thereof an act of interference, seizure or other wrongful exercise of control. The question of whether a particular act is lawful or unlawful is to be judged by the law of the State of registration of the aircraft or the law of the State in whose airspace the aircraft may be in flight. Paragraph 1 of Article 11 imposes on all the Contracting States the obligation to take appropriate measures to restore or to preserve the aircraft commander's control of the aircraft. The words "appropriate measures" are

---

[580] Mendelsohn (1967, p. 514).

intended to mean only those things which it is feasible for a Contracting State to do and also only those which it is lawful for a Contracting State to do. Thus, a Contracting State, which is situated thousands of miles away from the scene of the hijacking, is not under any obligation to take any action, because it would not be feasible for it do so.

Article 12 imposes another obligation on each Contracting State. This Article is a corollary to Articles 6 and 8 of the Convention. The latter two Articles authorize the aircraft commander to disembark any person who has committed, or is about to commit, an act of the type described in Article I of the Convention. Article 12 obliges a contracting State to allow the commander of an aircraft registered in another contracting State to disembark the alleged offender. Article 12 provides:

> Any Contracting State shall allow the commander of an aircraft registered in another contracting state to disembark any person pursuant to Article 8, paragraph 1.

Thus, it is clear that the obligation of a Contracting State to permit disembarkation of a hijacker, at the request of the aircraft commander, is an unqualified obligation.

Article 13 of the Convention deals with the obligation of a contracting State to take delivery of a person from the aircraft commander. This provision should be contrasted with the authority of the aircraft commander to disembark. The obligation of the contracting State under this Article is a corollary to the authority given to the aircraft commander under Articles 6, 7 and 9.

Paragraph 1 of Article 13 states the primary unqualified obligation of each Contracting State to "take delivery." Paragraph 2 addresses the obligation of a contracting State, after having taken delivery, to take custody. It provides that the Contracting State is under an obligation to take "custody" only if it is satisfied that the circumstances so warrant such action. Thus, the State is left free to judge for itself whether the act is of such a nature as to warrant such action on its part and whether it would be consistent with its law since under paragraph 2 any such custody is to be affected only pursuant to the law of the State taking custody. However, such custody may only be continued for that period of time which is reasonably necessary to enable criminal proceedings to be brought by the State taking custody, or for extradition proceedings to be instituted by another interested or affected State. On the other hand, any person taken into custody must be given assistance in communicating immediately with the nearest appropriate representative of the State of which he is a national (Article 13(3)).

IV. Extradition

Article 16 of the Convention provides that offenses committed on aircraft registered in a contracting State are to be treated, for the purpose of extradition, as if they had been not only in the place where the offence has occurred, but also in the territory of the State of registration of the aircraft. Without prejudice to this provision it is declared that "nothing in this Convention shall be deemed to create an obligation to grant extradition." A commentator observes:

> The Tokyo Convention does not oblige the Contracting State to punish an alleged offender upon his disembarkation or delivery. Ironically, the landing State must set him free and let him proceed to the destination of his choice as soon as is practicable if it does not wish to extradite or prosecute him. The Contracting States are obliged to extradite the offenders, if at all, only under provisions of other treaties between them.[581]

The failure to provide for a machinery of mandatory extradition if prosecution was not conducted was considered a major set-back of the Tokyo Convention. However, the above loopholes from which the Convention severely suffers are not the only ones:

> Looking for the vantage point today, it is obvious that the Tokyo convention left major gaps in the international legal system in attempting to cope with the scope of aircraft hijacking. there were no undertaking by anyone to make aircraft hijacking a crime under its national law, no undertaking to see to it the crime was one punishable y severe penalties and most important, no undertaking to either submit the case for prosecution or to extradite the offender to a State which would wish to prosecute.[582]

V. *Responsibilities of States*

As has been mentioned, all States party to the Convention undertake to permit disembarkation of any person when the commander considers that it is necessary to protect the safety of the flight or for the maintenance of good order and discipline on board. States also commit themselves to take delivery of any person the commander reasonably believes has committed a serious offence on board.[583] In this case, when they have taken delivery, States concerned must make an immediate inquiry into the facts of the matter and report the findings to both the State of Registration and to the State of which the person is a national.[584] Where the State considers the circumstances warrant such action, it shall take custodial or other measures, in accordance with its laws, to ensure that the person delivered to it remains available while the inquiry is conducted. Such measures may be continued for a reasonable time to permit criminal or extradition proceedings to be instituted when such proceedings follow from the inquiry.[585]

Although the Convention is unequivocal in providing clearly that all contracting States should ensure their legal competence in respect of aircraft on their register,

---

[581] Chung (1976, p. 150).
[582] Boyle (1964, p. 320).
[583] Article 13(1).
[584] Article 13(5).
[585] Article 13(2).

C. Concerted Action Under the Auspice of the ICAO 227

thus addressing jurisdictional issues with regard to crimes on bound aircraft, there are a number of lapses in the Convention which make it open for criticism.

Firstly, the Convention does not apply to "aircraft used in military, customs or police services."[586] This is a topical issue which requires clarity, as in modern exigencies of airlines, there are instances when civilian aircraft are called upon to carry military personnel or supplies, as much as military aircraft are sometimes deployed to execute civilian flights.

Problems concerning registration, particularly over when the Convention insists on registration as a pivotal issue may also change the circumstances, although commanders could be totally ignorant of the laws of the State in which the aircraft they are flying is registered. The commander may be required to determine whether a certain action on his aircraft does in fact constitute a crime and more particularly, a serious crime. Since at most, a commander may have some familiarity with the laws of the State of the operator. The United Kingdom,[587] has elected to incorporate the terms of the Convention into its domestic legislation, thereby widening its scope to cover any aircraft controlled by its own nationals.

The Convention could also be improved upon its terms of chronology of the offenses in that its applicability extends to the period from "the moment when power is applied for the purpose of take-off until the moment when the landing run ends,"[588] and in relation to the powers of the commander, who has authority for the purposes of the Convention, only from the time at which external doors closed following embarkation to the time when doors open for disembarkation.[589]

These parameters are far from satisfactory. In relation to the first, courts have been inconsistent in interpreting similar definitions of flights used in insurance policies. It has been contended that power is first applied "for the purpose of take-off" when the aircraft first begins to move under its own power to the take-off position.[590] In relation to the second, the terms "all its external doors" also leaves confusion worse confounded in that it makes unclear whether "all its external doors" includes, for example, cargo or baggage hold doors, or doors giving access to such areas as the electronic compartment of the aircraft. It is not difficult to envisage circumstances in which these areas could be of significance. The main problem, however, is that the Convention does not provide for the manner in which the offender should be dealt with after he has been removed from the aircraft. The somewhat poor and inadequate drafting in Articles 14 and 15 seems to suggest that it is only where the person disembarked or delivered cannot or does not wish to continue his journey, that the State of landing can take action.[591] They do not offer a

---

[586] Article 1(4).
[587] Civil Aviation Act 1982, Section 92.
[588] Article 1(3).
[589] Article 6(2).
[590] McNair, *International Law Opinions*, Cambridge University Press: 1856 at 224.
[591] Article 16(2).

State any guidance as to questions arising from requests for extradition of an offender or extradition by the state's own initiation.

The Convention also fails to identify the "offenses and certain other acts committed on board" which are its subject matter as extraditable offenses, and therefore all requests for extradition arising out of an offence under the Convention must be dealt with under existing extradition arrangements. Even where those agreements are existing between the two States concerned, this could often lead to confusion and delay. Furthermore, in any case, many "jeopardising" acts are unlikely to be recognized as forming a basis for extradition. A marked omission from the Convention is that while it creates and defines "jeopardising" acts, it does not require States to treat these as "serious crimes" although the Convention's procedures in respect of delivery and extradition are applicable only to serious crimes.

With respect to extradition, the State of Registration of a leased aircraft which is involved in an offence will have little interest in pursuing a matter in which none of its nationals have been involved. A dry lease can further complicate the issue of extradition, since often in these circumstances such a state which is not directly involved in the offence is unlikely to be enthusiastic about incurring the trouble and expense associated with extradition and subsequent trial.

VI. An Answer?

It was no less a personage than the Roman Emperor, *Marcus Aurelius*, who concluded sadly that the choice in most human issues was "educate, or endure."

The international community must take cognizance of the fact that the Tokyo Convention is relatively ineffective if States do not make provisions in their own laws to give legal effect to the concerted action that is required at international law to combat terrorism. They must be persuaded to ensure, for example, that their laws of custody are such as to permit the immediate inquiry prescribed by the Convention to be properly conducted, an essential requirement if the evidence required for a successful prosecution is to be gathered. For this reason there should also be a requirement that an inquiry should follow any disembarkation.

States must also ensure that their laws in respect of extradition are framed in such a manner as to facilitate the State of Registration in taking action against the perpetrators of crime or "jeopardising" acts on board its aircraft. These laws should also be at least receptive to the idea of the State of the operator exercising a jurisdiction in respect of events on board aircraft controlled by its nationals.

States must also be persuaded of the need to exercise the criminal jurisdiction they have in respect of their own aircraft in such a manner as to deter potential offenders. Finally, States might embark upon a process of education to make their airport immigration and police authorities aware of the existence of the Tokyo Convention and of its provisions for disembarkation and delivery.

The airlines must also embark on a programme of education within their own ranks. In general there is great uncertainty on the part of captains as to the extent

C. Concerted Action Under the Auspice of the ICAO

and limits of their authority and they are often in total ignorance of the Tokyo Convention. All airlines should use guidance material on the relevant sections of the Convention for carriage in the cockpit of any material of assistance together with a current list of the contracting States. This material can prove invaluable when a commander is confronted by officials whose first reaction is often to refuse to permit a requested disembarkation or delivery.

Airlines do need to inform their pilots on the contents of the Convention and to brief them on how to collect evidence, how to request an investigation and how to file a complete report of the incident. They also need to liaise with their own local authorities to ensure that they are aware of the extent and seriousness of the problem and of the measures which the international community has devised for dealing with it. There is much work to be accomplished by the security, legal and operations departments of the individual airlines. A further incentive is that such a programme of benign propaganda may have the collateral effect of persuading immigration authorities of the folly of insisting on putting potentially violent deportees on board our aircraft.

Finally, the airlines themselves can and must do more to deal with the problem themselves. Alcohol is the underlying cause of the majority of incidents. Yet too often obviously drunk and unruly passengers are boarded – regardless of laws which make it an offence to enter any aircraft when drunk or to be drunk in an aircraft, as in the United Kingdom,[592] or for a pilot to allow a person obviously under the influence of drink or drugs to be carried in his aircraft, as in the U.S.A. The airlines should be careful to include in their contract with their passengers a condition which permits them to refuse carriage for reasons of safety or if, in the exercise of its reasonable discretion, the carrier determines that:

> ... (b) the conduct, age or mental or physical state of the passenger is such as to ... (ii) cause discomfort or make himself or herself objectionable to other passengers or (iii) involve any hazard or risk to himself or herself or to other persons or property....[593]

Too often airlines fail to exercise reasonable discretion to avoid potential offenses from being committed. It is all too common an occurrence that, once airborne, cabin crew members, in the absence of clear instructions from their employer, continue to supply alcohol to passengers even when the signs of impending trouble are obvious.

Airlines are often strangely reluctant to impose the very effective sanction available to them of refusing to carry on the return leg, a passenger who has been troublesome on the outbound leg of his journey. This is a powerful measure of deterrent and each airline should explore the possibility of using it with their own legal adviser.

If potential troublemakers were aware that their disruptive behaviour was likely to be followed not only by effective action by the State authorities but also likely to result in their being blacklisted by airlines, it is probable that the aviation community would be advance a considerable distance towards at least preventing the problem of

---

[592] Air Navigation Order 1989, S.I. 1989 No. 2004 – Article 52.
[593] IATA General Conditions of Carriage (passenger and baggage), March 1988 – Article VIII.

crime and unruliness on our aircraft from spiralling out of control. Therefore, the airline industry must embark on a programme of education and persuasion.

D. The Hague Convention on Hijacking 1970

The vast increase in the number of aircraft hijackings and the growth of peril to international civil aviation posed by such incidents, together with the inadequacy of Tokyo Convention led the ICAO Assembly at its 15th Session held in Buenos Aires from 3 to 28 September 1968 to adopt resolution A16-37 on the subject. This Resolution reads as follows:

> WHEREAS unlawful seizure of civil aircraft has a serious adverse effect, on the safety, efficiency and regularity of air navigation.
>
> NOTING that Article 11 of the Tokyo Convention on Offenses and Certain Other Acts Committed on Board Aircraft provides certain remedies for the situation envisaged.
>
> BEING however of the opinion that this Article does not provide a complete remedy.
>
> THE ASSEMBLY
> (1) URGES all States to become parties as soon as possible to the Tokyo Convention on Offenses and Certain Other Acts Committed on Board Aircraft.
> (2) INVITES States, even before ratification of, or adherence to the Tokyo Convention, to give effect to the principles of Article 11 of that Convention.
> (3) REQUESTS the Council, at the earliest possible date, to institute study of other measures to cope with the problem of unlawful seizure.

In connection with Clause (3) above, the Council by its resolution of 16 December 1968, decided to refer the question of unlawful seizure to the Legal Committee of ICAO. Thus, the Legal Committee was once again ordered to draft a new Convention on the subject.

The Legal Committee held its first session from 10 to 22 February 1969 in Montreal. It considered that the basic objective in its search for a solution to the problem under study should be to deter persons from committing unlawful seizure of aircraft and, more specially, to ensure – as far as practicable – the prosecution and punishment of these persons. The most efficient way of attaining this objective would, in the opinion of the Sub-Committee of the Legal Committee, entrusted with the subject, be through an international agreement between States (either a protocol to Tokyo Convention or an independent convention) which would be capable of ratification or adherence independently of the Tokyo Convention.

On 1 December 1970, the draft Convention was submitted to an ICAO Conference at The Hague attended by 77 States, and there the Convention was adopted on 16 December without any alteration.

The Hague Convention, unlike the Tokyo Convention, makes hijacking a distinct offence and calls for severe punishment of any person found within the territory of a Contracting State who hijacked an aircraft. As one writer succinctly observes:

D. The Hague Convention on Hijacking 1970

The Hague Convention specifically defined the offence of unlawful seizure of aircraft as a model for individual national legislation, and provides ... that each Contracting State undertakes to make the offence punishable by severe penalties.[594]

Whereas Article I of the Tokyo Convention applied in respect of acts which, whether or not they are offences, the Hague Convention appears to provide the answer the first of the problems left by Tokyo Convention.

That offence as defined by Hague Convention reads as follows:

Any person who on board an aircraft in flight
(a) unlawfully, by force, or threat thereof, or by any other form of intimidation, seizes, or exercise control of, that aircraft, or attempts to perform any such act; or
(b) is an accomplice of a person who performs or attempt to perform any such act commit an offence.

Article 2 of the Convention provides that each Contracting State should make the offence punishable by severe penalties. However, the Convention does not list the exact penalties to be imposed by the Contracting State, other than describing them as severe penalties.

I. The Scope of the Convention

There are several limitations placed on the application of the Convention as expressed by the articles of the Convention. Under Article I, the act must be committed by a person on board an aircraft "in flight" and it thereby excluded offenses committed by persons not on board the aircraft such as saboteurs who remain on the ground. Thus, the Hague Convention seems to suffer in this respect from the same defects which his predecessor, the Tokyo Convention, has suffered from. D.Y. Chung has observed:

The question of hijacking has been pretty well covered by the Tokyo and Hague Conventions. However, the type of hijacking these two Conventions dealt with is only "on board hijacking", while "non-on board hijacking" is not included. It is possible that someone who is not on board but who has placed a bomb or some destructive device on an airliner, may practice extortion on the airline or divert the plane to another destination. In other words, it is possible to hijack the plane by remote seizure or remote control. Another possibility is that of sabotage. Such a situation is not also covered by the above two Conventions, *i.e.* Hague and Tokyo.[595]

Similarly, according to Article I, the Hague Convention only applies to accomplices who are on board an aircraft in flight, and not to those who may be on the ground aiding and abetting the unlawful act. The Representative of the Netherlands

---

[594]Feller (1972, p. 214).
[595]Chung (1976, p. 643).

on the ICAO Legal Committee once said in this respect that "it is obviously possible to be an accomplice without being on board an aircraft."[596]

Article 3 of the Convention provides that the aircraft is deemed to be "in flight" at any time from the moment when all its external doors are closed following embarkation until the moment when any such door is opened for disembarkation. Hence, any hijacking initiated or attempted before the closing of the doors of aircraft after embarkation or after the opening of the doors for disembarkation is not covered by the Convention. Rene Mankiewicz observes:

> This limitation leaves outside the scope of the Convention any hijacking initiated or attempted before the closing or after the opening of the aircraft doors. As a consequence, such acts are punishable only under the law of the State where committed; the jurisdictional articles of the new Convention do not apply thereto. Furthermore, it follows that such acts are punished merely by the general criminal or air law of the concerned State, unless special legislation is introduced for punishing unlawful seizure committed or attempted on the ground.[597]

A further limitation expressed by the Convention (Article 3(2)) is that it shall not apply to aircraft used in military, custom or police service, nor in the cases of joint air "transport" operating organizations or international operating agencies which operate aircrafts which are subject to joint or international registration (Article 5), if the place of take-off or landing of the aircraft on board which the offence is committed is situated in the State of registration of such aircraft (Article 3(4)). On the other hand, the Convention would apply if the place of take-off or that of actual landing is situated outside the territory of the State of registration of the aircraft, on the understanding that it is immaterial whether the aircraft is engaged in an international or a domestic flight.

II. Powers and Duties Imposed Upon States in Order to Combat Hijacking

Beside the obligation to make the offence of hijacking punishable by severe penalties, the Convention imposed upon the Contracting States a series of obligations that are geared towards stamping out hijacking, these obligations are that:

> Each State shall take measures as may be necessary to establish – apart from any existing national criminal jurisdiction (Article 4(3)) – its jurisdiction over the offence and any act of violence against a passenger or crew when (Article 4(1)):
> (a) the offence is committed on board an aircraft registered in that State;
> (b) the aircraft on board which the offence is committed lands in the territory with the alleged offender still on board;

---

[596]*ICAO Doc 9050 LC/169-2* at 72.

[597]Mankiewicz (1971, p. 201).

D. The Hague Convention on Hijacking 1970

(c) the offence is committed on board an aircraft leased without crew to lessee who has his principle place of business or, if he has no such place of business, his permanent residence in that State.

In addition, every Contracting State must take necessary measures to establish its jurisdiction over the offence in case where the alleged offender is present in its territory and it does not extradite him (Article 4(2)). Mankiewicz further observes:

> this provision is necessary in order to increase the effective punishment even if the hijacker is not prosecuted in, or escaped form, the State of landing or is not extradited to the State of registration of the aircraft. Thus, the alleged hijacker can be arrested no matter where the offence took place as long as he is present in a Contracting State. This provision seems to introduce the principle of universal jurisdiction into the Hague Convention.

The jurisdictional powers conferred upon States by paragraph 1(b) of Article 4 above, may be considered as an important factor in the attempts of the international community to stamp out and deter hijacking, in that it gives Contracting States a legal instrument, which they may otherwise lack, in view of absence of any link between them and the State of landing, to act in these situations. This is an acceptable situation, whereby contracting States can extend the basis of jurisdiction under international law.

On the other hand, according to Article 4(1) three States possess concurrent jurisdiction over an alleged offender: first, the State of registration of the aircraft; second, the State of landing if the offender is on board the aircraft, and third, any Party to the Convention within whose boundaries the alleged offender is present, once that State has chosen not to extradite him to the State of registration of the aircraft or to the State in which he landed while he is still on board the hijacked aircraft, or to the State described in subdivision 1(c). In addition, subsection (3) sanctions such bases of jurisdiction as "passive nationality" where the national law so provides. It is interesting to note that the jurisdiction of the State of registration of the aircraft is equal to the other States described in Article 4.

A third instance of concurrent jurisdiction was added to Article 4 of the Hague Convention during the Diplomatic Conference at The Hague. Jurisdiction was granted to the State where the carrier, who operates an aircraft but is not the owner of this aircraft, has his principal place of business, or his permanent residence. Article 4(c) of the Convention covers the case of the so-called "bare hull Charter agreements" or "dry lease," i.e., when an aircraft is hired without crew to an operator. Thus, when an offence is committed on board an aircraft which is registered in a contracting or non-contracting State, and which is "dry" leased to an operator having his head office or permanent residence in a contracting State, the latter shall take necessary measures to establish its jurisdiction over the offence. This has been a useful improvement of Article 4 of the Convention in view of the great frequency of the leased agreements that the air transport industry is using at present.

A very important point which worth mentioning is that although Article 4 requires the Contracting State to assume jurisdiction over the unlawful seizure of aircraft within the limit of Article 3, it does not provide for obligation on the part of any Stated to actually prosecute the alleged offender. However, a provision which may be of some relevance is found in Article 7 which reads:

> The Contracting State is whose territory the alleged offender is found shall, if it does not extradite him, be obliged, without exception whatsoever, and whether or not the offence is committed in the territory, to submit the case to its competent authorities, for the purpose of prosecution. Those authorities shall make their decision in the same manner as in the case of any ordinary offence of a serious nature under the law of that State.

Thus, Article 7 states that authorities having jurisdiction under Article 4 are at liberty not to prosecute the hijacker, or this accomplice if it is determined that the offenders would not be prosecuted. R.P. Boyle has observed that the Diplomatic Conference which discussed the draft of the Hague Convention rejected the contention to apply compulsory prosecution or alternatively extradition because:

> ... this obligation is only to submit the offender for prosecution. There is no obligation to prosecute. Many careful distinctions have been adduced. One obvious one is that in case of universal jurisdiction, the State having the hijacker may not have available to it proof of the crime since conceivably it was committed in a distant State and thus the witnesses and other necessary evidence to the State having custody of the hijacker.
>
> However, the reason for rejection of adopting compulsory prosecution appears to me to be a political one for some which States do not want any interference in their sovereign right to permit political asylum in some form for whatever purpose, despite the gravity of the offence. It is interesting to note that both U.S.A. and the Soviet Union have urged that States should be compelled to prosecute the alleged offender if extradition was not granted. It is submitted that this lack of either compulsory jurisdiction or extradition is a serious weakness in the Convention, and stands in the way of an effective international solution to hijacking.[598]

Another obligation which is imposed upon Contracting States is that each State is required to include the offence referred to in the Convention as an extradatable one in every new extradition treaty. Existing treaties are deemed to include it already. The Convention may in case of a request for extradition, and in absence of an extradition treaty, be given consideration by the States which make extradition conditional on the existence of an extradition treaty, as the necessary legal basis for extradition. For the purpose of extradition, the offence is treated as if it had been committed not in the place in which it occurred but in the territory of the States, required to establish their jurisdiction in accordance with Article 4 above. Article 8 of the Convention states:

1. The offence shall be deemed to be included as an extraditable offence in any extradition treaty existing between Contracting States. Contracting States undertake to include the offence as an extraditable offence in every extradition treaty to be concluded between them.
2. If a Contracting State which makes extradition conditional on the existence of a treaty receives a request for extradition from another Contracting State with which it has no extradition treaty, it may as its option consider this Convention

---

[598]Boyle (1972, p. 473).

D. The Hague Convention on Hijacking 1970

as the legal basis for extradition in respect of the offence. Extradition shall be subject to the other conditions provided by the law of the requested state.
3. Contracting states which do not make extradition conditional on the existence of a treaty shall recognize the offence between themselves subject to the conditions provided by the law of the requested state.
4. The offence shall be treated, for the purpose of extradition between Contracting States, as if it had been committed not only in the place in which it occurred, but also in the territories of the states required to establish their jurisdiction in accordance with Article 4, paragraph 1.

Thus according to Article 8, if a Contracting State receives a request for extradition from a State with which it has no extradition treaty, the Convention shall be considered as the legal basis for extradition. The effect of this provision is to enlarge the scope of existing international treaties on extradition to include hijacking. Where a State is usually prohibited by domestic law from extraditing a hijacker in the absence of a treaty, the State must extradite the offender under the provisions of the Convention.

The obligation to extradite an airline hijacker is subject to all other customary and conventional rules of law governing extraditable offenses. As a general rule, extradition is denied where an individual is accused of committing a political offence. Most States recognize the granting of political asylum as a right to be determined by the State from which it is requested. As the laws of a State may preclude extradition of an airline hijacker if the offence is regarded as political, the existence of hijacking in an extradition treaty may not result in mandatory extradition. However, if a State does not extradite the offender, according to Article 7, the case must be submitted to the proper authorities for prosecution. I.D. Johnston has stated the following in relation to Article 8:

> The Convention obliges the parties to include hijacking in extradition treaties to be concluded between them and insert it retrospectively into existing extradition treaties. Parties which have not concluded extradition treaties but which make extradition conditional on a treaty can regard the Convention itself as a legal basis for extradition. These provisions increase the possibility of extradition but by no means make it a certainty. The Russian Proposal, supported by the U.S.A., that hijackers be returned in all cases was rejected at the Conference. Automatic extradition, though probably the best deterrent, was considered too drastic a commitment by most of the negotiating States. What they are prepared to accept however, was the duty to prosecute offenders whom they did not extradite as provided for by Article 7.[599]

Be that as it may, so far as the extradition of nationals is concerned, there is no indication in the Convention as to what the position is. Shubber is of the view that even though there is no mention to the extradition of the States own national according to the Convention or to the term offender in Article 8 still such extradition is possible.

---

[599]*Aviation Security Legislation*, Aviation Security Legislation, Vol. 5, April 1973 at 307.

There is no reason to suppose that hijackers who happened to be nationals of the State requested to extradite him should be excluded from the scope of extradition under the Convention, provided that course of action is compatible with the national law of the State concerned. This interpretation is not incompatible with the intention of the drafters and the purpose for which the Convention has been created.[600]

III. Other Provisions

The Hague Convention imposed further obligations on the Contracting State to preserve the security and efficiency of air transport. States are obliged to take reasonable measures to restore control of aircraft to its lawful commander or to preserve his control over it and to facilitate the continuation of the journey of the passenger and the crew. In addition, States are obliged to return the aircraft and its cargo to those entitled without delay (Article 9) and report promptly as possible to the Council of ICAO any relevant information (Article 11).

Article 10 imposes an obligation on the Contracting States to give one another the greatest measure of assistance in connection with the criminal proceedings.

When comparing the contents of the Hague Convention with that of the Tokyo Convention, one observes that the two Conventions overlap and are even contradictory on some issues and their inter-relation is far from clear.

The Hague Convention may be considered as a significant step forward in the endeavour of the international community to suppress the hijacking of aircraft and remove the threat caused by it to international civil aviation. The Convention has enlarged the number of the States competent to exercise jurisdiction over a hijacker and included the introduction of new basis for the exercise of jurisdiction of the State where the charterer of an aircraft has his principal place of business or permanent residence.

Another encouraging fact is that the Hague Convention grants every Contracting State the power to exercise jurisdiction over a hijacker if such States are affected by an offence committed under the Convention, thus making it impossible for a hijacker to escape the normal process of the law.

The Hague Convention, despite its efficiency in some areas, is not without its weaknesses. Mankiewicz[601] comments:

> the Hague Convention deals only with "unlawful seizure committed on board aircraft" and does not apply to sabotage committed on ground, nor does it cover unlawful interference with air navigation, facilities and services such as airports, air control towers or radio communications. Attempts made further to extend the scope of the Convention were unsuccessful. Nevertheless, the Seventeenth session of the Assembly of ICAO, held in Montreal in June, 1970, adopted a Resolution directing the Council of ICAO to convene the Legal Committee, if possible not later than November, 1970, in order to prepare ... a draft

---

[600]Shubber (1973b, p. 725).

[601]See Mankiewicz (1971, p. 206).

E. The Montreal Convention (1971)

Convention on Acts of Unlawful Interference Against Civil Aviation with the view to its adoption ... as soon as practicable. Consequently, the draft Convention was prepared and was opened for signature at Montreal on September 23, 1971.

E. The Montreal Convention (1971)

Since both the Tokyo and the Hague Conventions dealt only with unlawful seizure committed on board aircraft, it did not cover sabotage committed on the ground, nor unlawful interference with air navigation facilities and services. The Montreal Convention was drafted *inter alia*, to remedy those lapses. The objectives of the Montreal Convention are best discussed as follows:

> The primary aim of the Montreal Convention was to arrive at a generally acceptable method of dealing with alleged perpetrators of acts of unlawful interference with aircraft. In general, the nations represented at the Montreal Conference agreed that acts of sabotage, or violence and related offenses interfering with the safety and development of international civil aviation constituted a global problem which had to be combated collectively by concerned nations of the international community. A multilateral international convention had to be adopted which extended both the scope and efficacy of national legislation and provided the legal framework for international co-operation in the apprehension, prosecution and punishment of alleged offender.[602]

I. Definition of In Service

To achieve the above objectives, the Montreal Convention first sought to expand the scope of the activity covered by the Convention in order to include a new series of offenses which can be committed without the offender being on board the aircraft. The same definition for an aircraft in flight as given in Article 3(1) of the Hague Convention applies but the Montreal Convention introduces a new term, "aircraft in service," which is defined as follows:

> Aircraft is considered to be in service from the beginning of the preflight preparation of the aircraft by ground personnel or by the crew for specific period until twenty-four hours after the landing. The period of the service shall, in any event, extend for the entire period during which the aircraft is in flight.[603]

The expression was deemed important as it covers a more extended period than that covered by the expression "in flight" as defined in Article 3(1) in the preceding Hague Convention. The term "in service" would cover such acts as the bombing of and discharge of weapons against aircraft on ground, as well as similar acts against

---

[602] Abramovsky (1975a, p. 278).
[603] Article 2(b).

aircraft in flight, whether or not the acts were performed by a person on board or outside the aircraft. Another significance of the term "in service" is that it serves to specify the physical position in which the aircraft must be if the offenses covered by the sub-paragraph of Article 2 are to come under the Convention. An extensive definition of the expression could encompass attacks against an aircraft while in the hangar or at a parking area. But the States at the Montreal Conference were not willing to go that far. This is because an extensive definition would mean that the States would, under another provision of the Convention, be bound either to extradite the suspected author of such attack, or if it did not extradite him, submit the case to its competent authorities for the purpose of prosecution. States are notoriously reluctant to enter into international arrangements on criminal matters if those arrangements markedly reduce domestic jurisdiction. Yet, too narrow a definition of the expression aircraft in service would compromise the utility of the Convention.[604]

The definition of the term in service posed a difficult problem during the deliberation of the Montreal Conference. Although the beginning of the in service period afforded few problems, the main difficulty was in the definition of the end of the "in service" period, when applied to lengthy stopovers or night stop in a country, and awaiting turn around before commencement of the homeward-bound journey. It was decided that the aircraft should be protected by the Convention, that is, it should be deemed to be "in service" when it makes a stopover or night stop in another country. The present wording of the term "in service" attempts to solve the problem by specifying that an aircraft shall be considered in service 24 hours after any landing. The expression "in service" as it stands include the term "in flight" under Article 2(a). Therefore, in the event of force landing occasioned by hijacking, the period "in service" is deemed to continue until the competent authorities take over the responsibility for the aircraft and for persons and property on board.

II. Definition of the Offence

Another approach adopted by the Montreal Convention in its endeavours to curb hostile acts against civil aviation is to define the offence broadly in order to embrace all the possible acts that might occur. The first issue which faced the drafters of the Convention in this respect elated to the provision of substantial coverage of serious offenses and at the same time avoiding the difficulties that may arise in connection with the listing of specific crimes in a convention intended for adoption by a great many States. After much debate and deliberation, this issue was settled and the final conclusion of the meeting is reflected in Article I. G.F. Fitzgerald described the method of enumerating the offenses in the Convention as being "novel":

---

[604]Fitzgerald (1971, p. 71).

E. The Montreal Convention (1971)

Article I is novel in that it describes a number of penal offenses within the framework of a multilateral convention.[605]

Article 1 of the Convention defines and enumerates the offenses of unlawful interference with aircraft as follows:

1. Any person commits an offence if he unlawfully and intentionally:

    (a) Performs an act of violence against a person on board an aircraft in flight if that act is likely to endanger the safety of that aircraft in flight, or
    (b) Destroys an aircraft in service or causes damage to such an aircraft in flight if that act is likely to endanger its safety in flight, or
    (c) Places or causes to be placed on board an aircraft in service, by any means whatsoever, a device or substance which is likely to destroy that aircraft, or to cause damage to it which renders it incapable of flight, or to cause damage to it which is likely to endanger its safety in flight, or
    (d) Destroys or damages air navigation facilities or interferes with their operation, if any such act is likely to endanger the safety or aircraft in flight, or
    (e) Communicates information which he knows to be false, thereby endangering the safety of an aircraft in flight.

2. Any person also commits an offense if he:

    (a) Attempts to commit any of the offenses mentioned in paragraph 1 of this Article;
    (b) Is an accomplice of a person who commits or attempts to commit any such offence.

It should be noted that while Article 1 delineates several different offenses, the dual requisites of unlawfulness and intent apply to act of the offenses enumerated. G.F. Fitzgerald observes:

> The introductory language of paragraph 1 makes it clear that the dual element of unlawfulness and intention must be present in all of the acts covered by sub-paragraphs (a) to (e); otherwise those acts will not be offenses. The dual element would also apply to attempts and complicity covered by sub-paragraph 2.[606]

Sub-paragraph (a) of Article 1 is designed to deter and punish acts of violence committed against person on board aircraft in flight. It should be noted that not all acts of violence come within the scope of the offence, but only those likely to endanger the safety of the aircraft. The notion of an act of violence referred to in this sub-paragraph includes armed attack and also attack against the lives of persons, the aircraft by other means, such as, by blows, strangling, poisoning or lethal injection.

The word "violence" used in sub-paragraph (a) can be interpreted as including not only an armed attack or physical assault, but also administration of poison through, for example, its introduction into the food or drink served on board aircraft.[607]

The act of violence perpetrated upon a person on board an aircraft according to sub-paragraph (b) may come from within or without the aircraft.

---

[605]Fitzgerald (1971, p. 67).
[606]Fitzgerald (1971, p. 68).
[607]Fitzgerald (1971, p. 68).

The manner in which sub-paragraph 1(a) is worded, when it is read with the opening language of Article 1, would lead one to conclude that the person performing the act of violence does not have to be on board the aircraft in order to come under the Convention. This means that the convention would apply to a person who, being outside the aircraft (for example a low flying and slow-moving helicopter or light aircraft) in flight or who, while on the ground has poisoned food which is later consumed by a person on board such aircraft.[608]

According to this sub-paragraph, the act of violence is not restricted to those acts which imperil the life of the victim. Any act of violence perpetrated against a person on board and which is likely to interfere with the safety of the aircraft falls within the scope of the offence. Hence, the standard for determining whether the Convention is applicable in a given situation does not hinge on the gravity or the heinousness of the act but rather on its effect on the safety of the aircraft in flight. It is to be noted that the same definition as given in Article 3 of the Hague Convention for an "aircraft in flight" applies in (Article 2(a) of the Montreal Convention).

The two offences which can be committed on board an aircraft in service are enclosed in sub-paragraphs (b) and (c) of Article 1 of the Montreal Convention.

Sub-paragraph (b) is designed to deter and penalise acts of sabotage perpetrated against the aircraft itself. The sub-paragraph encompasses attacks both from within and without the aircraft. The destruction and damage referred to in the sub-paragraph must occur while the aircraft is "in service" as the particular act, the consequence of which is the destruction of the aircraft, may be performed before the aircraft is "in service" as the particular act, the consequence of which is the destruction of the aircraft, may be performed before the aircraft is "in service." Destruction includes substantial destruction of the aircraft beyond the possibility of rendering it airworthy through repair while the concept "causing damage" is intended to cover the damaging of a vital but inexpensive piece of wiring, would render the aircraft incapable of flight. It could also cover any damage, whether caused to an aircraft on the ground or in the air, where there is a likelihood that the safety of the aircraft in flight would be endangered.[609]

Sub-paragraph (c) is an attempt by the Convention to encompass, through using the term 'by any means whatsoever,' all situations in which explosives or other devices are placed on board an aircraft.

> The word "by any means whatsoever" cover the placing of explosives on board an aircraft whether carried on board by the author of the act or any unwitting accomplice, sent on board in air cargo or by mail, or even attached to the outside of the aircraft before it undertakes its journey.[610]

Sub-paragraph (d) is intended address hostile acts against "air navigation facilities" which may include airports, towers, radio services and meteorological services used in international flights.

---

[608]Fitzgerald (1971, p. 68).
[609]Fitzgerald (1971, p. 68).
[610]Fitzgerald (1971, p. 70).

Sub-paragraph (e) is concerned with making it an offence for anyone to pass, or cause to pass false information relating to an offence, for example, the presence of an explosive device or would-be hijacker on board the aircraft. Although most national legislation may have already enacted legislation concerning this subject, it was felt that measures to restrain such acts could especially be included in this Convention, as it was intended to cover a type of offence which very definitely interferes with the orderly conduct of commercial air services. It must be noted that in order that the act may fall within the Convention, the offender who communicates the information must know that the information is false.

Article 1(2) covers the case of an attempt to commit an offence and the case of being an accomplice to commit one of the offenses listed in the sub-paragraphs of Article. During the debate on the Montreal Convention, there was an attempt to include conspiracy in the definition, but some delegations, including France, were of the view that since conspiracy was not an offence under their national systems of penal law, it should not be included in the convention. After long deliberations, it was decided by a vote that reference to conspiracy would not be made in the Convention.

III. Penalties and the Scope of the Convention

Like the Hague Convention, the Montreal Convention provides for the undertaking by each contracting State to make the offenses covered by the Convention punishable by "severe penalties." Article 3 of the Montreal Convention states that each contracting State undertakes to make the offenses mentioned in Article 1 punishable with severe penalties. Unlawful acts against the safety of civil aviation are thus considered to be serious crimes which the Contracting States must punish by severe penalties. The term "severe penalties" is, however, not defined.

The Delegate of France explained at the discussions leading to the adoption of The Hague Convention that in connection with Article 2 the Sub-Committee and the Committee relating to The Hague deliberations had been faced with the question as to whether or not the severity of the punishment to be imposed upon the offender should be stated. The Sub-committee had come to the conclusion that this could not be done, considering the diversity of criminal codes in different countries. A more general wording, i.e., "severe penalties," was therefore considered more appropriate. It was not customary for international conventions of this type to stipulate minimum penalties, and a number of States did not have any provisions for them in their national legislation.[611] This omission has been criticized as one of the weaknesses of the convention.[612]

---

[611] Legal Committee, 18th Session, London, 29 September – 22 October, Vol. 1, Minutes, See *ICAO Doc 8936 LC/164-1* at 39.

[612] See G.N. Horlick, Public and Private Responses to Aircraft Hijacking, 2 Vanderbilt Law Journal, 1976 at 21.

Article 4 of the Convention stipulates which flights are to be covered by the Convention. Paragraph 1 excludes from the operation of the Convention aircrafts used in military, customs, or police services.

According to Paragraph 2 of Article 4, the scope of the Convention is determined primarily in terms of the international element of aviation. In case of the offenses contemplated in clauses (a), (b), (c) and (e) of Article 1(1), the Convention applies irrespective of whether the aircraft is engaged in an international or domestic flight, only if, as stated in Article 4(2):

> (a) the place of take-off or landing, actual or intended, of the aircraft is situated outside the territory of the State of registration of that aircraft; or
> (b) the offence is committed in the territory of a State other than the State of registration of the aircraft.

The Convention also applies in cases of international flights if the offender or the alleged offender is found in the territory of the State other than the State of registration of the aircraft.

Paragraph 5 of Article 4 provided:

> In the case contemplated in sub-paragraph (d) of paragraph 1 of Article 1, this convention shall apply only if the air navigation facilities are used in international air navigation.

Hence, the Convention will apply only if the air navigation facilities are used in international air navigation, i.e., the sabotage of domestic air navigation facilities is outside the scope of the Convention, notwithstanding the fact that the saboteur of domestic facilities may be found in another State. G.F. Fitzgerald observes:

> In case of air navigation facilities mentioned in sub-paragraph (d) of Article 1(1), the Convention applies only if the facilities destroyed, damaged, or interfered with are used in international navigation.[613]

IV. Jurisdictional Powers Given to States Under the Montreal Convention (1971)

Article 5 of the Convention, which concerns jurisdiction, provides that each contracting State shall take such measures as may be necessary to establish its jurisdiction over offences in the same three instances as those contained in the Hague Convention, and a fourth instance of when the offence is committed in the territory of that State. This Convention as its predecessor[614] does not exclude any criminal jurisdiction exercised in accordance with national law. Fitzgerald states:

---

[613] G.F. Fitzgerald, *International Terrorism and Civil Aviation*, Unpublished Speech given to the Third Annual Conference of the Canadian Council of International Law, 2 October 1974.

[614] Hague Convention of 1970.

E. The Montreal Convention (1971)

A controversial topic in the Montreal Convention is that of jurisdiction, since, like The Hague Convention, this Convention attempts to establish a form of universal jurisdiction over the alleged offender.[615]

Article 5 of Montreal Convention provides that each contracting State shall take such measures as may be necessary to establish its jurisdiction over the offenders in the following cases:

1. (a) When the offence is committed in the territory of that State;
   (b) When the offence is committed against or on board an aircraft registered in that State;
   (c) When the aircraft on board which the offence is committed lands in its territory with the alleged offender still on board; and,
   (d) When the offence is committed against or on board an aircraft leased without crew to a lessee who has his principal place of business or, if the lessee has no such place of business, his permanent residence, in that State.
2. Each contracting State shall likewise take such measures as may be necessary to establish its jurisdiction over the offenses mentioned in Article 1, paragraphs 1 (a), (b) and (c) and in Article 1, paragraph 2, in so far as that paragraph relates to those offenses, in the case where the alleged offender is present in its territory and it does not extradite him pursuant to Article 8 to any of the States mentioned in paragraph 1 of this Article, i.e., Article 5.
3. This Convention does not exclude any criminal jurisdiction exercised in accordance with the national law.

An analysis of Article 5 would lead to the conclusion that at least four States are specifically empowered to exercise concurrent jurisdiction over an alleged offender. These States are: (1) the State within whose territorial boundaries the offence is committed (whether the offence takes place on its territory or within its airspace.) this reaffirming and codifying the traditional basis of territoriality; (2) the State of registration of the aircraft (hence, such State is empowered to exercise its jurisdiction over offenders who commit their crimes on board aircraft registered in those States); (3) the State of landing, if the offender is on board the aircraft; and (4) any party to the convention within whose boundaries the alleged offender is present, if that State refuses to extradite the offender to any of the States having jurisdiction under Article 5(1).

Article 5(2) adopts the interpretation of universal jurisdiction as contained in the Hague Convention. Furthermore, Article 5, paragraph 3, provides that the jurisdictional basis delineated by the Convention do not supersede any criminal jurisdiction that has derived from national laws of the parties to the Convention. Consequently, the jurisdictional relation to nationality may be asserted by the State of nationality of the alleged offender, and the States which are the targets of the offence or whose nationals are threatened, maimed or killed by the offender may invoke the

---

[615]Fitzgerald (1971, p. 73).

protective principle or the lesser recognized jurisdictional basis of passive nationality. These additional bases of jurisdiction are expected to further increase the possibility of suppressing the offenders.

In his concluding remarks, Professor Fitzgerald has observed:

> Thus, the Montreal Convention breaks new grounds and goes beyond codification in providing for the international legal action to be taken by States in respect of many acts ....[616]

By adopting the Montreal convention, concerned States attempted to provide a framework which would substantially widen the scope and application of national legislation and thereby both penalise and deter unlawful interference with aircraft.

1 Extradition or Prosecution

Article 7 of the Montreal Convention, like its predecessor, embodies the principle of *aut dedere aut judicare*, which is the basis of the whole draft. It reads as follows:

> The State party in whose territory the alleged offender is present shall, if it does not extradite him, submit without exception whatsoever and without undue delay, the case to its competent authorities for the purpose of prosecution, through proceeding in accordance with the laws of that State.

According to this provision, a contracting State, has an obligation either to extradite the alleged offender found in its territory or submit his case to the competent authorities for the purpose of prosecution. It appears from the overall reading of "without exception whatsoever" that the Convention makes prosecution mandatory. However, a deeper analysis of the Article brings to bear the fact that it does not mandate the actual prosecution of the offender but merely the submission of the case to the competent domestic prosecuting authorities. This contention is supported by the fact that during the Montreal conference the Israeli delegation proposed that the Convention includes a mandatory prosecution provision, although this proposal was defeated by vote of 35 to 2 with 6 abstentions.

The failure of the Montreal Convention to provide for an objection to prosecute, when the offender is not extradited, was considered a weakness regarding the system of sanctions *aut dedere aut punire*. A commentator observes:

> The lack of mandatory system of prosecution with respect to aerial terrorism must be emphasized. Despite the repeated efforts of some delegations during the Hague and Montreal Conferences, the existing texts on aerial terrorism do not recognize the system of mandatory prosecution in case of denial of extradition requests. On the contrary, the State authorities in charge of the handling of prosecution may well decide that according to their domestic law, the alleged offender should not be prosecuted at all.[617]

---

[616]Fitzgerald (1971, p. 75).
[617]Costello (1975, p. 488).

2 Extradition and Other Principles

As far as extradition is concerned, the Montreal Convention repeats *verbatim* The Hague provision regarding extradition. The Convention also repeats the Hague convention provision, discussed above, relating to: the taking of alleged offender into custody; joint air transport operating organizations or international operating agencies, which operate aircraft that are subject to joint international registration; continuation of the journey of the passengers, crew, and aircraft; assistance between States in connection with criminal proceeding; and, the reporting of the process to the ICAO Council.

Although the Montreal convention was considered a breakthrough in combating terrorism against air transport, it remains, like its predecessors, tenuous and destitute of real effect. It would be a platitude to state that the effectiveness of any convention, however well drafted and universally accepted, would depend on the willingness and ability of States to enforce within their own territory the rule of law.

> Even if it is widely ratified, a small number of States can undermine its (a treaty's) effectiveness by actively supporting or condoning acts of unlawful interference and by providing havens for the perpetrators of such acts. Because of conflicting ideologies and political exegesis, such events have in fact occurred.[618]

Another problem is that although all three Conventions have entered into force, barely half of the world community subscribes to either one or all of these agreements, and therefore their total impact has been less than inhibiting. Thus far, 153 States have ratified the Tokyo Convention, 153 States have ratified The Hague Convention and 155 States have ratified the Montreal Convention of 1971.[619] This low rate of ratification of the Conventions, when compared to the number of 183 member States of ICAO has drastically reduced their effectiveness:

> Whether the Convention will fulfil its aims is dependent upon the breadth of support it obtains, indicated in part by the number of ramifications it receives. For the Convention to be effective, it must be acceded to by almost all nations.[620]

Some States have not only failed to ratify the conventions, but have also undermined the Conventions' effectiveness by providing sanctuaries to alleged offenders. Motivated by political and economic interests, other States have granted tacit support, and occasionally even active aid, to the perpetrators.

As a direct effect of the failure of the international community to provide an effective machinery for combating terrorism against air transport, threats to international civil aviation have consistently become more alarming and grave. New facets and more spectacular types of offenses have evolved as a result.

In an effort to redress the situation, concerned actions, under the auspice of ICAO, have attempted to formulate and adopt multilateral international accord which

---

[618] Abramovsky (1975b, p. 300).

[619] See ICAO Doc *A31-WP/26, LE/2*, 4/7/95, at 11.

[620] See ICAO Doc *A31-WP/26, LE/2*, 4/7/95, at 11.

would compel recalcitrant States into adherence both of customary and international law and with the provisions of the Tokyo, Hague and Montreal Conventions.

F. The Bonn Declaration

At a close of a 2-day economic meeting held at Bonn, Germany, 16–17 July 1978, leaders of the Governments of Canada, France, the federal Republic of Germany, Italy, Japan, the United Kingdom of Great Britain and Northern Ireland, and the United States of America agreed to act jointly in common undertaking against countries failing to act swiftly against hijacking. The declaration on co-operative action reads:

> The heads of States and governments concerned about terrorism and the taking of hostages, declare that their governments will intensify their joint efforts to combat international terrorism.
> To this end, in cases where a country refuses extradition or prosecution of those who have hijacked an aircraft and/or do not return such aircraft, the heads of States and governments are jointly resolved that their governments should take immediate action to cease all flights to that country.
> At the same time, their governments will initiate action to halt all incoming flights from that country or from any country by the airlines of the country concerned. The heads of States and governments urge other governments to join in this commitment.

It is evident that the declaration was intended to create an international regime for preventing and deterring acts of unlawful interference with civil aviation by the imposition of stringent sanctions that would adversely affect the economic and political interests of a delinquent State. Mark E. Fingerman observes:

The Declaration focuses on sanctions designed to deter nations from encouraging the commission of the offence. In effect, the spirit of the Declaration is a recognition of the fact that States are frequently *de facto* accomplices to acts of skyjacking ... The rationale of the Declaration would appear to be the foreclosing of the possibility of a skyjacker finding refuge and thereby reducing the attractiveness of the offence.[621]

The object of the Bonn Declaration as is indicated in its preamble is to intensify the joint effort of States to combat international terrorism. In order to achieve this objective, the Declaration has set out respective obligations on third State in the event a hijacked aircraft ended in the territory of such State. If the third State failed to meet the obligations specified in the Declaration, the Declaration envisages that a definite sanction will be inflicted upon the State as a sort of punishment.

The Declaration refers to an act of hijacking, without actually defining the offence. It can be assumed that the act referred to would be interpreted in accordance with the definition in Article 1 of the Hague Convention. The Declaration seemingly refers to act that has been completed, which means that the hijackers

---

[621]Fingerman (1980, p. 142).

should have reached their final destination. Thus, a State in whose territory a hijacked aircraft lands only for the purpose of refuelling would not act contrary to the Declaration if it allows the landing without taking action against the hijacker.

The Declaration applies in instances where a State refuses to prosecute or extradite the hijackers and/or return the hijacked aircraft. The words "prosecution and extradition" as contained in the Declaration have the same meaning as used in Articles 8 and 7 of the Hague and Montreal Conventions respectively. Of course, for this provision to be applicable a State must be in a position to prosecute or extradite, i.e., the hijacker must stay in the country and be available for prosecution by the competent authorities. However, once a State is able to take appropriate action but does not act and the hijacker disappears, such omission would be regarded as defaulting according to the spirit of the declaration.

The sanctions which the Contracting States would impose are: (a) taking immediate action to cease all flights to that country, and (b) initiating suspension all incoming flights which arrive from the defaulting State or are operated by airlines of a defaulting State. These sanctions are in essence an economic boycott or a "reprisal" in international law and are meant as a deterrent. The Declaration is recognises in spirit that some States may act as *de facto* accomplices to acts of hijacking and may give refuge and safe haven to an offender.

I. The Legal Status of the Bonn Declaration

The suspension of aerial communication as envisaged in the Bonn Declaration has been considered a serious measure in the context of international relation:

> Naturally, the suspension of aerial communications was not an economic step... This was a political sanction, because the suspension of aerial communications meant in practice a deterioration in relations between States. It meant stoppage of the carriage of cargo and passengers, it would interfere with diplomatic communications, etc.[622]

Another contentious aspect of the Bonn sanction machinery is that the measure of boycotting of a delinquent State would not only affect the interests of the specific State that was violating the obligations specified in the Declaration, but also of those States which applied or agreed to apply such sanctions and third party States.

There is strong feeling among some jurists who consider that the imposition of sanctions against offending States fall exclusively within the domain of the Security Council of the United Nations, and thus, any independent convention or declaration permitting the use of sanctions by party States themselves would violate the United Nations Charter. In support of this argument, Articles 39 and 41 of the Charter of the United Nations have been cited.

Article 39 of the Charter provides:

---

[622]*ICAO Doc 9050-LC/169-1* at 41.

> The Security Council shall determine the existence of any threat to peace, breach of peace, or act of aggression and shall make recommendations, or decide what measures shall be taken in accordance with Articles 41 and 42 to maintain or restore international peace and security.

Article 41 provides:

> The Security Council may decide what measures not involving the use of armed force are to be employed to give effect to its decisions, and it may call upon the members of the United Nations to apply such measures. These may include complete or partial interruption of economic relations and of rail, sea, air, postal, telegraphic, radio and other means of communications and the severance of diplomatic relations.

It is customarily accepted therefore that States cannot take joint sanctions against another State unless such action was authorized by the Security Council of the United Nations.

The above view had also been voiced in the ICAO Legal Committee where delegates of France and U.S.S.R. expressed the opinion that to apply sanctions against States in the form of interruption of full or partial air services was within the exclusive jurisdiction of the Security Council. The French delegate observed:

> The sanctions approach had been very thoroughly discussed in the Special Sub-Committee and some rather serious objections to it had been raised. The first was whether the machinery for consideration of sanctions was compatible with Article 41 of the United Nations Charter, which empowered the Security Council to decide upon measures in the nature of sanctions, including the complete or partial interruption of air services, and called upon members of the United Nations to apply them.[623]

A similar opinion is voiced by the delegate of Soviet Union who argued that joint action in a form of suspension of flight, if implemented, would be in contradiction with the competence of the Security Council:

> Indeed, according to Article 41 of the United Nations Charter, one of the measures that the Security Council of the United Nations was empowered to apply included the suspension of air communications. The imposition of collective sanctions against States, outside the framework of the United Nations, would be precluded by U.N. Charter.[624]

Another approach which indicates the incompatibility of the measures adopted by the Bonn Declaration with international law is reflected in Article 2(3) and Chapter VII of the United Nations Charter. Under Article 2(3), all members of the United Nations pledge themselves to settle their international disputes by peaceful means in such manner that international peace and security, and justice are not endangered. Article 33 then enumerates various procedures for the settlement of such disputes, notably "negotiation, inquiry, mediation, arbitration, judicial settlement, resort to regional agencies or arrangements or other peaceful means (chosen by parties to the dispute)."

It has been observed by Brosche that:

---

[623] *ICAO Doc 9050-LC/169-1* at 10.
[624] *ICAO Doc 9050-LC/169-1* at 41.

even if this list is not considered to be exhaustive, it is quite clear that embargo, boycott, blockade, reprisal or other kinds of economic pressure do not constitute procedures of pacific settlement. They are not peaceful means and not appropriate for the solution of disputes. The use or imposition of such measures would constitute a violation of the obligation to settle international disputes by peaceful means. Due to these facts, it becomes evident that the use of any kind of [economic] pressure is contrary to the [Charter] principles of peaceful settlement of disputes.[625]

It is clear from the above that the aerial boycott adopted by the Bonn Declaration is not permissible according to international law as incorporated in the Charter of the United Nations. It is clear that States will be held responsible for a boycotting instituted directly by their governments if such measure is found by the international community to be *ultra vires* the established norms of international law.

Furthermore, it may also be relevant to view the Bonn Declaration by reference to the doctrine of non-intervention, as elaborated in various international instruments in recent years. Thus, paragraph 2 of its Declaration on the Inadmissibility of Intervention in Domestic Affairs of States and the Protection of their Independence and Sovereignty of 21 December 1965 [Resolution 2131(XX)], the United Nations Assembly decreed that:

> No State may use or encourage the use of economic, political or any other type of measures to coerce another State in order to obtain from it the subordination of the exercise of its sovereign rights and to secure from it an advantage of any kind.

II. Incompatibility of the Declaration with the Vienna Convention on the Law of Treaties

The Bonn Declaration was designed to be invoked by seven States against an allegedly defaulting State, whether or not the latter is a party to the Declaration. Mark E. Fingerman opines:

> The legal force of the Bonn Declaration upon non-parties is of critical importance; it is against these nations that the Declaration's sanctions were most intended to apply. The Declaration calls for the imposition of its sanctions upon any State that violates its provisions, whether or not the State in question is a party to the Declaration or any civil aviation convention.[626]

Articles 33 and 34 of the Vienna Convention on the Law of Treaties specifically state that States which do not become party to a treaty would not be bound by that treaty unless they expressly agree in writing to be bound by it. Only the State parties to a treaty (including a declaration) could be considered as being bound by the provisions of that treaty. Whilst it may be conceded that a treaty could create rights

---

[625]Brosche (1974, p. 2).
[626]Fingerman (1980, p. 144).

for third States which those States could accept, it cannot not impose obligations on third States in terms of requiring them to commit acts such as prosecuting or extraditing offenders against civil aviation or returning the aircraft against which such offence is committed. International law does not envisage the imposition of obligations on States which are not parties to a treaty. Therefore, the scope of application of the Declaration should be limited to States which are parties to it. The Note presented by the French Government on the Resolution adopted by the Council of ICAO on 19 June 1972 on the question of joint sanction stated:

> The Convention can establish obligations only for States parties to it and would permit imposing sanctions only on those parties, pursuant to Articles 34 and 35 of the Vienna Convention on the Law of Treaties. Therefore, the Convention could be effective only if it was universally accepted.[627]

Therefore, the Declaration will be ineffective as far as it intended to impose an obligation upon third parties to prosecute or extradite the hijacker and to release the aircraft.

III. The Incompatibility of the Declaration with the Convention on International Civil Aviation (Chicago Convention 1944) and the International Air Services Transit Agreement

Another difficulty emerging from the Bonn Declaration was the relationship of the Declaration to other international conventions. The problem of suspension of air transport services as a sanction under the Bonn Declaration becomes particularly relevant in this context. Article 5 of the Chicago Convention confers certain rights upon Contracting States:

> Each Contracting State agrees that all aircraft of the other Contracting States, being aircraft not engaged in scheduled international air services shall have the right subject to the observation of this Convention to make flights into or in transit non-stop across its territory and to make stops for non-traffic purposes without the necessity of obtaining prior permission and subject to the right of the States flown over to require landing.

The Transit Agreement or the so-called "Two Freedoms Agreement" also contains a reciprocal grant among ICAO members relating to their scheduled air services whereby their carriers could fly across the territory of a State without landing or land in its territory for non-traffic purposes.

Therefore, it appears that the imposition of a restriction upon the airline of a defaulting State to fly over or to land in the States subscribing to the Bonn Declaration is incompatible with the provisions of the Chicago Convention and the Transit Agreement. The rights to fly over or land belong to States parties to the Chicago Convention, and those rights could not be derogated by a contrary

---

[627]*ICAO Doc 9050-LC/169-2* at 42.

F. The Bonn Declaration

provision in another treaty, such as the Bonn Declaration. The Spanish delegate to the ICAO Legal Committee observed of sanctions against air services:

> One of the problems ... was the compatibility of the air services with the right of the States parties to the Chicago Convention and Air Transit Agreement.[628]

The French Government stated the following in regard to the question of suspension of air services as a sanction:

> decisions on the suspension of air services could not be taken without amending the bilateral agreement which grants traffic rights, and, perhaps, even the Chicago Convention itself.[629]

In negotiating air transport agreements, both parties will endeavour to promote safe commercial operations of the type contemplated by the Chicago Convention and seek the grant of rights for their carriers. Bilateral agreements on air traffic rights are usually not intended to cover the continuation of operation into and from victim States by aircraft of the States which are seen to promote the disruption of safe commercial aviation, in a manner specified in the Declaration. Failure by States to take practicable measures necessary to prevent the disruption of international aviation – which is caused by such acts of detention and seizure of aircraft as specified by the Declaration – would therefore not be consistent with the grant by peace-loving States, of rights necessary for the conduct of air traffic by another State. Therefore, it is not logical to say that bilateral air transport agreements can properly be interpreted as granting rights to airlines of States to continue air services to and from a delinquent State if such state detains passengers, crew or aircraft or fails to prosecute or extradite the perpetrators. Walter Schwenk is of the view:

> In interpreting the bilateral, the conclusion may be reached, however, that there seems to be sufficient justification for the suspension of air traffic rights under the bilateral itself without the need to resort to general principles of international law.[630]

The Chicago Convention established principles and arrangements designed to assure that international civil aviation would develop in safe and orderly manner. It imposes obligation upon each Contracting State "not to use civil aviation for any purpose inconsistent with" such aims[631] More directly, The Convention specifically requires each Contracting State "to adopt all practicable measures ... to facilitate and expedite navigation by aircraft between the territories of Contracting States, and to prevent unnecessary delays to aircraft crews, passengers and cargo."[632] Refusal by a State to adopt generally agreed procedures to eliminate the threat to international civil aviation posed by such acts of detention and unlawful seizure as

---

[628] *ICAO Doc 8936-LC/164-1* at 216.
[629] *ICAO Doc 9050-LC/169-2* at 42.
[630] Schwenk (1979, p. 317).
[631] Article 4.
[632] Article 22.

are specified in the Bonn Declaration, would constitute a failure by that State to carry out its obligation under Articles 2 and/or 44 of the Chicago Convention. Therefore, suspension of flights in the circumstances referred to in the Declaration would not be incompatible with the Chicago Convention or the Transit Agreement, contrary to views of some scholars.

Moreover, the sanction adopted by the Bonn Declaration involving suspension of air services in no way deprives a State of a fair opportunity to operate an international airline pursuant to Article 44(f). On the contrary, a State found to be in default would not be giving a fair opportunity to other States. The U.S. Representative in the Legal Committee held in Montreal in 1973 said:

> Defaulting States had no longer had the privilege of Article 44(f) until such time as it provided a fair opportunity to the rest of the community. Article 44(f) would have to be read in context with Articles 44(d) and (h), and with the directive as stated in Article 44(a) that the Organization shall ensure safe and orderly growth of international civil aviation throughout the world.[633]

The U.S. Representative said that the power to suspend air services as a sanction is not only compatible with the Chicago Convention but also with international law.

> If a party to the Chicago Convention committed a material breach of the obligation to ensure the safety of civil aviation, then other parties individually had the right to suspend the operation of the Convention in whole or in part with respect to the defaulting State in accordance with customary international law, as specified in Article 60 of the Vienna Convention on the Law of Treaties.[634]

When they ratify or adhere to the Chicago Convention, they not only become members of ICAO, but also, they undertake to take appropriate steps to ensure the safety and security of international civil aviation. Hence, the harbouring of perpetrators by way of failure to prosecute or extradite may be said to constitute a violation of the basic rationale of the Chicago Convention. To breach the obligation set forth by the Chicago Convention is to impliedly denounce the Convention. Therefore, the defaulting State in such an instance cannot claim the rights conferred upon it by the Convention.

IV. *Problem of Prosecution or Extradition*

The fourth problem of the Bonn Declaration is the issue of prosecution or extradition. Sanctions under the Declaration are expected to follow the failure of the delinquent State to prosecute, extradite and/or return the aircraft. However, public international law provides no rule which imposes a duty to extradite, or prosecute. Hence extradition or prosecution becomes either a matter of comity or treaty

---

[633] *ICAO Doc 8936-LC/164-1* at 228.
[634] *ICAO Doc 9050-LC/169-1* at 39.

between States. Even when a treaty exists, extradition may be refused in certain circumstances. Therefore, surrender of an alleged criminal cannot be demanded as of a right in the absence of a binding treaty between the respective parties. Any attempt to bind States to extradite or prosecute offenders in the absence of a treaty to that effect would definitely be an encroachment on State-sovereignty, making such act a violation of customary international law.

Besides the above problems surfacing from the Bonn Declaration, there exist also certain gaps with respect to the application of the Declaration:

1. How would the decision to suspend air services be taken by the members of the Declaration, would it be by majority or unanimously?
2. Who will judge that a State is no longer in default, and when will the services be resumed? Should there be disagreement among the seven States parties on these points and should a procedure be needed to be laid down in order to regulate these matters?
3. Did the Declaration take into account the diversity of violations attributable to the defaulting State and that whether there would be the same penalty automatically applicable to every case?

The Bonn Declaration, unlike the three Conventions discussed above, represents a fragmented attempt on the part of the international community to control terrorism or unlawful interference with international civil aviation. This, however, by no means confirms the fact that the three international conventions were comprehensive attempts by the entirety of the international community. While the conventions lacked a certain compulsion in their requirements, the Bonn Declaration, which seemingly had a punitive flavour, lacked respectable representation by the international community. It is time to view both these approaches with a view to coalescing them to form a synthesis of action. The element of sanction as introduced by the Bonn Declaration should be fused with the international flavour of the three conventions. The international community may be able to work out a workable, effective and enforceable instrument on this basis.

G. A New Convention on the Marking of Plastic Explosives for the Purpose of Detection

Following the Resolution of the ICAO of the Council and the adoption of a United Nations General Assembly Resolution, the International Conference on Air Law was held under the auspices of the ICAO in Montreal from 12 February to 1 March 1991, which unanimously adopted a Convention on the Marking of Plastic Explosives for the Purpose of Detection. The Convention was opened for signature on 1 March 1991 and on that day was signed on behalf of 41 States.[635]

---

[635]ICAO Doc 9571.

The events that led to he Conference are well known.[636]

The ill effects of the disaster of flight Pan Am 103 over Lockerbie in Scotland on 21 December 1988 were more felt when the world learned that it was determined to be caused by the explosion of a lethal charge of SEMTEX, which is a high performance a plastic explosive manufactured commercially in Czechoslovakia, which was placed aboard the Pan Am flight hidden in a radio cassette player.

The adoption of a new Convention on the marking of Explosives was further encouraged by a unanimously adopted United Nations Security Council Resolution on 14 June 1989, which expressed concern 'at the ease with which plastic and sheet explosives can be used in acts of terrorism with little risk of detection' and urged ICAO to intensify its work aimed at preventing all acts of terrorism against international civil aviation, and in particular, its work on devising an international regime for the marking of plastic or sheet explosives for the purpose of detection.[637]

At the 27th Session of the ICAO Assembly in September–October 1989, the Delegations of the United Kingdom and Czechoslovakia presented a draft Resolution which was unanimously adopted as Resolution A27-8. The Assembly calls upon the Council:

> To convene a meeting of the Legal Committee, if possible in the first half of 1990, to prepare a draft international instrument 'on the marking of plastic and sheet explosives for the purpose of detection,' with a view to its adoption at a diplomatic conference as soon as practicable thereafter in accordance with the ICAO procedures set out in Assembly Resolution A7-6.[638]

At the broadest international level, the 44th Session of the United Nations General Assembly urged ICAO 'to intensify its work on devising an international regime for the marking of plastic or sheet explosives for the purpose of detection.'[639]

In June 1989, the Council of ICAO considering the overall preference of the international community decided to include in the General Work Programme of the ICAO Legal Committee with the highest and overriding priority the subject 'Preparation of a new legal instrument regarding the marking of explosives for detectability.'[640]

Work started quickly after the completion of the 27th Session of the ICAO Assembly. The preparations for the new instrument passed through the following stages:

A Rapporteur (Mr. A.W.G. Kean, CBE – United Kingdom) was invited to study and prepare the subject; he presented his report in September 1989. The report was considered by a Special Sub-Committee of the Legal Committee which met at

---

[636] For the thorough analysis of the history of the Convention see: Milde (1990, pp. 155–179).
[637] United Nations, Security Council, SC/RES/635 (1989).
[638] Doc 9551, A27-RES.
[639] Resolution 44/29 of 13 December 1989.
[640] C-DEC 127/20.

G. A New Convention on the Marking of Plastic Explosives   255

Montreal from 9 to 19 January 1990 and prepared its Report for the 27th Session of the Legal Committee which met at Montreal from 27 March to 12 April 1990.

The Legal Committee devoted almost all of its time to the new instrument and presented its Report which contained the text of a 'Draft Convention on the Marking of Plastic (and Sheet) Explosives for the Purpose of Detection' which the Committee considered to be "a final draft" under the terms of Assembly Resolution A7-6, to the Council. Consequently, a report was drafted by the Rapporteur appointed to study and prepare the issue with regard to the adoption of a Convention.

The Council considered the Report of the Legal Committee together with the Report of the Third meeting of the Ad Hoc Group of specialists on 4 July 1990 and decided to circulate the draft text of the Convention to States and international organizations for comments.

At the same time the Council convened an International Conference on Air Law to meet at Montreal from 12 February to 1 March 1991.

The major issues to be resolved by the Conference relate to:

(a) The scope of the Convention – in particular whether it should be confined to plastic explosives;
(b) The obligations of States, especially those related to the prohibition and effective prevention of the manufacture in their territories of such explosives and of the movement in and out of their territories of unmarked explosives;
(c) The exceptions that should be created and the extent to which they should be created in relation to activities by military or police authorities that were not inconsistent with the purposes and objectives of the convention;
(d) The manner of and the timing for the disposal of existing stocks;
(e) The function of the Annex to the Convention as a flexible instrument to address further development of technology, its amendment and the role of the Explosives Technical Commission.

I. *Scope of the Convention*

The Convention, which is titled: 'Convention on the Marking of Plastic Explosives for the Purpose of Detection.' It has been drawn as authentic in the English, French, Russian, Spanish and Arabic languages.

From the early beginning, there has been some confusion with respect to the proper reference to 'plastic explosives.' The relevant UN resolutions as well as Assembly Resolutions A27-8 refer to 'plastic or sheet explosives,' the Legal Committee put in its definition of explosives the wording 'and sheet' between square brackets and referred the matter to the Ad Hoc Group of Specialists suggested in its report,[641] that the problem was more of a linguistic than of a technical nature.

---
[641] AH-DE/3, Report, Restricted.

The Conference deleted the words 'or sheet' in the understanding that they are superfluous and confusing in most languages, except for the French language. Hence, in the French version the words 'et en feuille' when referring to plastic explosives have been retained.

In the preambular clauses reference is made to other means of transport and targets other than aircraft, indicating that the scope of this Convention is clearly wider than the Civil Aviation sector. However, as aviation has been so far the major victim of the criminal use of plastic explosives, the Conference upheld the tasks entrusted to ICAO and its Council in the draft Convention by both the Legal Sub-Committee and the Legal Committee.[642]

Both in the preparatory stage and during the Conference, ICAO's 'Sister Organization' the International Maritime Organization, participated actively in the deliberations.

The Legal Committee draft in Article I defined 'Explosives', 'Detection Agent' and the verb "Mark," while referring to the Annex for further description or clarification of these terms.

The Conference chose to broaden the definition of Explosives which now refers to '... explosive products, commonly known as "plastic explosives," including explosives in flexible or elastic sheet form, as described in the Technical Annex to this Convention.'

Thus an elegant if somewhat superfluous solution for the 'Plastic and Sheet' problem was attained.[643]

Furthermore, definitions of the verb 'Manufacture,' of 'Duly authorized military devices' and 'Producer State' were added. The Conference discussed at some length the desirability of expanding the scope of the Convention to explosives other than Plastic Explosives. The supporters of such an approach argued that, by simply deleting any reference to 'plastic or sheet' in the definition, the Convention would cover more than plastic explosives alone. While the description of the explosives only, it could be amended easily in the future to cover other explosives as well. As the Annex would have its own procedure of amendment, no new Convention or Protocol to the Convention would be needed.

The Conference ultimately decided against this suggestion. The 'Don't-rock-the-boat' approach prevailed, while many delegations also felt, that such action would be outside their mandate and not in accordance with the relevant UN and ICAO Council decisions which all mentioned plastic and/or sheet explosives. As a compromise, the Conference included in its Final Act a Resolution which, among other things, 'Requests the Council of the International Civil Aviation Organization to initiate, as a matter of high priority, studies into methods of detecting explosives

---

[642]The last preambular clause to the Convention explicitly confirms this by stating 'Noting with satisfaction the role played by the Council of the International Civil Aviation Organization in the preparation of the Convention as well as its willingness to assume functions related to its implementation' – See ICAO Doc 9571.

[643]Superfluous, because the Technical Annex to the Convention provides in: PART 1: DESCRIPTION OF EXPLOSIVES, a detailed definition of Plastic Explosives.

G. A New Convention on the Marking of Plastic Explosives   257

or explosive materials, especially into the marking of those explosives of concern, other than plastic explosives, whose detection would be aided by the use of marking agents, with a view of the evolution, if needed, of an appropriate comprehensive legal regime.'

II. *Obligations of States*

Already at an early stage of the preparations for the new Convention a majority in both the Legal Sub-Committee and the Legal Committee opposed initiatives to create an international offense, e.g., the act of manufacture or movement of unmarked explosives. Most delegations field – that the 1971 Montreal Convention and the additional 1988 Montreal Protocol adequately covered the offenses that needed to be universally punishable in this context and that the new Convention should concentrate on the prevention of the use of plastic explosives by the perpetrators of those offences.

Article II expresses the obligation of States to take the necessary and effective measures to prohibit and prevent the manufacture in their territories of unmarked explosives, while Article III formulates the same obligation towards the movement into or out of the territory of a State Party of unmarked explosives.

It is noteworthy, that Article II and III simply refer to 'necessary and effective measures' to be taken by States as a fulfilment of their obligations. It will be left to the individual States and their national legislation to provide for the necessary prohibitions and sanctions. While these provisions may facilitate the ratification of the Convention, they probably will not contribute greatly to a uniform and strict approach to the problem of containment and destruction of unmarked explosives.

Article IV, which contains provisions for strict control by States over unmarked plastic explosives within their territory and their eventual destruction, was the subject of extensive discussions during the Conference. In the Legal Committee draft, States were under the obligation to consume or destroy their military and commercial stocks of unmarked explosives within 15 and 3 years respectively.

The Conference agreed on the 3-year period of commercial stocks proposed in the Legal Committee draft, but modified the regime applicable to military explosives. It was felt, that it was not realistic to require unmarked explosives, that had already been incorporated into duly authorized military devices,[644] to be destroyed within any time frame. The chances of illegal use of the plastic explosives incorporated into those devices were judged to be remote as they were likely to be under tight control. Moreover, the destruction of these devices would pose grave technical, economical and environmental problems. Consequently, paragraph 3 of Article IV has been redrafted and now contains the obligation for States to

---

[644]Duly authorized military devices refers to: devices such as shells, bombs, projectiles, mines, missiles, rockets, shaped charges, grenades, etc.

consume, destroy, mark or render permanently ineffective within a period of 15 years only those military or police stocks which are not permanently incorporated in duly authorized military devices, in other words, the bulk, raw stock of unmarked plastic explosives.

Another modification included the addition of three new paragraphs to Article IV as a result of a change in the Technical Annex. During its fourth meeting, from 26 to 30 November in Montreal, the Ad Hoc Group of Specialists concluded, that it was necessary to exclude in certain cases unmarked plastic explosives need to be manufactured or held for research and development, testing and training and forensic science purposes. This has been effected, in the Annex, by excluding them from the definition of explosives referred to in Article 1, paragraph 1 of the Convention.

As a consequence and in order not to create a possible loophole in the Convention, the obligation for States to exercise strict and effective control over these exempted explosives and to destroy them when they are no longer used for those purposes has been added in Article IV.

III. Technical Annex

One of the difficulties facing the Conference (and the other preparatory bodies) was the fact that the issue at hand was a complicated one. Not only the 'visual' legal and political problems associated with international instruments aiming at preventing certain criminal activities arose, but parties were also confronted by new, highly technical problems such as the development of detection techniques, chemical additives and marking methods.

It may be assumed that the ICAO Council and the 27th Assembly foresaw in the new Convention an instrument with two major components:

1. Strict obligations for States with regard to unmarked plastic explosives pending the eventual extinction of those forms of plastic explosives
2. Strict technical rules on the detectability of plastic explosives through obligatory marking techniques

From the beginning, these conditions led to a twofold, almost completely separated approach. On the one hand the customary ICOA sequence of rapporteur, Subcommittee, Legal Committee prepared essentially the legal part of he desired international instrument, solving questions and finding solutions on issues like transport and possession of legal and illegal unmarked explosives, depletion of existing military and civil stocks, States obligations, enforcement and international penal aspects, etc.

On the other hand, the Ad Hoc Group of Specialists already established by the Council on 30 January 1989, set out to find one or more acceptable methods of marking plastic explosives for the purpose of detection. This was not an easy task, as a acceptable has many connotations in the field of safe handling and production,

G. A New Convention on the Marking of Plastic Explosives

effectiveness of the proper explosive an of course effective and reliable detection. As a result of its work, culminating in four meetings in Montreal the Ad Hoc Group presented the Council on 3 December 1990 with its product,[645] which contained a proposal for the Annex to the Conventions.

When the two products were joined together by the Conference, there were some anxious moments. Due to amendments proposed by a number of Delegations, the Annex appeared to be developing into a something with the appearance of a mini-convention, where, e.g., directives and obligations towards States Parties were formulated. This process was unacceptable to other Delegations, who stressed the need for straightforward technical Annex. After some frantic redrafting a compromise was reached and parts of the proposed draft Annex were relocated in Article IV.

The Conference further decided to refer to the Annex as 'Technical Annex' so as to underline its essentially technical nature. A new Article X was added to the Convention, stipulating that the Annex shall form an integral part of the Convention. The legal status of the Annex to the Convention had already been the subject of lengthy discussions in the Legal Sub-committee, Legal Committee and Council. It had been conceived as an integral part of the Convention. Because of its nature and purpose, the Convention would be meaningless and indeed could not exist without the Annex. Therefore it must be subject to the same consensus as the Convention itself at the time of its adoption. On the other hand, the Annex had been given a special status in Article VII with respect to its amendment. The Legal Sub-committee and Legal Committee had realised, that the Annex would contain strictly technical provisions which were subject to evolution and which might require adjustments to be made promptly and with greater flexibility than would be possible under the traditional procedure of amending the Convention through a diplomatic conference.

The Annex consists of two Parts:

Part 1 provides a technical description of plastic explosives. This part also contains the reference to the aforementioned explosives used for training and laboratory purposes.

Part 2 contains, *inter alia*, the chemical and molecular formulas of four[646] different detection agents as determined earlier by the Ad Hoc Group of Specialists on the Detection of Explosives.

The four selected compounds are:

- Ethylene glycol dinitrate (EGDN)
- 2,3-Dimethyl-2,3-dinitrobutane (DMNB)
- *para*-Mononitrotoluene (*p*-MNT)
- *ortho*-Mononitrotoluene (*o*-MNT)

---

[645] C-WP/9209, Restricted.

[646] The Ad Hoc Group of Specialists concluded after its 4th meeting, that all four additives which had been selected earlier, should be included in the Technical Annex. As the compounds meet the same criteria in respect of detectability and useability, their mutual inclusion may offer producer States some flexibility in selecting a particular additive.

IV. *International Explosives Technical Commission*

The Convention provides in Article V for the establishment of an International Explosives Technical Commission (IETC), which is to evaluate the technical developments relating to the manufacture, marking and detection of explosives. The Commission shall also make recommendations to the Council for amendments to the Technical Annex. (Article VI)

The Conference decided to put the membership of the Commission at no less than 15 and no more than 19 members, who will be appointed by the ICAO Council from among persons nominated by States parties to the Convention.[647]

In the discussions on this subject it was reaffirmed that the Council in its appointment policy would be guided by the need for the presence of experts from both producer and user countries, while the long standing ICAO principle of equitable geographical representation would also be applied. Para 5 of Article V stipulates that the Commission shall adopt its rules of procedure, subject to the approval of the Council.

The amending procedure of the Annex in the Legal Committee draft was upheld by the Conference.[648] Consequently, Article VII provides for an exhaustive consultation procedure on proposed amendments of the Technical Annex: States Parties would be notified of a proposed amendment and invited to communicate their views; after consideration of the comments the Council would formally propose the amendment to all States within 90 days, the Amendment would be deemed to have been adopted and would enter into force after 180 days or after such period as specified in the proposed amendment.

If five or more States Parties have objected to the proposed amendment, the proposal will be referred back to the IETC for further consideration. In such a situation the Council may also decide to convene a conference of all States Parties.

A new Article VIII invited States Parties to provide the Council with information that would assist the IETC in its work and to keep the Council informed of measures they have taken to implement the provisions of the Convention. The Council in its turn is to communicate such information to all States Parties and international organizations concerned.

The Conference took the decision to insert a new Article IX which is directed at facilitating the implementation of the Convention. The Council of ICAO shall, in co-operation with States Parties and International Organizations concerned, take

---

[647] The Conference found an analogy with Article 56 of the Chicago Convention which deals with the composition of the Air Navigation Commission – the most important advisory body to the ICAO Council. Its present membership is limited to 15, but a decision was taken by the 27th ICAO Assembly to amend Article 56 in order to increase its membership to 19. The amendment will require 108 ratifications for its entry into force.

[648] Proposals by several Delegations to introduce in the Convention, *expressis verbis*, the possibility to amend the Convention itself were decided against by the Conference in the understanding that the main body of the Convention could be amended in the manner provided for and codified in the Vienna Convention on the Law of Treaties.

G. A New Convention on the Marking of Plastic Explosives

appropriate measures to facilitate the implementation of the Convention, including the provision of technical assistance and measures for the exchange of information relating to technical developments in the marking and detection of explosives.[649]

V. *Final Clauses and Final Act*

ICAO has been designated the Depositary of the Convention. The Convention is subject to ratification, acceptance, approval or accession by States and shall enter into force 60 days after the deposit of the 35th such instrument with ICAO, provided that at least five of the ratifying or acceding States are Producer States.

Thus a qualifier for the entry into force of this Convention has been added. As a result, Article XIII para. 2. of the Convention requires States when depositing their instrument of ratification, acceptance, approval or accession, to declare whether or not they are a producer State.

Article I, para. 6 defines Producer State as 'any State in whose territory explosives are manufactured.' No further qualifications have been added, so ultimately it is left to States Parties to decide whether they wish to be considered a Producer State.

In the Final Act[650] of the Conference the text of a Resolution is included which was adopted by consensus by the Conference. That Resolution addresses the importance of the marking of explosives to prevent unlawful acts against, *inter alia*, civil aviation, maritime navigation and other modes of transportation and urges States to become Party to the Convention as soon as possible.

The Resolution also urges the international community to consider increasing technical, financial and material assistance to States in need of such assistance in order to benefit from the achievement of the objectives of the Convention.

In Resolving Clause 5 it invites the Council to assume the functions assigned to it in the Convention and to maintain, pending the entry into force of the Convention, the existence of the Ad Hoc Group of Specialists and to respect the principle of equitable geographical representation in the appointment of he members of the IETC. Finally, the Resolution requests the Council to initiate, as a matter of high priority, studies regarding the marking of explosives other than those referred to in the Convention.

In itself the new Convention represents an impressive achievement of States. The speed of its conception is truly remarkable and reflects positively on the

---

[649] An important decision. The effective detection of marked plastic explosives is of course as important as the marking itself, as the one would be completely useless without the other. As civil aviation security is by definition a global concept, mutual assistance and co-operation is a must.

[650] Final Act of the International Conference on Air Law held under the auspices of the International Civil Aviation Organization in February–March 1991. Adopted by the Conference on 1 March 1991.

capabilities of ICAO in these matters, while it may be considered through its scope and purpose a genuine multilateral instrument.

Although the Conference went through several critical phases, the determination of its participants to get results never wavered and that led to the unanimous adoption of a new instrument of international law. Again, as in the case of the additional 1988 Protocol to the Montreal Convention, this result has been achieved in record time.

It is too early to judge the practical value of the Convention, but it is submitted that the possibilities of he Convention are unnecessarily limited by its strict adherence to the term 'plastic' in its definition of explosives. This restriction may impede further possible developments aimed at broadening its impact.

Much will depend on the concrete implementation of the measures proposed in the Convention through a hopefully quick and widespread ratification and the further work of the Ad Hoc Group of Specialists and its successor, the International Explosives Technical Commission.

Interference with civil aviation should be viewed as an extortion oriented act committed against the international order and world peace and which is calculated to take advantage of the most susceptible human quality of seeking personal security as a priority. The offence is an immediate threat to world peace and should be treated with the utmost care. It is needless to say that any nation which views the offence differently encourages world discord. Any wilful act calculated to endanger the safety of an aircraft, its passengers or any aviation related property should be collectively regarded as an offence against the safety of air travel.

As for the need for a more flexible approach to the extradition of offenders the establishment and recognition of a universal offence against the safety of aircraft would almost automatically nurture mutual cooperation between nations. Often, if an offender imputes politics to the offence committed by him, he is granted political, asylum by the host nation merely because the latter sympathises with the alleged motive for the offence as represented by the offender. Once this takes place it no longer remains the commission of an offence universally condemned but becomes an altercation between nations on political beliefs and convictions.

References

Abramovsky A (1974) The Hague Convention. Columbia Journal of Transnational Law 3(3):89
Abramovsky A (1975a) The Montreal Convention. Columbia Journal of International Law 14 (2):278
Abramovsky A (1975b) Multilateral Convention for the suppression of unlawful seizure and interference with aircraft. Part II of the Montreal Convention. Columbia Journal of Transnational Law 14:300
Boyle RP (1964) The Tokyo Convention on offenses and certain other acts committed on board aircraft. Journal of Air Law and Commerce 30:305–328
Boyle RP (1972) International action to combat aircraft hijacking. Lawyers of the Americas 4:460–473

References

Brosche H (1974) The Arab oil embargo and the United States pressures against Chile. Case Western Reserve Journal of International Law 7:3

Chung DY (1976) Some legal aspects of aircraft hijacking in international law. Ph.D. dissertation, University of Tennessee

Costello D (1975) International terrorism and the development of the principle *Aut Dedere Aut Judicare*. Journal of International Law and Economics 10:483

Evans AE (1969) Hijacking: its cause and cure. The American Journal of International Law 63:708

Feller SZ (1972) Comment on criminal jurisdiction over aircraft hijacking. Israel Law Review 7:212

Fingerman ME (1980) Skyjacking and the Bonn Declaration of 1978: sanctions applicable to recalcitrant nations. California Western International Law Journal 10:142

Fitzgerald GF (1971) Toward legal suppression of acts against civil aviation. International Conciliation 585:42–78

Hudson MO (1941) International legislation, vol VII. Carnegie Endowment for International Peace, Washington, DC

Kunz JL (1948) The inter-American treaty of reciprocal assistance. The American Journal of International Law 42:111

Lauterpact H (1950) International law and human rights. Stevens & Sons, London

Mankiewicz RH (1971) The 1970 Hague Convention. Journal of Air Law and Commerce 37(2):201

Margo RD (1989) Aviation insurance, 2nd edn. Butterworths, London

McWhinney E (1987) Aerial piracy and international terrorism, the illegal diversion of aircraft and international law, 2nd revised edn. Martinus Nijhoff, Dordrecht

Mendelsohn AI (1967) In flight crime, the international and domestic picture under the Tokyo Convention. Virginia Law Review 53:509

Milde M (1986) Interception of civil aviation vs. misuse of civil aviation (background of Amendment 27 to Annex 2). Annals of Air and Space Law 11:122

Milde M (1990) Draft convention on the marking of explosives. Annals of Air and Space Law 15:155–179

Oppenheim L (1958) International law, vol 1, 8th edn. Longmans, Green & Co., London

Reiff H (1959) The United States and the treaty law of the sea. University of Minnesota Press, Minneapolis, MN

Sarkar AK (1972) International air law and safety of civil aviation. Indian Journal of International Law 12:200

Schwenk W (1979) The Bonn declaration on hijacking. Annals Air and Space Law 4:307

Shubber S (1973a) Jurisdiction over crimes on board aircraft. Martinus Nijhoff, The Hague

Shubber S (1973b) Aircraft hijacking under the Hague Convention. International and Comparative Law Quarterly 22:725

Van Panhuys HF (1970) Aircraft hijacking and international law. Columbia Journal of Transnational Law 9:13

Vlasic IA (1982) Casebook on international air law. McGill, Montreal

Chapter 6
Aviation Security Audits

As discussed earlier, on the basis of Assembly Resolution A33-1 adopted in 2001 and the recommendations of the High-level, Ministerial Conference on Aviation Security (Montreal, February 2002), the Council adopted in June 2002 the Aviation Security Plan of Action, which included the establishment of a comprehensive programme of regular, mandatory, systematic and harmonized audits to be carried out by ICAO in all Contracting States. The ICAO Universal Security Audit Programme (USAP) was subsequently launched, with the objective of all Contracting States having benefited from an initial audit by the end of 2007.

The ICAO USAP has been implemented on schedule and within its budget allocation. The audits have proven to be instrumental in the identification of aviation security concerns and in providing recommendations for their resolution. From its inception, the USAP has enjoyed the support of Contracting States and is promoting positive change as States become increasingly sensitized to the international requirements. The USAP follow-up missions have validated a markedly increased level of implementation of ICAO security Standards, thereby attesting to States' commitment to achieving the objective of the USAP to strengthen aviation security worldwide.

A security culture, if such were to exist among ICAO's member States, would mean that the States would be aware of their rights and duties, and, more importantly, assert them. Those who belong to a security culture also know which conduct would compromises security and they are quick to educate and caution those who, out of ignorance, forgetfulness, or personal weakness, partake in insecure conduct. This security consciousness becomes a "culture" when all the 190 member States as a whole makes security violations socially and morally unacceptable within the group.

All ICAO Member States are to have been successfully audited by the end of 2007, with strengths and weaknesses identified, regional and global trends tracked, and recommendations made to States for improving their security regimes. However, there remains a small number of States that have made little or no progress in implementing the ICAO recommendations to correct the deficiencies identified through the audits. Although security audit information has been restricted in the past, steps should be taken to increase the transparency of the audit programme and ensure that the global aviation network remain protected. It is therefore proposed that, in addition to a review of deficiencies by the Audit Results Review Board, consideration be given to the development of a process that will notify all Member States when deficiencies identified during the course of a USAP audit remain unaddressed for a sustained period. A notification process could involve the use of information which does not divulge specific vulnerabilities but enables States to initiate consultations with the State of interest to ensure the continued protection of aviation assets on a bilateral basis.

Upon completion of a USAP audit, States are required to submit a corrective action plan addressing deficiencies and schedule a follow-up visit. Audit follow-up visits were initiated in mid-2005 in order to validate the implementation of States' corrective action plans and to provide support to States in remedying identified deficiencies. These visits are normally conducted in the second year following the date of a State's audit. According to USAP reports, follow-up visit results have shown that the majority of States have made progress in the implementation of their corrective action plans. The average implementation rate of Annex 17 Standards in visited States increased significantly when compared with the initial audit results. At the same time, however, it is significant to note that follow-up visits have also revealed that there remains a small number of States that have made little or no progress in implementing the ICAO recommendations to correct the deficiencies identified through the audits.

Comprehensive statistical analysis of audit results and levels of compliance (globally, by region, and by subject matter) is available on the USAP secure website. Key findings are presented at both the national and airport levels. According to the progress report submitted to the ICAO Council in 2006 the ICAO Secretariat advised that in the case of States that are demonstrating little or no progress by the time of the follow-up visit, a cross analysis of the USAP audit results with those of the USOAP reveals that generally, States that have difficulty in implementing the safety-related SARPs are also experiencing difficulties with the implementation of the Annex provisions on the security side. Certain contributing factors have been identified. These often include a lack of financial and/or suitably qualified human resources as well as frequent changes in key personnel within a State's Appropriate Authority. In certain cases, there also appears to be a certain complacency and general lack of interest in implementing the ICAO recommendations.

In order to address the issue of States that are not responding effectively to the ICAO audit process, a high-level Secretariat Audit Results Review Board has recently been established for the purpose of examining both the safety and security histories of specific States brought to its attention by either ICAO's USOAP or USAP. The objective would be is to highlight or raise the profile of these States within the system in order to encourage them to take responsible actions in a measured and timely manner.

The Committee on Unlawful Interference of the ICAO Council has recommended to the Council that these data and trends be made public at the Assembly. Although such information has been restricted in the past, the Committee believes all States and the public should be aware of the areas needing improvement without identifying specific States or vulnerabilities. Further, the Council has been discussing with the Secretariat ways in which it can most effectively exercise its oversight responsibilities with respect to States that do not comply with their responsibilities under the Convention and its Annexes.

While reports show that many ICAO Member States have actively used information gathered from USAP audits to improve their security systems, reports also demonstrate that other States cannot or will not make necessary changes. For those States that lack resources to improve their security systems, new mechanisms such as ICAO's Coordinated Assistance and Development (CAD) Programme are in place to assist in directing longer-term attention to problems. For those States that remain unable to improve their security systems, bringing such challenges before the Audit Results Review Board, and possibly the Council, for consideration are valuable steps toward addressing the deficiencies in the longer term. However, the vulnerabilities presented by unresolved and sustained issues represents a significant weakness in the global protective network and a possible critical or urgent area of vulnerability for other Member States with air carrier service at the airport of interest, particularly when combined with indications of a heightened threat.

In building a security culture within ICAO member States it is imperative that Consideration should also be given to the development of a process for ensuring that all Member States are notified when deficiencies identified during the course of a USAP audit remain unaddressed for a sustained period of time. A notification process could involve the use of information which does not divulge specific vulnerabilities but enables States to initiate consultations with the State of interest to ensure the continued protection of aviation assets on a bilateral basis. Such a notification process may result in a strengthened ability on the part of ICAO to ensure that States unwilling to meet basic security standards will be held accountable and allow for a limited amount of transparency in the security audit programme without divulging specific potential security vulnerabilities.

One of the significant results of the 36th Session of the ICAO[651] Assembly, held in September 2007, was the adoption of a Resolution calling for the sharing of information through the ICAO Council pertaining to security audits conducted by ICAO. This brings to bear a certain shift of focus from the original confidentiality of the audits to one of limited transparency. It also raises the more compelling issue as to what the legal principles applicable are that would attribute to the Council the ability to divulge information and the limitations if any, on carrying out the instructions of the Assembly, which is one of the mandatory functions of the Council. The question also arises as to whether such a function could be sustained in the face of other overriding factors, one of which is the extent to which ICAO

---

[651] The International Civil Aviation Organization, a specialized agency of the United Nations, was established by Article 43 of the *Convention on International Civil Aviation* (Chicago Convention), signed at Chicago on 7 December 1944 (ICAO Doc 7300/9, Ninth Edition, 2006). The main objectives of ICAO are to develop the principles and techniques of international air navigation and to foster the planning and development of air transport. ICAO has 190 Member States.

stands empowered by its constituent member States to divulge information pertaining to aviation activities in their territories.

A. Security Oversight

ICAO has a security oversight programme called the Universal Security Audit Programme (UASP). The ICAO Universal Security Audit Programme (USAP), launched in June 2002, represents an important initiative in ICAO's strategy for strengthening aviation security worldwide and for attaining commitment from States in a collaborative effort to establish a global aviation security system.

The programme, which is part of the Aviation Security Plan of Action, provides for the conduct of universal, mandatory and regular audits of the aviation security systems in all ICAO member States. The objective of the USAP is to promote global aviation security through the auditing of States on a regular basis to assist States in their efforts to fulfil their aviation security responsibilities. The audits identify deficiencies in each State's aviation security system, and provide recommendations for their mitigation or resolution.

Implementation of the programme commenced with the first aviation security audit taking place in November 2002 and between three and four audits continue to be conducted around the world each month. The 35th Session of the Assembly held from 28 September to 8 October 2004 mandated ICAO to maintain strict confidentiality of all State-specific information derived from audits conducted under the Universal Security Audit Programme (USAP). However, in order to promote mutual confidence in the level of aviation security between States, the Assembly urged all Contracting States to "share, as appropriate and consistent with their sovereignty, the results of the audit carried out by ICAO and the corrective actions taken by the audited State, if requested by another State".[652]

While noting the importance of continuing bilateral exchanges of information between States, the 36th Session of the Assembly, held from 18 to 28 September 2007, also recognized the value of proposals presented by the Council and Contracting States for the introduction of a limited level of transparency with respect to ICAO aviation security audit results.[653] The Assembly directed the Council to consider such an introduction of a limited level of transparency, balancing the need for States to be aware of unresolved security concerns with the need to keep sensitive security information out of the public realm. In doing so, the Assembly emphasized that it was essential that any methodology developed to provide for increased transparency also ensure the appropriate safeguarding of a State's

---

[652]A35-9, Appendix E, Resolving Clause 4; and Recommended Practice 2.4.5 of Annex 17 – *Security*.

[653]Resolution A36-20, A36-WP/336 and Plenary Action Sheet No. 3.

A. Security Oversight

security information in order to prevent specific information that could be used to exploit existing vulnerabilities from being divulged.

The 36th Session of the ICAO Assembly adopted Resolution A36-20,[654] *Appendix E* of which addresses the USAP. As mentioned earlier, it must be emphasized that the Resolution *inter alia* directs the Council to consider the introduction of a limited level of transparency with respect to ICAO aviation security audits, balancing the need for States to be aware of unresolved security concerns with the need to keep sensitive security information out of the public realm and requests the Council to report to the next ordinary session of the Assembly (in 2010) on the overall implementation of the USAP.

Since the launch of the USAP in 2002, 169 aviation security audits and 77 follow-up missions have been conducted.[655] The audits have proven to be instrumental in the ongoing identification and resolution of aviation security concerns, and analysis reveals that the average implementation rate of Annex 17 Standards in most States has increased markedly between the period of the initial audit and the follow-up mission.

A critical part of the audit process is the requirement that all audited States submit a corrective action plan to address deficiencies identified during an audit. As directed by the Council, all States are notified (by State letter and on the USAP secure website) of those states that are more than 60 days late in submitting a corrective action plan. As of 31 July 2007, there were seven States that were more than 60 days late. In the case of late corrective action plans, repeated reminders are sent to States, including at the level of the Secretary General and with the involvement of the applicable Regional Office, and ICAO assistance is offered should the State require advice or support in the preparation of its action plan. Extensive feedback is provided to each audited State on the adequacy of its corrective action plan, and an ongoing dialogue is maintained where necessary to provide support in the implementation of proposed actions.

ICAO performs comprehensive analyses of audit results on levels of compliance with Annex 17 – *Security* Standards on an ongoing basis (globally, by region and by subject matter). This statistical data is made available to authorized users on the USAP secure website and is shared with other relevant ICAO offices as a basis for prioritizing training and remedial assistance projects. As of 31 July 2007, 77 follow-up missions had been conducted. These missions take place two years after the initial audit with the purpose of validating the implementation of State corrective action plans and providing support to States in remedying deficiencies. These missions are normally conducted by the applicable Regional Office, with close

---

[654] Resolution A36-20, Consolidated statement on the continuing CA policies related to the safeguarding of international civil aviation against acts of unlawful interference, Report of the Executive Committee (Report Folder) Assembly, 36th Session, A36 – WP/336, p/46, at 16-2.

[655] The 36th Session of the ICAO Assembly was informed that there are some 150 certified auditors on the USAP roster, from 59 States in all ICAO regions. The participation of certified national experts in the audits under the guidance of an ICAO team leader has permitted the programme to be implemented in a cost-effective manner while allowing for a valuable interchange of expertise.

coordination through Headquarters. The results of the follow-up visits indicate that the majority of States have made significant progress in the implementation of their corrective action plans.

A high-level ICAO Secretariat Audit Results Review Board (ARRB) has been established as part of an overall coordinated strategy for working with States that are found to have significant compliance shortcomings with respect to ICAO Standards and Recommended Practices (SARPs). The ARRB examines both the safety and security histories of specific States and provides an internal advisory forum for coordination among ICAO's safety, security and assistance programmes.

B. The Role of the ICAO Council

As future measures in the audit programme of ICAO, the ICAO Council approved in 2007, the practice that not all States need to be audited at the same frequency, although the USAP should always preserve the principle of universality. The Council was of the view that, with a solid baseline of audit results established for all States by the end of 2007, a more effective use of resources can be achieved by developing an appropriate scheduling/frequency model to determine the priority of future audits and frequency of visits to States. It remains as a requirement however that the principle of universality will be maintained with all States audited at least once within a 6-year period.

Another decision of the Council was that future audits under the USAP should be expanded to include relevant security-related provisions of Annex 9 – *Facilitation*. With the recent expansion of the Universal Safety Oversight Audit P to a comprehensive systems approach covering all safety-related Annexes, Annex 9 is currently the only Annex which is not included in either of ICAO's two audit programmes. There are a number of security-related provisions contained in Annex 9, particularly as related to the security and integrity of travel documentation, which can be audited under the USAP along with the related Standards of Annex 17.

The Council also decided that wherever possible, ICAO aviation security audits should be focused on a State's capability to provide appropriate national oversight of its aviation security activities. Using the results of the initial audits and follow-up visits, the scope of future ICAO audits should be adjusted to the prevailing situation in each audited State. Those States that have demonstrated the requisite national infrastructure necessary to oversee security activities at their airports may undergo a targeted oversight audit to verify adequate implementation of the State's national quality control programme. Such oversight audits would continue to include a verification of the implementation of ICAO provisions through spot checks at the airport level.

The decision of the 36th Session of the Assembly – that a "limited level of transparency" with respect to ICAO aviation security audit results be maintained, ensures a balance between the need to divulge certain information while protecting the interests of States. As such the Council has to draw a fine line between

B. The Role of the ICAO Council

potentially conflicting interests. As for safety, the 35th Session of the Assembly, when it addressed the issue of expanding the audits from a limited Annex basis to a comprehensive systems approach, instructed the Secretary General to make the final safety audit reports available to all Contracting States and also to provide access to all relevant information derived from the Audit Findings and Differences Database (AFDD) maintained by ICAO.[656] Furthermore, in Resolution A36-2 (Unified Strategy to Resolve Safety Related Deficiencies) the Assembly, in operative Clause 6 of the Resolution, has directed the Council to apply and review, as necessary, the procedures to inform Contracting States, within the scope of Article 54(j) of the Chicago Convention, in the case of a State having significant shortcomings with respect to ICAO safety related SARPs in order for other Contracting States to take action in an adequate and timely manner.

Article 54(j) makes it a mandatory function of the Council to report to any Contracting State any infraction of the Chicago Convention as well as any failure to carry out recommendations and determinations of the Council.[657] There are various dimensions to this provision in the context of Resolution A36-2. Firstly, it is surprising that the Assembly Resolution does not also request the Council to perform its mandatory function in Article 54(k), which is to report to the Assembly any infraction of the Convention where a Contracting State fails to take appropriate action within a reasonable time after notice of the infraction. This would have arguably been a more coercive and effective tool than the measure prescribed in Article 54(j) in that States would be quite concerned if their shortcomings were to be aired out in front of 190 Contracting States at an ICAO Assembly.

The second dimension to the Resolution is that it the function of the Council in this case, to use the words of operative clause 6 of Resolution A36-2 to "apply and review... the procedure to inform Contracting States within the scope of Article 54(j) of the Chicago Convention, in the case of a State having significant shortcomings with respect to ICAO safety related SARPs in order for other Contracting States to take action in an adequate and timely manner". Surprisingly the Council is asked by the Assembly to restrict itself to determining the adherence to SARPs and report its findings thereof, which is already a function handed down in the Convention to the Council in Article 38.[658] Again, it is not clear as to why the Assembly refrained from applying the rest of Article 54(j) to its Resolution, which makes it incumbent upon the Council to report the failure to carry out recommendations or determinations of the Council. This application would have served the purpose of the Assembly better than the mere restriction to the SARPs in the Annexes.

---

[656] Resolution A35-6, Operative Clause 7.

[657] *Convention on International Civil Aviation* (Chicago Convention), signed at Chicago on 7 December 1944 (ICAO Doc 7300/9, Ninth Edition, 2006).

[658] Article 38 provides: *inter alia* that any Contracting State can file a difference to a standard and notify the Council which in turn is required to make immediate notification to all other States of the difference which exists between one or more features of an international standard and the corresponding national practice of that State.

The third dimension is that the Council, under the Convention, has only functions (which are in essence duties) and no powers.[659] On the other hand the Assembly has powers and duties accorded to it in the Chicago Convention,[660] one of which is to delegate to the Council the powers and authority necessary or desirable for the discharge of the duties of the Organization and revoke or modify the delegations of authority at any time.[661] However, in this instance there is no indication that the Assembly exercised its powers to delegate its authority or power to the Council to apply and review the procedure in Article 54(j). If this had been the case, the Council would have had the same right and the authority of the Assembly to take appropriate action as deemed necessary in the manner in which the information derived from safety and security audits would be disseminated and reported to other States.

A power is the capacity to direct the decisions and actions of others. A function on the other hand is to perform, execute or administer.[662] A power is also defined as an ability on the part of a person to produce a change in a given legal relation by doing or not doing a certain act.[663] In this context the Council only has a duty or function to report to States shortcomings of other States detected during the course of safety and security audits with regard to adherence by the ICAO member States of SARPs. It is therefore incontrovertible that Assembly Resolution A36-2 merely hands over to the Council the function to report an infraction of the Chicago Convention as well as shortcomings with regard to SARPs and recommendations and determinations of the Council in that regard.

The Chicago Convention bestows neither the ability nor the power on the Council to investigate and determine on its own initiative whether there has been an infraction of the Convention. There is also no specific provision which entitles the Council to notify the State concerned that an infraction has taken place. However, Article 54(n) provides that the Council can consider any matter relating to the Convention which any Contracting State refers to it, giving the Council the capacity to make its own determination and recommendations pertaining to a matter referred to it. It is also noteworthy that both Article 15 of the Convention, which allows the Council to report and make recommendations resulting from a review by the Council of charges imposed for the airports and other facilities, and Article 69, which gives the Council competency to make recommendations to member States for the improvement of air navigation facilities, are two instances of specific provision being made within the Convention where the Council can make recommendations for the consideration of ICAO member States.

---

[659] Although Jacob Schenkman, in his well documented and logically reasoned treatise on ICAO states that "The Council has been entrusted with duties, powers and functions..." he does not give a single example of such a power. See Capt. Schenkman (1955, p. 158).
[660] Article 49 of the Convention.
[661] Article 49(h).
[662] Black (1990, p. 673).
[663] Black (1990, p. 1189).

B. The Role of the ICAO Council

Clearly, non compliance with SARPs and shortcomings or deficiencies in security cannot be classified as infractions of the Convention. An infraction is a violation and arguably applicable to the Chicago Convention itself and not to the Annexes which only contain SARPs that are not strictly legally binding so as to constitute a violation if not followed. Therefore, the Assembly, in A36-2 quite clearly meant the reportage of failure to carry out recommendations and determinations of the Council with regard to SARPs. This is clearly an administrative function and not a judicial function, since an administrative act is usually referred to as similar or related activities regarding the handling and processing of information.

Another important dimension to the Council's role as per A36-2 in divulging security information is that ICAO has already entered into memoranda of understanding with the States audited that audit reports will be confidential and made available to the State audited and relevant ICAO staff on a need-to-know basis. These agreements also require that, concurrently with the preparation of the report, a non-confidential audit activity report limited to the name of the audited State, the identity of airports visited during the audit, and the completion date of the audit will be developed for release to all Contracting States. Reports to the Council are required to be in a form that maintains the confidentiality of the audit report in relation to the State concerned. Accordingly, ICAO has restricted itself for purposes of confidentiality to giving only limited and non specific details of audits to its member States. This raises a legal issue as to ICAO's right to contravene its agreement with member States in deference to an Assembly Resolution. This issue also seemingly goes to the root of ICAO's empowerment by its member States and ICAO's accreditation to such States.

International organizations can generally only work on the basis of legal powers that are attributed to them. Presumably, these powers emanate from the sovereign States that form the membership of such organizations.[664] Therefore, the logical conclusion is that if international organizations were to act beyond the powers accorded to them, they would be presumed to act *ultra vires*.[665] It should be noted that ICAO does not only derive implied authority from its Contracting States based on universality but it also has attribution from States to exercise certain powers. The doctrine of attribution of powers comes directly from the will of the founders, and in ICAO's case, powers were attributed to ICAO when it was established as an international technical organization and a permanent civil aviation agency to administer the provisions of the Chicago Convention. In addition, ICAO could lay claims to what are now called "inherent powers" which give ICAO power to perform all acts that the Organization needs to perform to attain its aims not due to any specific source of organizational power but simply because ICAO inheres in organization hood. Therefore, as long as acts are not prohibited in ICAO's constituent document (the Chicago Convention), they must be considered legally valid.[666]

---

[664] See de Witte (1998, pp. 277–304).
[665] Klabbers (2002, p. 60).
[666] Seyersted (1963, p. 28).

Over the past two decades the inherent powers doctrine has been attributed to the United Nations Organization and its specialized agencies on the basis that such organizations could be stultified if they were to be bogged down in a quagmire of interpretation and judicial determination in the exercise of their duties. The advantages of the inherent powers doctrine are twofold. Firstly, inherent powers are functional and help the organization concerned to reach its aims without being tied by legal niceties. Secondly, it relieves the organization of legal controls that might otherwise effectively preclude that organization from achieving its aims and objectives. The ability to exercise its inherent powers has enabled ICAO to address issues on aviation insurance and establish an insurance mechanism; perform mandatory audits on States in the fields of aviation safety and security; and establish a funding mechanism to finance aviation safety projects, all of which are not provided for in the Chicago Convention but are not expressly prohibited.

With regard to the conferral of powers by States to ICAO, States have followed the classic approach of doing so through an international treaty. However, neither is there explicit mention of such a conferral on ICAO in the Chicago Convention nor is there any description of ICAO's powers, except for an exposition of ICAO's aims and objectives. The Council of ICAO is designated both mandatory and permissive "functions", although the Council could impose certain measures when provisions of the Convention are not followed. Therefore States have not followed the usual style of conferral of powers in the case of ICAO, which, along the lines of the decision of the International Court of Justice in the 1996 *WHO Advisory Opinion* case[667] was that the powers conferred on international organizations are normally the subject of express statement in their constituent instruments.[668] This notwithstanding, it cannot be disputed that ICAO Contracting States have conferred certain powers on ICAO to perform its functions independently. For example, ICAO is a legal entity having the power to enter into legal agreements with legal entities including other international organizations with regard to the performance of its functions.

Conversely, an international organization must accept conferred powers on the basis of Article 34 of the Vienna Convention on the Law of Treaties which stipulates that a treaty does not create rights or obligations of a third State without its consent. This principle can be applied *mutatis mutandis* to an international organization such as ICAO. The conferral of powers on an international organization does not *ipso facto* curtail the powers of a State to act outside the purview of that organization unless a State has willingly limited its powers in that respect. This principle was recognized in the *Lotus* Case[669] where the Provisional International Court of Justice held that a State can exercise powers on a unilateral basis even while the conferral to the Organization remains in force. The Court held that restrictions upon the independence of States cannot be presumed.[670]

---

[667]Legality of the Threat or Use of Nuclear Weapons, Advisory Opinion, ICJ Reports, 1996, p. 64.
[668]Legality of the Threat or Use of Nuclear Weapons, Advisory Opinion, ICJ Reports, 1996, p. 79.
[669]*PCIJ Reports* Series A, No. 10, p. 4.
[670]*PCIJ Reports* Series A, No. 10, p. 18.

B. The Role of the ICAO Council

ICAO's conferred powers enable the Organization to adopt binding regulations by majority decision (which is usually unnecessary as most of ICAO policy is adopted through consensus). However, States could opt out of these policies or make reservations thereto, usually before such policy enters into force. This is because States have delegated power to ICAO to make decisions on the basis that they accept such decisions on the international plane. In such cases States could contract out and enter into binding agreements outside the purview of ICAO even on subjects on which ICAO has adopted policy. The only exception to this rule lies in the adoption of Standards in Annex 2 to the Chicago Convention on Rules of the Air, in particular navigation over the high seas and other overflight areas where freedom of flight prevails which all Contracting States are bound to follow in order to maintain global safety.

The 35th Session of the Assembly, when it addressed the issue of expanding the audits from a limited Annex basis to a comprehensive systems approach, instructed the Secretary General to make the final safety audit reports available to all Contracting States and also to provide access to all relevant information derived from the Audit Findings and Differences Database (AFDD) maintained by ICAO.[671] Furthermore, in Resolution A36-2 (Unified Strategy to Resolve Safety Related Deficiencies) the Assembly, in operative Clause 6 of the Resolution, directs the Council to apply and review, as necessary, the procedures to inform Contracting States, within the scope of Article 54(j) of the Chicago Convention, in the case of a State having significant shortcomings with respect to ICAO safety related SARPs in order for other Contracting States to take action in an adequate and timely manner.

From the above discussion it becomes clear that, while on the one hand the ICAO Assembly, which in essence is the representative voice of the 190 member States comprising ICAO, has directed the Council to apply and review procedures to inform member States within the scope of Article 54(j) of shortcomings, on the other hand, the overriding separate and individual memoranda signed by ICAO with its member States in the area of security would have to be revised in terms of the confidentiality clause. Additionally, the Council would have to set in place an understanding with States and appropriate mutually agreed guidelines on the content of such information and the manner in which it is to be divulged.

States retain the powers to act unilaterally and they are not bound to comply with obligations flowing from the Organization's exercise of conferred powers. States which have delegated powers on ICAO have the legal right under public international law to take measures against a particular exercise by ICAO of conferred powers which is considered to be *détournement de pouvoir, ultra vires* or an internationally wrongful act with which the objecting States do not wish to be associated. A State could also distance itself from the State practice of other Contracting States within the Council if such activity is calculated to form customary international law that could in turn bind the objecting State if it does not persist in its objections.[672]

---

[671] Resolution A35-6, Operative Clause 7.
[672] See Sarooshi (2005, p. 110).

The above notwithstanding, a significant issue in the determination of ICAO's effectiveness as an international organization is the overriding principle of universality and global participation of all its 190 Contracting States in the implementation of ICAO policy. This principle, which has its genesis in the Chicago Conference of 1944, has flowed on gaining express recognition of legal scholars. This is what makes ICAO unique as a specialized agency of the United Nations and establishes without any doubt that ICAO is not just a tool of cooperation among States.

References

Black HC (1990) Deluxe Black's law dictionary, 6th edn. West group, St. Paul, MN
de Witte B (1998) Sovereignty and European integration: the weight of tradition. In: Slaughter A-M et al (eds) The European court and national courts: doctrine and jurisprudence. Hart, Oxford
Klabbers J (2002) An introduction to international institutional law. Cambridge University Press, Cambridge
Sarooshi D (2005) International organizations and their exercise of sovereign powers. Oxford University Press, Oxford
Schenkman J (1955) International Civil Aviation Organization. Librairie E. Droz, Geneva
Seyersted F (1963) Objective international personality of intergovernmental organizations: do their capacities really depend upon the conventions establishing them? Krohns Bogtrykkeri, Copenhagen

Chapter 7
Conclusion

A perceived inadequacy of the global framework of aviation security is the lack of an implementation arm. ICAO has taken extensive measures to introduce relevant international conventions as well as Standards and Recommended Practices (SARPs) in Annex 17 to the Chicago Convention. There is also a highly classified *Aviation Security Manual* developed by ICAO which is provided to States. Additionally, the Organization provides focused security training courses to its member States. However, ICAO's role is largely confined to rule making and the provision of guidance, bringing to bear the need for an aviation security crisis management team on a global scale that could work towards effectively precluding acts of terrorism.

Another measure that could proactively facilitate the arrest of terrorism is the global curbing of the financing of terrorism. The United Nations General Assembly, on 9 December 1999, adopted the International Convention for the Suppression of the Financing of Terrorism, aimed at enhancing international co-operation among States in devising and adopting effective measures for the prevention of the financing of terrorism, as well as for its suppression through the prosecution and punishment of its perpetrators.

The Convention, in its Article 2 recognizes that any person who by any means directly or indirectly, unlawfully or wilfully, provides or collects funds with the intention that they should be used or in the knowledge that they are to be used, in full or in part, in order to carry out any act which constitutes an offence under certain named treaties, commits an offence. The treaties listed are those that are already adopted and in force and which address acts of unlawful interference with such activities as deal with air transport and maritime transport. Also cited is the International Convention for the Suppression of Terrorist Bombings, adopted by the General Assembly of the United Nations on 15 December 1997.

The *Convention for the Suppression of the Financing of Terrorism* also provides that, over and above the acts mentioned, providing or collecting funds toward any other act intended to cause death or serious bodily injury to a civilian, or to any other person not taking an active part in the hostilities in the situation of armed conflict, when the purpose of such act, by its nature or context, is to intimidate a population, or to compel a government or an international organization to do or to abstain from doing any act, would be deemed an offence under the Convention.

The use of the word "terrorism" in the title of the Convention brings to bear the need to examine in greater detail both the etymology and the connotations of the word in modern parlance. The term "terrorism" is seemingly of French origin and is believed to have been first used in 1798. "Terrorism," which originally had connotations of criminality to one's conduct is now generally considered a system of coercive intimidation brought about by the infliction of terror or fear. The most frustrating obstacle to the control of unlawful acts against international peace is the paucity of clear definition of the offence itself. Many attempts at defining the offence have often resulted in the offence being shrouded in political or national barriers.

In 1980 the Central Intelligence Agency of the United States of America adopted a definition of terrorism which states that terrorism is the threat or use of violence for political purposes by individuals or groups, whether acting for or in opposition to established governmental authority, when such actions are intended to shock, stun or intimidate victims. Terrorism, has involved groups seeking to overthrow specific regimes, to rectify perceived national or group grievances, or to undermine international order as an end in itself.

This all embracing definition underscores the misapprehension that certain groups which are etched in history such as the French Resistance of Nazi occupied France during World War II and the Contras in Nicaragua would broadly fall within the definitive parameters of terrorism. In fact, this formula labels every act of violence as being "terrorist" engulfing in its broad spectrum such diverse groups as the Seikigunha of Japan and the Mujahedeen of Afghanistan, although their aims, modus operandi and ideologies are different. A narrower definition which is that a terrorist is an individual or member of a group that wishes to achieve political ends using violent means, often at the cost of casualties to innocent civilians and with the support of only a minority of the people they claim to represent.

Even this definition although narrower than the 1980 definition cited above is not sufficiently comprehensive to cover for instance the terrorist who hijacks an air plane for his own personal gain. The difficulty in defining the term seems to lie in its association with political aims of the terrorist as is found in the definition that terrorism is really terror inspired by violence, containing an international element that is committed by individuals or groups against non-combatants, civilians, States or internationally protected persons or entities in order to achieve political ends.

The offence of terrorism has also been defined as one caused by any serious act of violence or threat thereof by an individual. Whether acting alone or in association with other persons which is directed against internationally protected persons, organizations, places, transportation or communication systems or against members of the general public for the purpose of intimidating such persons, causing injury to or the death of such persons, disrupting the activities of such international organizations, of causing loss, detriment or damage to such places or property, or of interfering with such transportation and communications systems in order to undermine friendly relations among States or among the nationals of different States or to extort concessions from States.

7 Conclusion

It is time that terrorism is recognised as an offence that is *sui generis* and one that is not always international in nature and motivated by the political aims of the perpetrator. For the moment, if terrorism were to be regarded as the use of fear, subjugation and intimidation to disrupt the normal operations of humanity, a more specific and accurate definition could be sought, once more analysis is carried out on the subject. One must always be mindful however, that without a proper and universally acceptable definition, international cooperation in combating terrorism would be impossible.

A terrorist act is one which is *mala in se* or evil by nature and has been associated with the political repression of the French Revolution era where, it is said, the word terrorism was coined. A terrorist is a *hostis humani generis* or common enemy of humanity.

International terrorism has so far not been defined comprehensively largely due to the fact that owing to its diversity of nature the concept itself has defied precise definition. However, this does not preclude the conclusion that international terrorism involves two factors. They are:

1. The commission of a terrorist act by a terrorist or terrorists.
2. The "international" element involved in the act or acts in question, i.e., that the motivation for the commission of such act or acts or the eventual goal of the terrorist should inextricably be linked with a country other than that in which the act or acts are committed.

Perhaps the oldest paradigm of international terrorism is piracy which has been recognized as an offence against the law of nations and which is seen commonly today in the offense of aerial piracy or hijacking.

Acts of international terrorism that have been committed over the past two decades are too numerous to mention. Suffice it to say, that the most deleterious effect of the offense is that it exacerbates international relations and endangers international security. From the isolated incidents of the sixties, international terrorism has progressed to becoming a concentrated assault on nations and organizations that are usually susceptible to political conflict, although politics is not always the motivation of the international terrorist. International terrorism has been recognized to engulf acts of aggression by one State on another as well as by an individual or a group of individuals of one State on another State. The former typifies such acts as invasion, while the latter relates to such individual acts of violence as hijacking and the murder of civilians in isolated instances. In both instances, the duties of the offender-State have been emphatically recognized. Such duties are to condemn such acts and take necessary action.

The United Nations has given effect to this principle in 1970 when it proclaimed that:

> Every State has the duty to refrain from organizing or encouraging the organization of irregular forces or armed bands, including mercenaries, for incursion into the territory of another State. Every State has the duty to refrain from organizing, instigating, assisting or participating in acts of civil strife or terrorist acts in another State or acquiescing in organized activities within its territory directed towards the commission of such acts, when the acts referred to in the present paragraph involve a threat or use of force.

The most pragmatic approach to the problem lies in identifying the parameters of the offense of international terrorism and seeking a solution to the various categories of the offense. To obtain a precise definition would be unwise, if not impossible. Once the offense and its parasitic qualities are clearly identified, it would become necessary to discuss briefly its harmful effects on the international community. It is only then that a solution can be discussed that would obviate the fear and apprehension we suffer in the face of this threat.

Index

A

Advance passenger information (API)
  Australia Immigration and Customs, 147
  guidelines, 138–140
  Mexican approach, 149
  passenger name record, 121–122
  public authorities, 135
  recommendation, 136–137
  United States legislation, PNR, 140–142
Air Transport Committee (ATC), 63, 64
American Transportation & Security Act, 140, 141
Audit Results Review Board (ARRB), 270
Audits
  aviation assets, 266
  corrective action plan, 269
  follow-up missions, 269
  ICAO Council
    Chicago Convention, Article 54(j), 271
    inherent powers, 273
    Resolution A36-2, 271
    Vienna Convention, Article 34, 274
  identified deficiencies, 266
  scheduling/frequency model, 270
  security culture, 265
  security oversight
    follow-up visits, 270
    limited level of transparency, 268–269
  Universal Security Audit Programme (USAP), 268
Aviation Security Manual, 277
Aviation Working Group (AWG), 98

B

Biometric identification, innovative security tools
  challenges, 112
  identification process, 111
  machine readable travel document, 110
  physiological, 111
  quantum cryptography, 109
Bonn declaration
  aerial communication, 247
  French delegate, 248
  incompatibility, Vienna Convention, 249
  international civil aviation
    Contracting States, 250
    transit agreement, 250–251
  legal status, 247–249
  prosecution or extradition, 252–253
  Soviet Union delegate, 248
  transit agreement, 250–252

C

Canadian Immigration and Refugee Act, 142
Chicago Convention
  aviation security, 57
  carriage of documents, 160
  civil aircraft, 159
  civil aviation, 210–212, 251
  Contracting State, 56, 128, 250
  customs, 195–196
  scope in Article 3, 211–212
Complicity theory, 42–43
Condonation theory, State responsibility
  categories, 50
  causality principle, 50
  Jane case, 48
  separate delict theory, 50
  United Nations General Assembly, 49
  war crime, 51
Conference of Parties (COP), 102–104
Contracting States
  Australian position, 147–148

281

Contracting States (*cont.*)
   Canadian position
     accuracy of the data, 143
     Canadian Immigration and Refugee Act, 142
   German and Swiss positions, 148–149
   Mexican approach, 149–151
   Safe Harbour principles, 146–147
   United Kingdom position
     British Airways, 145
     UK Data Protection Act, 146
   United States legislation, API and PNR
     American Transportation & Security Act, 140, 141
     WCO/IATA guideline, 141
Convention for the Suppression of the Financing of Terrorism, 277
Convention on International Civil Aviation
   advance passenger information, 131
   Contracting State, 56
Corporate Manslaughter and Corporate Homicide Act, 100–101

D
Departure control systems (DCS), 124, 125

E
Early twenty-first century initiatives
   advance passenger information, 121–123, 139–141
   innovative security tools
     biometric identification, 109–112
     public key directory, 112–121
   liability Conventions
     General Risks Convention, 93–98
     Unlawful Interference Compensation Convention, 98–109
   machine readable travel documents
     capacity-building strategy, 153
     ICBWG, 152, 153, 155
     UN Security Council's Resolution 1373, 152
     vocational training, 155
   passenger name record
     advanced passenger information guidelines, 138–140
     advantages of unified guidelines, 126–138
     Contracting States' positions, 140–151
     definition and application, 123–125
     importance, 125–126
   unmanned aerial vehicles
     air traffic services, 166–167
     high seas, operations, 163–166
     legal and regulatory issues, 162–163
     as state aircraft, 168–174
Emerging threats
   anti-missile system installation, 32
   AVSEC panel
     aviation security regime, 19, 20
     European Civil Aviation Conference, 19
   bioterrorism
     international air transport regulations, 22
     real-time outbreak disease surveillance, 23
     SARS virus, 21, 22
     Spanish flu virus, 21
   carriage of liquids, aerosols and gels, 38
   cyber-terrorism, 24–25
   deterrence, 13–16
   high level ministerial conference, 2–7
   ICAO, threat assessment, 16–19
   international accord, 33–36
   Man Portable Air Defense System (MANPADS)
     aircraft security, 25
     El Al airliners, 28
     ex post facto, 26
     PFLP group, 30
     Rand corporation, 26
     RPG-7 portable rocket launch, 31
     surface-to-air missiles, 27
     surface-to-surface missiles, 27
   missile attacks, 29–31
   perimeter guard, 32–33
   post conference work, 7–9
   probability, 9–13
Enhanced Border Security and Visa Reform Act, 132

G
General Risk Convention
   damage by aircraft, 95
   Exclusive Remedy provision, 97
   ICAO initiative, 94
   International Air Transport Association, 98
   mental and bodily injury, 96–97
   weight of the aircraft, 97, 98
Geneva Convention on the High Seas (1958)
   aircraft piracy, 217
   hijacking and piracy, 215–218
   illegal violence, 216
   international law, 214
   offence, 213–214
   piracy jure gentium, 26

Index 283

H
Hague Convention on hijacking 1970, 230
  efficiency, 236–237
  powers and duties
    extradition treaty, 234–235
    jurisdictional powers, 233–234
    obligations, 232–233
  scope, 231–232
Hostis humani generis, 279

I
Implementation and Capacity Building Working Group (ICBWG), 152
Innovative security tools
  biometric identification
    challenges, 112
    identification process, 111
    machine readable travel document, 110, 111
    physiological biometrics, 111
    quantum cryptography, 109
  public key directory
    encryption and decryption process, 113, 114
    ICAO, 115–120
    PKI scheme, 113
    waiver of immunity, 120–121
International Air Services Transit Agreement (IASTA), 129, 250–252
International Air Transport Association (IATA)
  State responsibility, 40, 41
  WCO, 139, 141
International Civil Aviation Compensation Fund, 102
International Civil Aviation Organization (ICAO)
  Air Transport Committee, 63
  ARRB, 270
  capacity-building strategy, 153
  capacity to conduct business, 117–119
  Chicago Convention, 57, 271
  Contracting States, 88, 265
  environmental protection, 88, 89
  Facilitation Division, 135
  high level ministerial conference, 2–7
  immunities and liabilities, 119–120
  inherent powers, 273
  initiatives
    action against opiate carriage, 192–193
    Annex 9, 196
    consideration, 192
    drug smuggling, 197
    narcotic drugs and psychotropic substances, 194–195
    opiates carriage, first-aid kits, 189
    recommendation, 190–191
  legal liability, 117
  market-based measures, 90
  member States, 267
  operational measures, 89
  organizational matters, 116
  post conference work, 7–9
  protagonist, 117
  recommendations, 265
  Resolution A36-2, 271
  TAG/MRTD, 115, 116
  technology and standards, 89
  USAP follow-up missions, 265
  Vienna Convention, Article 34, 274
International Convention for the Suppression of the Financing of Terrorism, 49
International Conventions
  Chicago Convention of 1944
    civil aviation, 210–212
    scope in Article 3, 211–212
  Geneva Convention on the High Seas (1958)
    aircraft piracy, 217
    hijacking and piracy, 215–216
    illegal violence, 216
    international law, 214
    offence, 217–218
    piracy jure gentium, 216
  Convention for the Prevention and Punishment of Terrorism (1937), 210
  United Nations Charter, 212–213
International Court of Justice (ICJ)
  Corfu Channel case, 52
  Lockerbie case, 44
International Criminal Police Organization (ICPO), 204
International drug control system, 178
International Explosives Technical Commission (IETC), 260–261
International Law Commission
  Report to General Assembly, 52–53
  United Nations Charter, Article 51, 50
International Maritime Organization, 185

L
Liability Conventions
  General Risk Convention
    damage by aircraft, 95
    Exclusive Remedy provision, 97
    ICAO initiative, 94
    International Air Transport Association, 98

mental and bodily injury, 96–97
weight of the aircraft, 97, 98
Unlawful Interference Compensation Convention
commercial airline pilots, 101–102
Conference of Parties, 103, 104
Diplomatic Conference, 105
domestic opt-in declaration, 104
"gross breach," 101
ICAO Security Panel, 108
International Civil Aviation Compensation Fund, 102
offence, 101
operators, 100
"relevant duty of care," 101
victims of violent crime, 106–107
Lockerbie case, State responsibility
ICJ, 44
Montreal Convention, 44, 46
Security Council Resolution 748 (1992), 45, 46
United Nations Charter, Article 24, 46

M
Machine readable travel documents (MRTD)
capacity-building strategy, 153
ICBWG, 152, 153
passports complaint, 156
UN Security Council's Counter-Terrorism Committee, 155
UN Security Council's Resolution 1373, 152
vocational training, 155
Man Portable Air Defense System (MANPADS), 25–29
Mexico–United States General Claims Commission, 43
Montreal Convention (1971)
aircraft in service, 237–238
jurisdictional powers of States, 242–246
offence
aircraft in flight, 240
definition, 238
violence, 239–240
severe penalties
air navigation facilities, 242
Hague Convention, 241
Montreal Protocol of 1978, 80–81

N
Nairobi Convention, 197
Narco-terrorism
Article 3 *bis*

civil aircraft, 200
foreign aircraft, 200–201
Convention on International Civil Aviation, Article 4, 198–199
definition, 179
drug trade, 177–178
facets, 177
global drug production, 178–179
ICAO assembly resolution A 27-12
Contracting States, 202
misuse of civil aviation, 203
ICAO initiatives
action against opiate carriage, 192–193
Annex 9, 196
consideration, 192
Nairobi Convention, 197
narcotic drugs and psychotropic substances, 194–195
opiates carriage, first-aid kits, 189
recommendation, 190–191
smuggling, 197
international drug control system, 178
offence, 177
Tokyo Convention, 201–202
United Nations initiatives
council decision 1985/131 of 28 May 1985, 180
draft Convention, 182
drug trafficking and drug abuse, 187
efforts, 187–188
ICAO, 184
against illicit traffic in narcotic drugs and psychotropic substances, 186–187
international co-operation, 183, 185
ratification, 184
resolution 36/168, 181
resolution 40/122, 179–180
sea transport, 185
Single Convention on Narcotic Drugs of 1961, 180
State co-operation, 185–186
National Aeronautics and Space Administration, 24

O
Office for Victims of Crime (OVC), 41–42

P
Passenger name record (PNR)
advanced passenger information guidelines, 138–140
advantages of unified guidelines

concept, 134–135
Fourth Panel Meeting Facilitation Panel, 133
history, 135–138
IASTA, 129
international air transport, 131
international law, 127, 128
Simplifying Travel Organization, 132
Chicago Convention, Article 13, 123
Contracting States
  Australian position, 147–148
  Canadian position, 142–143
  German and Swiss positions, 148–149
  Mexican approach, 149–151
  Safe Harbour principles, 146–147
  United Kingdom position, 144–146
  United States legislation, 140–142
data to States, importance, 125–126
definition and application, 123–125
Plastic explosives
  air law, 253, 255
  final clauses and act
    ICAO, 261
    producer state, 261
  international explosives technical commission, 260–261
  scope of the Convention, 255–257
  State obligations, 257–258
  technical annex
    ad hoc group, 258–259
    components, 258
Principles of responsibility
  Rome Convention of 1952
    Air Transport Committee, 63, 64
    background, 70–71
    high contracting party, 62
    ICAO Assembly, 65
    insurance, 71–76
    modernization, 81–90
    Montreal Protocol of 1978, 80–81
    provisions, 77–80
    sub-committees, 65
  State responsibility
    airport profiling, 55–58
    complicity theory, 42–43
    condonation theory, 48–51
    extradition of offenders, 43–48
    IATA, 40, 41
    knowledge, role of, 51–54
    OVC, 41–42
    principles, 42
    profiling of passengers, 54–55
    right of privacy, 58–61

Public key directory (PKD)
  encryption and decryption process, 113, 114
  ICAO
    capacity to conduct business, 117–119
    immunities and liabilities, 119–120
    legal liability, 117
    organizational matters, 116
    protagonist, 117
    TAG/MRTD, 115, 116
  PKI scheme, 113
  waiver of immunity, 120–212
Public key infrastructure (PKI), 113

R
Racial profiling, 55–58
Rome Convention of 1952
  Air Transport Committee, 63, 64
  background, 70–71
  high contracting party, 62
  ICAO Assembly, 65
  insurance
    aircraft accident, 74
    cost of third party, 72, 73
    coverage, 62
    liability limits, 71–72
    proposals, 76
    types of aircraft, 75
  modernization
    advance payments, 86
    catastrophic losses, 87
    contributory negligence, 84–85
    environmental protection, 88, 89
    ICAO's work, 89
    issue of proximity, 84
    operative systems of liability, 81
  Montreal Protocol of 1978, 80–81
  provisions
    Contracting State, 79
    liability for compensation, 77, 78
    purpose, 77
    United States delegation, 79, 80
    weight, 78
sub-committees, 65

S
Security culture
  emerging threats
    anti-missile system installation, 32
    AVSEC panel, 19–20
    bioterrorism, 21–24
    carriage of liquids, aerosols and gels, 36

cyber-terrorism, 24–25
deterrence, 13–16
international accord, 33–36
Man Portable Air Defense System (MANPADS), 25–29
missile attacks, 29–31
perimeter guard, 32–33
probability, 9–13
threat assessment, 16–19
ICAO response
ICAO high-level ministerial conference, 2–7
post conference work, 7–9
risk-based approach, 1–2
Severe acute respiratory syndrome (SARS), 22
Standards and Recommended Practices (SARPs), 2, 266, 270, 273–276, 281
State responsibility
airport profiling
Chicago Convention, 57
Convention on International Civil Aviation, 56
and racial profiling, 56, 58
complicity theory, 42–43
condonation theory
categories and causality principle, 50
Jane case, 48
separate delict theory, 50
United Nations General Assembly, 49
war crime, 51
IATA, 40, 41
ICAO Council, 40
knowledge, role of
immutability, 52
International Law Commission, 52–53
Island of Palmas case, 51
Pan Am case, 53
private acts, 54
wrongful acts, 53
Lockerbie case
ICJ, 44
Montreal Convention, 44, 46
Security Council Resolution 748 (1992), 45, 46
United Nations Charter, Article 24, 46
OVC, 41–42
principles, 42
profiling of passengers, 54–55
rights of privacy
human and minority rights, 60
personal data, 59
Sui generis, 279

T
Technical Advisory Group on Machine Readable Travel Documents, 115, 116
Tokyo Convention (1963)
aircraft commander, powers of geographic scope, 224
traditional concepts, 223
airlines, 229
combating international terrorism, 217
extradition, 225–226
ICAO, 218
jurisdiction, terrorist punishment, 223–224
legal committee, 219
objectives, 220
powers and duties of States, 224–225
State responsibilities, 226–228

U
United Nations Convention on the Law of the Sea (UNCLOS), 165
United Nations initiatives
Convention adoption, 183–184
draft convention, 182
drug trafficking and drug abuse, 187
efforts, 187–188
ICAO, 184
Convention Against Illicit Traffic in Narcotic Drugs and Psychotropic Substances, 186–187
international co-operation, 183, 185
ratification, 184
sea transport, 185
Single Convention on Narcotic Drugs of 1961, 180
State co-operation, 185–186
Universal Declaration of Human Rights, 61
Universal Security Audit Programme (USAP), 265–272
Unlawful Interference Conventions
Bonn declaration
incompatibility, Vienna Convention, 249–250
legal status, 247–249
prosecution or extradition, 252–253
transit agreement, 250–252
compensation
commercial airline pilots, 101–102
Conference of Parties, 103, 104
Diplomatic Conference, 105
domestic opt-in declaration, 104
"gross breach," 101
ICAO Security Panel, 108

Index

International Civil Aviation
  Compensation Fund, 102
  offence, 101
  operators, 100
  "relevant duty of care," 101
  victims of violent crime, 106–107
Hague Convention (1970)
  efficiency, 236–237
  powers and duties, 232–236
  scope, 231–232
International Conventions
  Chicago Convention of 1944, 210–212
  Convention for the Prevention and
    Punishment of Terrorism (1937), 210
  Geneva Convention on the High Seas
    (1958), 213–217
  United Nations Charter, 212–213
Montreal Convention (1971)
  aircraft in service, 237–238
  jurisdictional powers of States, 242–246
  offence, definition, 239–241
  severe penalties, 241–242
plastic explosives marking
  final clauses and final act, 261–262
  International Explosives Technical
    Commission, 260–261
  States obligations, 257–258
  technical annex, 258–259
Tokyo Convention (1963)
  aircraft commander, powers of, 221–223
  airlines, 229
  alcohol, 229
  combating terrorism, 228
  extradition, 225–226
  jurisdiction, terrorist punishment, 223–224
  powers and duties of States, 224–225
  State responsibilities, 226–228
United Nations General Assembly
  Resolutions
  civil aviation, 205–206
  ICAO Conventions, 208–209
  international terrorism, 207–208
Unmanned aerial vehicles (UAV)
  air traffic services, 166–167
  high seas, operations
    air navigation services, 164
    United Nations Convention on the Law
      of the Sea, 165
  legal and regulatory issues, 162–163
  personnel licensing, 159, 160
  as state aircraft
    Hague Protocol of 1955, 170
    ICAO Assembly, 168, 169
    International Law Commission, 173
    Montreal Convention of 1999, 171
    United States Department of Defence, 157

W

Wassenaar Arrangement, 33–36
World Customs Organization (WCO), 196, 204
  IATA, 139, 141
  objectives, 138
  Permanent Technical Committee, 137

Breinigsville, PA USA
14 February 2011
255270BV00007B/67/P